T0256379

FOREIGN DIRECT INVESTMENT AND REGIONAL DEVELOPMENT IN EAST CENTRAL EUROPE AND THE FORMER SOVIET UNION

Ashgate Economic Geography Series

Series Editors:
Michael Taylor, Peter Nijkamp and Tom Leinbach

Innovative and stimulating, this quality series enlivens the field of economic geography and regional development, providing key volumes for academic use across a variety of disciplines. Exploring a broad range of interrelated topics, the series enhances our understanding of the dynamics of modern economies in developed and developing countries, as well as the dynamics of transition economies. It embraces both cutting edge research monographs and strongly themed edited volumes, thus offering significant added value to the field and to the individual topics addressed.

Other titles in the series:

The Emerging Economic Geography in EU Accession Countries
Edited by Iulia Traistaru, Peter Nijkamp and Laura Resmini
ISBN 0 7546 3318 7

Urban Growth and Innovation
Spatially Bounded Externalities in the Netherlands
Frank G. van Oort
ISBN 0 7546 3867 7

The Influence of the World Bank on National Housing and Urban Policies
The Case of Mexico and Argentina During the 1990s
Cecilia Zanetta
ISBN 0 7546 3491 4

The Sharing Economy
Solidarity Networks Transforming Globalisation
Lorna Gold
ISBN 0 7546 3345 4

The Future of Europe's Rural Peripheries
Edited by Lois Labrianidis
ISBN 0 7546 4054 X

China's Rural Market Development in the Reform Era
Him Chung
ISBN 0 7546 3764 6

Foreign Direct Investment and Regional Development in East Central Europe and the Former Soviet Union

A Collection of Essays in Memory of
Professor Francis 'Frank' Carter

Edited by
DAVID TURNOCK
University of Leicester, UK

Routledge
Taylor & Francis Group

LONDON AND NEW YORK

First published 2005 by Ashgate Publishing

Reissued 2018 by Routledge
2 Park Square, Milton Park, Abingdon, Oxon OX14 4RN
711 Third Avenue, New York, NY 10017, USA

Routledge is an imprint of the Taylor & Francis Group, an informa business

© David Turnock 2005

David Turnock has asserted his right under the Copyright, Designs and Patents Act, 1988,
to be identified as the editor of this work.

All rights reserved. No part of this book may be reprinted or reproduced or utilised in any
form or by any electronic, mechanical, or other means, now known or hereafter invented,
including photocopying and recording, or in any information storage or retrieval system,
without permission in writing from the publishers.

A Library of Congress record exists under LC control number: 2004054431

Notice:
Product or corporate names may be trademarks or registered trademarks, and are used only
for identification and explanation without intent to infringe.

Publisher's Note
The publisher has gone to great lengths to ensure the quality of this reprint but points out
that some imperfections in the original copies may be apparent.

Disclaimer
The publisher has made every effort to trace copyright holders and welcomes correspondence
from those they have been unable to contact.

ISBN 13: 978-0-815-38905-7 (hbk)
ISBN 13: 978-1-351-15812-1 (ebk)

Contents

PART ONE: THEMATIC STUDIES

PART TWO: REGIONAL STUDIES

List of Figures and Tables

Figures

Tables

List of Contributors

Michael J. Bradshaw, Professor, Department of Geography, University of Leicester, University Road, Leicester LE1 7RH, U.K. <mjb41@le.ac.uk> Professor of Human Geography with interests in the geography of Russia (especially the economic development of Siberia and the Far East), transition studies and developing areas in a global context. His many books include *Regional Patterns of Foreign Investment in Russia* (1995) and two edited works: *East Central Europe and the Former Soviet Union: The Post-Socialist States* (2004, with Alison Stenning) and *The Russian Far East and Pacific Asia: Unfulfilled Potential* (2001).

✚ *Francis W. Carter*, Professor, Department of Social Sciences, School of Slavonic and East European Studies, University College London <cartom@clara.co.uk> for family contact. Professor of Geography with research interests in East Central Europe – extending over four decades – covering both historical perspectives and such contemporary issues as environmental problems, ethnicity and foreign investment. His main works were on *Dubrovnik: A Classic City State* (1972) and *An Economic Geography of Cracow* (1994). He also edited several volumes including *Shock-Shift in an Enlarged Europe* (1999); also *The Changing Shape of the Balkans* (with H. Norris) and *Central Europe after the Fall of the Iron Curtain* (with P. Jordan and V. Rey) both in 1996.

Hugh Clout, Professor of Geography, University College London, 26 Bedford Way, London WC1H 0AP <h.clout@geography.ucl.ac.uk> Currently Dean of the Faculty of Social and Historical Sciences with research interests in the historical geography of France, the reconstruction of settlements after the two world wars, the practice of academic geography in France and the UK, and the historical geography of London. He has written many books including edited volumes on *Europe's Cities in the Late Twentieth Century* (1994) and *Regional Development in Western Europe* (1987).

Remus Crețan, Dr., Departamentul de Geografie, Facultatea de Chimie-Biologie-Geografie, Universitatea de Vest din Timişoara, Str. Pârvan 5, 1900 Timişoara, Romania <cretan_remus@yahoo.com> As 'Conferenţiar' (Senior Lecturer) and member of the Faculty's Scientific Board, Remus researches historical and contemporary issues in human geography with particular reference to his home region of Banat. Recent books include *Toponomie Geografică* and (with S.Truţi and C. Ancuta) *Geografie Umană a României* (both in 2000). He has developed his research experience in Sweden and the UK where he has taken a particular interest in emigration.

Andrew Dawson, Dr., School of Geography and Geology, University of St Andrews, Purdie Building, North Haugh, St Andrews, Fife KY16 9ST, U.K. <ahd@st-andrews.ac.uk> Senior Lecturer in Geography and author of many papers and several books dealing with the economic development of the region, and especially Poland, both under communism and since the fall of the Berlin Wall. He is co-editor of *Geopolitics in East Central Europe* (2001) and editor of the influential volume on *Planning in Eastern Europe* (1987).

Liliana Guran-Nica, Dr., Universitatea Spiru Haret, Facultatea de Geografie, Str. Ion Ghica 13, 030044 Bucharest Sector 3, Romania <liliana_gn@hotmail.com> Lecturer in Geography and Senior Researcher at the Romanian Academy's Geography Institute. She has researched a range of rural problems and published papers on foreign investment and social risk. In 2002 she produced a book based on her doctorate thesis: *Investiţii Străine Directe: Dezvoltarea Sistemului de Aşezări din România*.

Derek Hall, recently retired Professor of Regional Development and Head of the Department of Leisure & Tourism Management, Scottish Agricultural College, Auchincruive, Ayr KA6 5HW, U.K. <derekhall.seabank@virgin.net> He has research interests in Balkan development issues, tourism, transport, and the impacts of EU enlargement. Edited volumes include *Reconstructing the Balkans* (1996) and *Europe goes East* (2000) (both with Darrick Danta), and *Tourism and Transition* (2004). He has researched the geography of Albania for over thirty years and published *Albania and the Albanians* in 1994.

Peter Jordan, Associate Professor, Österreichisches Ost- und Sudosteurupa-Institut, Josefsplatz 6, 1010 Wien, Austria <peter.jordan@osi.ac.at> A native of Hermagor (Carinthia), Peter was educated in geography and ethnology at the University of Vienna (D. Phil) with habilitation at the University of Klagenfurt. He is currently Director of his institute (where he has been based since 1977) and also Head of the Geography Department. He specialises in cultural and tourism geography and is editor-in-chief of the *Atlas of Eastern and Southeastern Europe*.

Włodzimierz Kurek, Associate Professor, Instytut Geografii, Uniwersytetu Jagiellonskiego, 31-044 Krákow, ul. Grodska 64, Poland <w.kurek@interia.pl> Włodzimierz is Head of the Department of Tourism and Health Resort Management, with additional interests in agriculture, rural geography and the geography of mountains. In 2003 he edited a major work on *Issues of Tourism and Health Resort Management* in the Institute's prestigious series 'Prace Geograficzne'.

Erika Nagy, Dr., Alfold Institute, 5600 Békéscsaba, Szabo Dezso Str. 42, Hungary

<nagye@rkk.hu> Senior Research Fellow at the Centre for Regional Studies, Hungarian Academy of Sciences, researching on retailing, advanced producer services and the transformation of urban space in East Central Europe. She has previously taught at the University of Szeged and Széchenyi István College (Györ) and edited the Hungarian professional journal of regional science 'Tér és Társadalom'.

Jonathan Oldfield, Dr., School of Geography, Earth and Environmental Sciencies, University of Birmingham, Edgbaston, Birmingham, B15 2TT <j.d.oldfield@ bham. ac.uk> As Lecturer in the Human Geography of Post-Communist States in Birmingham (after earlier work in London and Nottingham universities), Jon has published widely on Russian environmental issues in the journals and edited volumes. He is currently engaged on projects concerned with Russian understandings of sustainable development and Russia-EU environmental cooperation.

Petr Pavlínek, Associate Professor, Department of Geography and Geology, University of Nebraska, Omaha NE 68182 0199, U.S.A. <pavlinek@mail.unomaha.edu> Petr's interests lie in the geographical and environmental implications of post-communist transformations in Central & Eastern Europe. He has written extensively on environmental matters, car industry restructuring and the regional effects of FDI. He has recently published *Environmental Transitions: Post-Communist Transformations and Ecological Defense in CEE* (2000), with John Pickles, and *Economic Restructuring and Local Environmental Management in the Czech Republic* (1997).

John Pickles, Professor, Department of Geography, University of North Carolina, Chapel Hill, NC 27599-3220, U.S.A. <jpickles@email.unc.edu> Earl N. Phillips Distinguished Professor of International Studies. His recent publications include *A History of Spaces: Cartographic Reason, Mapping, and the Geo-Coded World* (2004) and (with Adrian Smith) *Theorising Transition: The Political Economy of Post-Communist Transition* (1998). He is currently working with Robert Begg, Milan Bucek, Poli Roukova, and Adrian Smith on a National Science Foundation supported project on the new geographies of production in the East European clothing industry.

Dan Platon, Dr., Agenţia pentru Dezvoltare Regională Bucureşti-Ilfov, Calea Victoriei 16-20, Scara A, Etaj 2, 030027 Bucharest Sector 3 <dplaton@home.ro> With a doctorate from the Romanian Academy's Institute for Economic Research Dan acts as coordinator for the Bucharest-Ilfov regional plan and for the working group on technological research, IT and the information society. He writes on matters relating to Romania's adoption of EU regional policy.

Adrian Smith, Dr., Department of Geography, Queen Mary, University of London, Mile End Road, London E1 4NS, U.K. <a.m.smith@qmul.ac.uk> Professor of Human Geography, with a research focus on the reconfiguration of political and economic power in post-socialist states. He is currently working on a project researching the dynamics of garment production and outward processing in Europe. Author of *Reconstructing the Regional Economy: Industrial Transformation and Regional Development in Slovakia* and joint editor (with John Pickles) of *Theorising Transition: The Political Economy of Post-Communist Transition* (both 1998).

Alan Smith, Professor, Department of Social Sciences, School of Slavonic and East European Studies, University College London, Senate House, Malet Street, London WC1E 7HU <a.smith@ssees.ac.uk> Professor of Political Economy and a specialist on the political economy of East Central Europe. He has produced several major works including *The Reintegration of Eastern Europe into the European Economy* (2000), *Challenges for Russian Economic Reform* (1995), and *Russia and the World Economy* (1993).

Caedmon Staddon, Dr., School of Geography and Environmental Management, Faculty of the Built Environment, University of the West of England, Frenchay Campus, Coldharbour Lane, Bristol BS16 1QY, U.K. <Caedmon.Staddon @uwe. ac. uk> Chad has conducted research on local-level environmental politics and economics, focusing on post-communist change in East Central Europe since the early 1990s. He has also made a special study of the Bulgarian forestry sector and has researched ecological problems across the region in the light of restructuring and privatisation.

Andrew Tickle, Dr., Council for the Preservation of Rural England, 22a Endcliffe Crescent, Sheffied S10 3EF, U.K. <andytickle@cprepeakandsyorks.org.uk> Formerly Lecturer in Environmental Policy & Management at Birkbeck, University of London and Visiting Fellow at the School of Slavonic & East European Studies, he is currently working for the Campaign to Protect Rural England. He has written extensively on environmental issues including *Environment and Society in Eastern Europe* (1998), co-edited with Ian Welsh. In 1988–9 he was British Council visiting researcher to the Czechoslovak Academy of Sciences and he has subsequently researched transition impacts on nature and landscape conservation.

David Turnock, Emeritus Professor, Department of Geography, University of Leicester, University Road, Leicester LE1 7RH, U.K. <dt8@le.ac.uk> Has research interests in the rural and regional development of East Central Europe, with particular reference to Romania. He established a partnership with Frank Carter in the late 1980s which eventually produced several editions of *Environmental Problems in Eastern Europe* (1993-2002) and *The States of East Central Europe* (2001). He also cooperates with the WWF over the Carpathian ecoregion initiative.

Craig Young, Dr., Department of Environmental and Geographical Sciences, Manchester Metropolitan University, John Dalton Building, Chester Street, Manchester M1 5GD, U.K. <C.Young@mmu.ac.uk> Senior Lecturer in Human Geography with research interests in post-socialist transformation in East Central Europe, particularly economic and social restructuring, place marketing and the cultural geographies of post-socialism. He has contributed widely to the literature through journals and edited collections.

Preface

This book is a judicious mix of the planned and the spontaneous. As any observer of the transition states of ECE and FSU will appreciate the last decade has produced massive transformations and it is now difficult to recall the days when the old communist bloc – having opted out of globalisation by design – was peddling a distinctive brand of autarky which was as idiosyncratic for its economic management and location policy as it was in its ambivalent approach to human rights and democracy. But Mikhail Gorbachev tried to reform the unreformable and, as they say, the rest is history. While there is nostalgia for the old system, opening-up to the world has brought massive changes especially evident in the high streets and suburban fringes of the larger towns and also on the roads where the number of vehicles has increased considerably and Western makes are now much more prominent. In Hungary where the number of cars has increased from 1.95mln in 1990 to 2.63mln, the ratio 12.8 vehicles of manufactured in communist countries – Dacias, Ladas, Polski Fiats, Škodas, Trabants, Wartburgs and Zastavas – for every Western car has switched to 1.9 Western cars (23 different makes) for every one surviving from the communist past. The change has been even more striking for lorries and vans where numbers have increased in Hungary from 224 to 369th but with the ratio transformed from 132:1 in favour Eastern makes (dominated by the East German IFA) to 6.4:1 in favour of Western makes led by Volkswagen, Mercedes and IVECO.

These changes owe much to foreign capital since domestic private enterprise has made a comparatively slow start and is additionally inconspicuous through the takeover of former state enterprises with limited impact on the landscape with relatively few greenfield projects. Foreign capital has been more prominent in this sense and the impact reflects a revised locational agenda. Governments have opened their economies to global competition and have come to accept foreign investment as a fact of life, though not without a measure of political soul-searching after the polemics of the communist era when it seemed preferable to do without new resources rarther than share them with external capital. And there is still a lingering insensitivity over the damage that can arise from unstable fiscal environments: at the time of writing, car manufacturing by Volkswagen in Sarajevo (Bosnia and Hercegovina) is threatened by disagreement over customs-free import of disassembled vehicles between the Trade Ministry and the Federation's Customs Department. The key importance of foreign investment is brought out in all general geographical studies of the region and seemed a likely candidate for a full-length study in its own right. However the untimely death of Frank Carter, who made such an indelible impression on the geographical landscape of what we now call East Central Europe, effectively promoted a latent project into one with greater immediacy and also endowed it with the blessing of authors drawn from several overlapping research networks. As one of Frank's old associates privy to works that 'might have been' it has been a privilege to canvass support which has found a universally positive response. Indeed, the only disappointment has arisen from the passing of Ian Hamilton, one Frank's former colleagues at the School of Slavonic and East European Studies and a major contributor to the literature as the following reading list demonstrates. However every good literary idea needs a publisher and we have been fortunate to have the support of Ashgate who have welcomed the book to help launch a series of books on Economic Geography.

So this is a memorial volume but one with a clear theme which we hope will increase its relevance to the present generation of students of 'transition' which continues to claim its own domain within our global world. We have tried to deal with the broad concepts and themes which apply throughout the region once subject to Soviet-inspired communism and the early essays also explore the variations at national level with the aid of United Nations' statistics. Then in a second section there is a focus on the national scale with reference to specific themes where it is can be shown how foreign investment – or the lack of it – has relevance for the prospects of ordinary people. In view of the difficulty of collating all the various national breakdowns on a comparable basis, we have not managed to produce a map of foreign investment by regions and cities across the whole study area – a much-needed guide for students and policy-makers – but it is clear that investment is spatially uneven and this has clear implications which can be seen in migration: including substantial movement into Western Europe, where concern is arising over the consequences of EU enlargement that is an immediate prospeect as these words are written. Regional policy is still in its infancy but more must be done to entice investors beyond the handfull of capital and provincial cities that have been relatively privileged to date. However, the modest progress that the book hopefully represents would not have been possible without the support of all the authors who have managed to deliver according to a tight schedule. My wife Marion has made a massive contribution and coped with a steep learning curve to cope with the mysteries of camera-ready copy and East European diacriticals. And we have had generous support from Carolyn Court, Amanda Richardson and Valerie Rose at Ashgate.

We have tried to be consistent in our use of jargon and I would draw attention to the rather lengthy list of abbreviations which are used throughout the book. We have stuck to short country names – including the use of 'Macedonia' rather that the more unwieldy 'Former Yugoslav Republic of Macedonia' (adopted officially to allay Greek sensitivities) – with the one exception of 'Czech Republic' where we follow the near-universal disinclination to use to simpler form of 'Czechia'. 'Serbia and Montenegro' is used retrospectively for the whole period of the former Yugoslavia's demise while Bosnia and Hercegovina (the English rather than the official spelling) may be abbreviated BiH. We also use 'East Central Europe' (ECE) to distinguish the transition states outside the former Soviet Union (FSU) from 'Central and Eastern Europe' (CEE) which covers a wider group of countries extending to the FSU's European states. There is however a slight confusion in that the FSU's Baltic States may be regarded as part of ECE – featuring prominently in the EU enlargement process – in contrast with the other former Soviet countries that comprise the Commonwealth of Independent States (CIS). It is also difficult to find unanimity over other possible subdivisions for ECE and there are some variations between chapters. 'South East Europe' – abbreviated SEE – is widely used for the southern group (although Croatia and Slovenia would see themselves – as indeed would Romania – as Central European) but the Balkan label is also used, especially in Chapter 9. The project was initially conceived with an East Central European thrust and although the FSU is included to allow the discourse to consider transition states more widely, the single country studies maintain the original focus.

Leicester, April 2004   David Turnock

References

This is a concise list, which refers to key authors, texts and journals.

Andrusz, G., Harloe, M. and Szelenyi, I. (eds) (1996), *Cities after Socialism: Urban and Regional Change and Conflict in Post-Socialist Societies*, Blackwell, Oxford.

Artisien, P., Rojec, M. and Svetlicic. M. (eds) (1993), *Foreign Investment in Central and Eastern Europe*, Palgrave, Basingstoke.

Artisien-Maksimenko, P. (ed) (2000), *Multinationals in Eastern Europe*, Macmillan, Basingstoke.

Artisien-Maksimenko, P. and Adjubei, Y. (eds) (1996), *Foreign Investment in Russia and Other Soviet Successor States*, Palgrave, Basingstoke.

Artisien-Maksimenko, P. and Rojec, M. (eds) (2001), *Foreign Investment and Privatisation in Eastern Europe*, Palgrave, Basingstoke.

Bevan, A.A. and Estrin, S. (2000), *The Determinants of Foreign Direct Investment in Transition Economies*, London Business School Centre for New and Emerging Markets Discussion Paper 9, London.

Beyer, J. (2002), '"Please Invest in Our Country": How Successful were Tax Incentives for Foreign Investment in Transition Countries?', *Communist and Post Communist Studies*, 35, 191–211.

Bradshaw, M. (1995), *Regional Patterns of Foreign Investment in Russia*, Royal Institute of International Affairs Post-Soviet Business Forum, London.

Bradshaw, M. and Swan, A. (2004), 'Foreign Investment and Regional Development', in M. Bradshaw and A. Stenning (eds), *East Central Europe and the Former Soviet Union: The Post-Socialist States*, Harlow, Pearson Education, 59–86.

Brock, M. (1998), 'Foreign Direct Investment in Russia's Regions 1993–1995: Why so Little and Where has it Gone?', *Economics of Transition*, 6, 349–60.

Chen, J.R. (ed) (2000), *Foreign Direct Investment*, Macmillan, London.

Crocker, J.E. and Berentsen, W.H. (1998), 'Accessibility between the EU and ECE: Analysis of Market Potential and Aggregate Travel', *Post-Soviet Geography and Economics*, 39, 19–44.

Deichmann, J. (2001), 'Distribution of Foreign Direct Investment among Transition Economies in Central and Eastern Europe', *Post-Soviet Geography and Economics*, 42, 142–52.

Deichmann, J.I. (2001), 'Distribution of FDI among Transition Economies in CEE', *Post-Soviet Geography and Economics*, 42, 142–52.

Dicken, P. and Quevit, M. (eds) (1994), *Transnational Corporations and European Regional Restructuring*, Utrecht, Royal Dutch Geographical Society, Netherlands Geographical Studies 181.

Dobosiewicz, Z. (1992), *Foreign Investment in Eastern Europe*, Routledge, London.

Dyker, D.A. (ed) (1999), *Foreign Direct Investment and Technology Transfer in the Former Soviet Union*, Edward Elgar, Cheltenham.

Estrin, S. (1994), *Privatisation in Central and Eastern Europe*, Longman, Harlow.

Estrin, S., Hughes, K. and Todd, S. (1997), *Foreign Direct Investment in Central and Eastern Europe: Multinationals in Transition*, Pinter, London.

Estrin, S., Richet, X. and Brada, J.C. (eds) (2000), *Foreign Direct Investment in Central Eastern Europe: Case Studies of Firms in Transition*, Sharpe, New York.

European Bank for Reconstruction and Development (Various Years), *Transition Report*, EBRD, London.

Fabry, N. and Zeghni, S. (2002), 'Foreign Direct Investment in Russia: How the Investment Climate Matters', *Communist and Post-Communist Studies*, 35, 289–393.

Fischer. P. (2000), *Foreign Direct Investment in Russia*, Palgrave, Basingstoke.

Gacs, J., Holzmann, R. and Wyzan, M.L. (eds) (1999), *The Mixed Blessing of Financial Inflows: Transition Countries in Comparative Perspective*, Edward Elgar, Cheltenham.

Gorzelak, G. (1996), *The Regional Dimension of Transformation in Central Europe*, Regional Studies Association, London.

Grabher, G. (1995), 'The Disembedded Regional Economy: The Transformation of East German Industrial Complexes into Western Enclaves', in A. Amin and N. Thrift (eds), *Globalization Institutions and Regional Development in Europe*, Oxford: Oxford University Press, 177–95.

Hamilton, F.E.I. (1999), 'Transformation of Space in Central and Eastern Europe', *Geographical Journal*, 165, 135–44.

Hamilton, F.E.I. (2001), 'Industrial Restructuring between Magdeburg and Magadan: On the Road from Marx to Market', in D. Turnock (ed), *East Central Europe and the Former Soviet Union: Environment and Society*, Arnold, London, 118–29.

Hanson, P. and Bradshaw, M. (eds) (2000), *Regional Economic Change in Russia*, Edward Elgar, Cheltenham.

Hardy, J. and Rainnie, A. (1996), *Restructuring Krakow: Desperately Seeking Capitalism*, Mansell, London.

Havas, A. (1997), 'Foreign Direct Investment and Intra-Industry Trade: The Case of the Automotive Industry in Central Europe', in D.A. Dyker (ed), *The Technology of Transition*, Central European University Press, Budapest, 211–40.

Heinrich, A., Kusznir, J. and Pleines, H. (2002), 'Foreign Investment and National Interests in the Russian Oil and Gas Industry', *Communist Economies and Economic Transformation*, 14, 495–507.

Hunya, G. (ed) (2000), *Integration through Foreign Direct Investment*, Edward Elgar, Cheltenham.

Hunya, G. (2002), *Recent Impacts of Growth and Restructuring in Central European Transition Countries*, WIIW Research Report 284, Vienna.

Kaminski, B. (2000), *How Accession to the European Union has Affected External Trade and Foreign Direct Investment in Central European Economies*, World Bank Working Paper WPS2578, Washington, D.C.

Kushnirsky, F.I. (1997), 'Post-Soviet Attempts to Establish Free Economic Zones', *Post-Soviet Geography and Economics*, 38, 144–62.

Kuznetsov, A. (1994), *Foreign Investment in Contemporary Russia*, Palgrave, Basingstoke.

Lankes, H-P. and Venables, A. (1997), 'Foreign Direct Investment in Eastern Europe and the Former Soviet Union: Results from a Survey of Investors', in S.Zecchini (ed), *Lessons from the Economic Transition: Central and Eastern Europe in the 1990s*, Kluwer/OECD, London, 555–65.

Levigne, M. (1999), *The Economics of Transition: From Socialist Economy to Market Economy*, Macmillan, Basingstoke.

Marinov, M. and Marinova, S. (1999), 'Foreign Direct Investment: Motives and Marketing Strategies in Central and Eastern Europe', *Journal of East-West Business*, 5(1–2), 25–55.

Meyer, K.E. (ed) (1999), *Direct Investment in Economies in Transition*, Edward Elgar, Cheltenham.

Meyer, K.E. and Pind, C. (1999), 'The Slow Growth of Foreign Direct Investment in the Soviet Union Successor States', *Economics of Transition*, 7, 201–14.

Parker, D. and Saal, D. (eds) (2004), *International Handbook on Privatisation*, Edward Elgar, Cheltenham (especially chapters 16–21 on transition economies).

Pavlínek, P. (2004), 'Regional Development Implications of FDI in ECE', *European Urban & Regional Studies*, 11, 47–70.

Pavlínek, P. and Pickles, J. (2000), *Environmental Transitions: Transformation and Ecological Defence in Central and Eastern Europe*, Routledge, London.

Peck, A.E. (1999), 'Foreign Investment in Kazakhstan's Mineral Industries', *Post-Soviet Geography and Economics*, 40, 471–518.

Petrakos, G. and Totev, S. (2001), *The Development of the Balkan Region*, Ashgate, Aldershot.

Phelps, N.A. and Alden, J. (eds) (1999), *Foreign Direct Investment and the Global Economy*, Regional Studies Association, London.

Pickles, J. and Smith, A. (eds) (1998), *Theorising Transition: The Political Economy of Post-Communist Transformations*, Routledge, London.

Popov, V. (1998), 'Investment in Transition Economies: Factors of Change and Implications for Performance', *Journal of East-West Business Studies*, 4(1–2), 47–97.

Resmini, L. (2000), 'The Determinants of Foreign Direct Investment in the Central and East European Countries: New Evidence from Sectoral Patterns', *Economics of Transition*, 8, 865–90.

Ruble, B.A. and Popson, N. (1998), 'The Westernization of a Russian Province: The Case of Novgorod', *Post-Soviet Geography and Economics*, 39, 433–46.

Smith, A. (1998), *Reconstructing the Regional Economy: Industrial Transformation and Regional Development in Slovakia*, Edward Elgar, Cheltenham.

Smith, A. and Ferenčíková, S. (1998), 'Inward Investment Regional Transformations and Uneven Development in Eastern and Central Europe', *European Urban and Regional Studies*, 5, 155–73.

Smith, A. and Pavlínek, P. (2000), 'Inward Investment Cohesion and "Wealth of Regions"', in J. Bachtler, R. Downes and G. Gorzelak (eds), *Transition Cohesion and Regional Policy in Central and Eastern Europe*, Ashgate, Aldershot, 227–42.

Smith, A.H. (2000), *The Reintegration of Eastern Europe into the European Economy*, Macmillan, Basingstoke.

Sokol, M. (2001), 'Central and Eastern Europe a Decade after the Fall of State Socialism: Regional Dimensions of Transition Processes', *Regional Studies*, 35, 645–55.

Swain, A. (1998), 'Governing the workplace: the workplace and regional development implications of automotive foreign direct investment in Hungary', *Regional Studies*, 32, 653–71.

Tickell, A. and Welsh, I. (eds) (1998), *Environment and Society in Eastern Europe*, Longman, Harlow.

United Conference on Trade and Development (Various Years), *World Investment Report*, UNCTAD, New York/Geneva.

Veremis, T. and Dianu, D. (eds) (2001), *Balkan Reconstruction*, Frank Cass, London.

Von Zon, H. (1998), 'The Mismanaged Integration of Zaporizhzhaya into the World Economy: Implications for Regional Development in Peripheral Regions', *Regional Studies*, 32, 607–18.

Weresa, M. (2001), 'The Impact of Foreign Direct Investment on Poland's Trade with the EU', *Communist Economies and Economic Transformation*, 13, 71–8.

World Bank (2002), *Transition: The First Ten Years – Analysis and Lessons for Eastern Europe and the Former Soviet Union*, World Bank/Oxford University Press, Oxford.

Websites

1. Sites of general relevance to the area as a whole

Alexander's Gas & Oil Connections: www.gasandoil.com
Bankwatch: www.bankwatch.org
Black Sea Regional Energy Centre: www.bsrec.bg
Central Europe Online: www.europeaninternet.com/centraleurope/
EC/Security Pact Danube Cooperation Process: Danube Portal:
www.danubecooperation.org
European Bank for Reconstruction & Development: www.ebrd.org/
European Commission: www.europa.eu.net
European Governments: www.gksoft.com/govt/en/europa.html
European Union Enlargement: www.europa.eu.int/comm/enlargement/index.htm
Food and Agriculture Organisation: www.fao.org/regional/SEUR/
Governments on the World Wide Web: www.gksoft.com
#National Statistical Offices: www.stat.gov.pl/english/stale/odnosniki.htm
NGO Networking Resource: www.ngonet.org/
Open Media Research Institute News Site: www.tol/cz/
Organisation for Security and Cooperation in Europe: www.osceprag.cz
Radio Free Europe/Radio Liberty: www.rferl.org/
Regional Environment Centre for CEE: www.rec.org
United Nations Development Programme (Europe and CIS): www.undp.org/rbec/
*University of Birmingham, Centre for Russian and East European Studies:
www.crees.bham.ac.uk/
*University of London, School of Slavonic and East European Studies:
www.ssees.ac.uk/
*University of Pittsburgh, Centre for Russian and East European Studies:
www.ucis.pitt.edu/reesweb/
World Bank (Europe & Central Asia): http://Inweb18.worldbank.org/eca/eca.nsf

access is through the Polish Statistical Office website.
*these sites are listed largely because of the comprehensive links offered to other
sites relevant to individual countries.

2. Sites dealing with English language newspapers and agencies

Baltic Times: www.baltictimes.com
Bucharest Business Week: www.bbw.ro
Central Asia-Caucasus News: www.eurasianet.org
@Warsaw Voice: www.warsawvoice.pl/

@ one of the best-known in a range of English language newspapers which also includes: *Baku Sun* <www.bakusun.az>; *Budapest Sun* ; *Georgia Times* ; *#Central Europe Business News* ; *Kyiv Post* ; *Moscow Times* <www.moscowtimes.ru>; *Prague Post* <www.praguepost.com>; *St. Petersburg Times* <www.sptimes.ru/index/htm>; *Slovak Spectator* ; *Sofia Echo* ; *Times of Central Asia* <www.times.kg>. *#Central Europe Business News* provides links to the *Budapest Business Journal* and *Prague Business Journal.*

3. Sites dealing with investment in individual states

Albania: www.albaniabiz.org; www.fiaalbania.org
Armenia: www.armavir.am; www.armenia.am; www.armeniadiaspora.com
Azerbaijan: www.invest-in-azerbaijan.com
Belarus: www.belarus.net; www.e-belarus.org
Bosnia & Hercegovina: www.apf.com.ba; www.cbbh.gov.ba
Bulgaria: www.investbg.government.bg; www.bulgaria.com
Croatia: www.croatia-info.net; www.croatiainvest.8m.com
Czech Republic: www.czechinvest.cz; www.investfor.cz; www.ifo.cz; www.mpo.cz; www.fleet.cz; www.nmunion.cz
Estonia: www.eu-invest.ee; www.eia.ee; www.riik.ee
Hungary: www.bbj.hu; www.info.gov.hu; www.fsz.bme.hu; www.aries.hu
Kazakhstan: www.kazecon.kz; www.kazakhstan.com; www.herald.kz
Kyrgyzstan: www.kyrgyzstan.org
Latvia: www.search.lv; www.latinst.lv; www.lda.gov.lv
Lithuania: www.lda.lt; www.litnet.lt
Macedonia: www.gov.mk; www.macedonia.com; www.invest-in-macedonia.com
Moldova: www.invest.molodova.md; www.moldovainvest.com; www.naai.moldova.md
Poland: www.paiz.gov.pl; www.poland.pl
Romania: www.arisinvest.ro; www.evenimentulzilei.ro; www.investromania.ro
Russia: www.invest2russia.com; www.investinpoland.pl; www.polishworld.com; www.trans-siberiangold.com; www.russiajournal.com; www.ibc.valuehost.ru
Serbia &Montenegro: www.invest-in-serbia.com; www.invest-import.co.yu; www.economy.co.yu
Slovakia: www.slovakia.org; www.slovakiaguide.com; www.slovensko.com
Slovenia: www.ijs.si; www.sigov.si; www.uvi.si; www.cmsr.si
Tajikistan: www.traveltajikistan.com
Turkmenistan: www.turkmenistanembassy.org; www.usemb-ashgabat.usia.co.at
Ukraine: www.ukraine.com; www.ukraine.org
Uzbekistan: www.freenet.uz; www.uzland.uz

List of Abbreviations

AEBR	Association of European Border Regions
BEK	Beskid Euroregion
BFIA	Bulgarian Foreign Investment Agency
BGL	Bulgarian Leva
BiH	Bosnia and Hercegovina
bln	billion
BP	British Petroleum
BSP	Bulgarian Socialist Party
BSR	Baltic Sea Region
CADSES	Central Adriatic Danubian and South East European Space
CARDS (European)	Community Assistance for Reconstruction, Development and Stabilisation
CBC	cross-border cooperation
CCI	Chamber of Commerce and Industry
CCIA	Chamber of Commerce, Industry and Agriculture
CEE	Central and Eastern Europe
CEFTA	Central European Free Trade Area/Agreement
CEM	European Conference of Ministers Responsible for Spatial Policy
CER	Carpathian Euroregion
cif	Cost Insurance and Freight (Shipping)
CIS	Commonwealth of Independent States
CMB	Construcţii Metalice Bocşa
CoE	Council of Europe
Comecon	Council for Mutual Economic Assistance (CMEA)
CPEs	centrally planned economies
CREDO	East-East Cross-Border Cooperation Programme
cu.m	cubic meter
DCTMER	Danube Criş Tisa Mureş Euroregion
DGE	Directorate General Environment (EC)
DGER	Directorate General External Relations (EC)
DM	Deutschmark
EBRD	European Bank for Reconstruction and Development
EC	European Commission
ECE	East Central Europe
EDA	European Development Association

EIB	European Investment Bank
EIU	Economist Intelligence Unit
ELER	Elbe Labe Euroregion
EMBO	employee-management buy-out
EPCE	Environmental Partnership for Central Europe
ER	Euroregion
ESI	European Stability Initiative
EU	European Union
FAO	Food and Agriculture Organisation
FDI	foreign direct investment
FIAS	Foreign Investment Advisory Service
FIEs	foreign investments enterprises
fob	free on board
FSC	Forestry Stewardship Council
FSU	Former Soviet Union
FTAs	free trade agreements
FTOs	foreign trade organisations
FYROM	Former Yugoslav Republic of Macedonia
G8	Group of eight 'leading' economies (Canada, France, Germany, Italy, Japan, Russia, UK and USA)
GDP	gross domestic product
GEF	Global Environmental Facility
GM	General Motors
GNI	gross national income
GNP	gross national product
ha	hectare
HRK	Croatian kuna
HUF	Hungarian forint
ICPDR	International Commission for the Protection of the Danube River
IFC	International Finance Corporation
IMF	International Monetary Fund
Interreg	Community Initiative Concerning Border Areas
IRA	Industrial Restructuring Area
ISPA	Instrument for Social Policies for Pre-Accession Aid
IT	information technology
JIT	just in time
LACE	Linkage Assistance and Cooperation for European Border Regions
LFA	less-favoured area
LVC	Łódź Voivodship Council
m	meter
MD	managing director
MIG	minimum income guarantee

MIGA	Multilateral Investment Guarantee Agency
mln	million
MNE	multi-national enterprise
NATO	North Atlantic Treaty Organisation
nd	no data/date
NDP	National Development Plan
NER	Neisse Euroregion
np	no publisher/place of publication
NGO	non-governmental organisation
NRDF	National Regional Development Fund
NUTS	National Units for Territorial Statistics
OECD	Organisation for Economic Co-operation and Development
OSCE	Organisation for Security and Co-operation in Europe
pc	per capita
PCF	Prototype Carbon Fund
PHARE	Poland and Hungary: Actions to Restructure their Economies
PLA	Protected Landscape Area
ptp	per thousand of the population
PZL	Polish złoty
Rb	rouble
RDA	Regional Development Agency
REReP	Regional Environmental Reconstruction Programme
RICOP	Industrial Restructuring and Professional Reconversion Programme
ROL	Romanian lei
Ro-Ro	roll on-roll off ferryboat
RPP	Rila Pulp and Paper
SAP	Stabilisation and Association Process
SAPARD	Special Action Programme for Agriculture and Rural Development
SCSP	Special Coordinator of the Stability Pact for South Eastern Europe
SECI	South East European Cooperative Initiative
SEE	South Eastern Europe
SEZ	special economic zone
SKK	Slovak crown
SME	small and/or medium size enterprise
SNBER	Spree Neisse Bóbr Euroregion
SOE	state-owned enterprise
SP	Stability Pact for Southeastern Europe
sq.km	square kilometer
sq.m	square meter
STEG	Staatseisenbahngesellschaft

SUV	sports utility vehicles
SWOT	strengths, weaknesses, opportunities and threats
TACIS	Technical Assistance for the Confederation of Independent States
TER	Tatra Euroregion
UDF	Union of Democratic Forces (Bulgaria)
UDR	Uzinele şi Domeniile din Reşiţa
UK	United Kingdom
UN	United Nations
UNCTAD	United Nations Conference on Trade and Development
UNDP	United Nations Development Programme
UNECE	United Nations High Commission for Europe
UNHCR	United Nations High Commission for Refugees
USA	United States of America
USAID	United States Agency for International Development
VW	Volkswagen
WBG	World Bank Group
WIIW	Vienna Institute for International Economic Studies
WTO	World Trade Organisation
WWF	Worldwide Fund for Nature
$	United States dollar (all dollar references in the book relate to US dollars)

In Memoriam

Francis William Carter 1938–2001: An Appreciation

Hugh Clout

Professor Francis William 'Frank' Carter, who died at the age of 62 on 4 May 2001, was an exceptional geographer and a remarkable character. He cherished his linguistic abilities and his erudition but made light of them in public. As a teacher he was blessed with the ability to enthuse, to make learning fun. Never was he at a loss to enliven an index of production or a trade statistic with some personal anecdote to help his students and colleagues appreciate that these numerical abstractions were aggregate reflections of the lives and hopes of real people, some of whom Frank had met, and had even visited their homes. His fieldtrips to Yugoslavia were legendary; his seminars memorable for the alcohol that was served; and his escape from Prague at the time of the Soviet reoccupation in 1968 was a narrow one.

Frank's language skills, his photographic memory, and his love of people and places formed the key to his achievement. I still remember the look of amazement on the face of a distinguished Western expert on the physical resources of Romania whom Frank greeted in fluent Romanian tinged, I am sure, with more than a hint of Black Country accent. Certainly Frank's Polish, Czech, Serbo-Croat, Russian, Bulgarian, German, French, Italian, Arabic and (I suspect) Spanish were not the silver-tongued language of diplomacy but rather the language of the ordinary people. He had a remarkable knack of absorbing vocabulary and linguistic ability, just as he absorbed musical sounds and rhythms enabling him to play the piano without sheet music, and could memorise material to deliver lecture after lecture without a single note. He greatly admired the compendious writings of Fernand Braudel both for their power of synthesis and for the diversity of source material interrogated. Braudel's *The Mediterranean and the Mediterranean World in the Age of Philip II* was a long-standing source of inspiration to him. Doubtless, some of the rough edges of Frank's Polish were smoothed away after his marriage to Krystyna Tomaszewska in 1977.

Frank was, of course, a 'rough diamond'. He could be stubborn, awkward and in some ways unreliable, but at the same time he was a remarkably hard researcher, a very productive writer, and a much-loved teacher and colleague. His sense of humour was both wicked and inimitable; deflating egos was one of his specialities.

Some colleagues in London will recall how a burst of laughter at the same time as a mouthful of cream cake was being eaten could change social relations in seconds – and for ever. Quite a lot of his tricks involved cakes, cats and carpets as I remember. Frank's family background was far from scholarly, with his father fully intending that he would follow him into the haulage and odd-job business and make a fortune. But as a child, Frank learned the rudiments of Polish from a lodger in the family home, showed promise at school (even though there was a rebellious streak, and he failed his '11 plus' examination), and responded to encouragement from perceptive teachers, especially John Inch at Wolverhampton. He then studied at the University of Sheffield for his first degree under the watchful eye of Professor Alice Garnett (who had written about Yugoslavia for the Naval Intelligence Handbooks during World War Two), to Cambridge for his postgraduate Diploma in Education, and to the London School of Economics for research on the historical geography of Dubrovnik, with guidance from Dr Audrey Lambert and Professor Michael Wise.

After a short while lecturing at Kings College London, Frank moved to University College in October 1966, holding a Hayter Lectureship jointly at the School of Slavonic and East European Studies. This was eventually transformed into a full-time post and a Readership at SSEES. Throughout his early years in London, the term 'rough diamond' still held true. Some of his contemporaries at UCL, all now professors thinking of retirement, will recall the amazing exploits Frank got up to during long vacations spent in Czechoslovakia, Greece, Poland or Yugoslavia. He always returned home laden with books, peasant crafts, cheap tapes, maps, samovars, carpets, leather bags, rough but potent alcohol – the list was seemingly endless. His office and his home resembled Aladdin's Cave. We never ceased to be amazed at the number of times his return to UCL could be delayed by the intricacies of 'Red Army manoeuvres' in some part or another of what was then regarded as Eastern Europe. Inevitably this meant that his start of session did not go down well with successive heads of department.

Yet all this time Frank was reading, collecting, analysing, synthesising and writing. I well remember the fun we all had with his computation of eigenvalues to delineate the ecumene of medieval Serbia. The publications flooded forth in a torrent year after year, decade after decade. Initially they focused on historical geography since current information was either unobtainable or of doubtful reliability. Also it made good political sense for scholars who wished to return to Eastern Europe to avoid sensitive issues. A monograph on *Dubrovnik: a classic city state* (1972) expanded his London MA; suites of articles on the industrialisation and urban structure of Prague echoed his London PhD and his doctorate from the Charles University in Prague; his massive tome on *Trade and Urban Development in Poland: an economic geography of Kraków from its origins to 1795* (1994) earned a doctorate in Poland; and his long-running series of articles on the historic settlement of Hvar Island in the Dalmatian Adriatic eventually came together for a doctorate from Zagreb.

As a boy from a home where there were no books and whose initial education was at Willenhall Secondary Modern School, Wednesbury Technical School and then Wulfrun College of Education – rather than at a grammar school – Frank had proved himself time and time again. Yet that was not all, since he edited a truly remarkable collection of original essays by geographers, archaeologists and historians that appeared as *An Historical Geography of the Balkans* (1977) and required translation from various languages and some measure of internal standardisation – or so I thought. I well remember spending day after day of a freezing Christmas vacation, in the Map Room of the old Geography Department housed in Foster Court at UCL, going through page after page of typescript with Frank in order to help standardise place names and knock chapter bibliographies into a common format. And then he changed his mind. The stubborn streak came through, with the book displaying an amazing diversity of detailed presentation that was certainly true to each author's typescript but must have driven the copy-editor to distraction.

Frank was canny enough to know that the historical geography of Balkans and even Eastern Europe was a rather recherché field even in a large and pluralistic geography department like that at UCL. Hence he developed a second portfolio of work that was of direct relevance to teaching and reported on current developments in the constituent countries, embracing agriculture, urban planning, transport provision, human migration, political geography, recreation, heritage, environmental change, pollution, and countless other topics. His repertoire was enormous and whilst these articles and chapters were arguably less 'original' than his work in historical geography, they conveyed a great deal of otherwise inaccessible information to scholars and students in Western Europe and North America. Frank's thematic and regional box files, and his card indexes and bibliographies, relating to items in apparently every conceivable European language, were second to none.

With the fall of the Berlin Wall and the real opening up of the region to Western scholars, Frank's skills and knowledge came into great demand. He found himself summoned to give radio and television interviews, to present keynote addresses at international conferences, to draft position papers, to undertake research on current economic changes, and to collaborate with colleagues in various countries in editing book after book. Frank was a man who rarely said 'no' and 'Professor Carter', with Krystyna at his side, became a well-known academic in Western Europe as well as further east. Whenever I visited a European university for the first time, I seemed to be greeted with the words: 'you are from UCL, then you must know Professor Carter'. In 1997 he received the Edward Heath Award of the Royal Geographical Society for his research on Eastern Europe over three decades. Medals, prizes and diplomas from former Iron Curtain countries were legion.

Tragically as Frank's productivity mounted, so his health faltered. He battled on with teaching, writing and travel, only to collapse time after time. Guest lectures in his final years were predominantly on the theme of foreign direct investment and regional development in what was now being called East Central Europe. Everyone knew – but preferred not to acknowledge – that he could not sustain this punishing routine, but he soldiered on with advising students, reading recent publications, and drafting new articles and chapters to the very last. His promotion to Full Professor of the University of London was arranged a little ahead of the normal round of official meetings so that he could be aware that – in his own words – he had 'made it' for the last five weeks of his life. By a cruel quirk of fate, he died half an hour before the final decisions of the annual promotion round were taken and results could be conveyed informally to applicants.

This volume of essays is dedicated to the memory of Professor Frank Carter by friends and colleagues who share his passion for the societies and economies of a region that he made his own. They have not been written in sorrow, since that time has passed, but rather in celebration of the life of a remarkable man who contributed so much to advance geographical scholarship, and whose personality and sense of humour enriched the lives of those who had the privilege to know him. 'Diamond' he certainly was.

List of Publications

(All single author works except for: * which denotes editor and contributor; # joint editor and contributor; @ joint author.)

1968:
'Population Migration to Greater Athens', *Tijdschrift voor Economische en Sociale Geografie*, 59, 100–5.
'The Decline of the Dubrovnik City State', *Balkan Studies*, 9, 127–38.
1969:
'Dubrovnik: The Early Development of a Pre-Industrial City', *Slavonic and East European Review*, 47, 355–68.
'Balkan Exports Through Dubrovnik 1358–1500: A Geographical Analysis', *Journal of Croatian Studies*, 9–10, 133–59.
'The Trading Organization of the Dubrovnik Republic', *Historická Geografie* (Prague) 3, 12–24.
'An Analysis of the Medieval Serbian Oecumene: A Theoretical Approach', *Geografiska Annaler*, 51B, 355–68.
'La Coopération économique dans les Balkans', *L'Information Géographique*, 33(4), 157–165.
1970:
'Natural Gas in Romania', *Geography*, 55, 214–20.
'Bulgaria's Economic Ties with Her Immediate Neighbours and Prospects for Future Development', *East European Quarterly*, 4, 209–24.
'Czechoslovakia's North Moravian Region: A Geographical Appraisal', *Revue Géographique de l'Est*, 10, 65–86.
'Prague: Some Contemporary Growth Problems', *Société Royale de Géographie d'Anvers: Bulletin*, 81, 197–218.

1971:
'The Commerce of the Dubrovnik Republic 1500–1700', *Economic History Review*, 24, 370–94.
'Rozvoj historické geografie se zvlastním zretelem k Anglii', *Historická Geografie*, (Prague) 6, 54–76.
'The Woollen Industry of Ragusa (Dubrovnik) 1450–1550: Problems of a Balkan Textile Centre', *Textile History*, 2, 3–27.
1972:
Dubrovnik(Ragusa): A Classic City State, Seminar Press, London/New York.
'Bulgariens wirtschaftsbeziehungen zu den Nachbarstaaten', *Österreichische Osthefte*, 14,181–97.
'Dubrovnik at her Zenith', *Geographical Magazine*, 44, 827–8.
1973:
'Prague et la Bohème Centrale: Quelques Problèmes de Croissance', *Annales de Géographie*, 82, 165–92.
'The Industrial Development of Prague 1800–1850', *Slavonic and East European Review*, 11, 243–75.
'Changements Fonctionnel et Structurel de l'Après-Guerre dans la Conurbation de Sofia', *Géographie et Recherche*, 8, 25–39.
'Public Transport Development in Nineteenth Century Prague', *Transport History*, 6, 205–26.
'Albania: Some Problems of a Developing Balkan State', *Revue Géographique de l'Est*, 4, 351–76.
'Post-War Functional and Structural Changes within the Sofia Conurbation', *Department of Geography University College London Occasional Papers*, 21.
1974:
'Concentrated Prague', *Geographical Magazine*, 41, 537–44.
1975:
'Č-K-D Employees Prague 1871–1920: Some Aspects of their Geographical Distribution', *Journal of Historical Geography*, 1, 69–97.
'Bulgaria's New Towns', *Geography*, 60, 133–6.
@'New Era in Slovenia', *Geographical Magazine*, 48, 556–60.
'The Cotton Printing Industry in Prague 1766–1873', *Textile History*, 6, 132–55.
1976:
'Post-War Internal Migration in South-Eastern Europe', in L.A.Kosinski (ed), *Demographic Developments in Eastern Europe*, Praeger, New York, 260–86.
'Four Countries Develop their own Energy (Albania, Bulgaria, Romania and Yugoslavia)', *Geographical Magazine*, 49, 10–16.
1977:
An Historical Geography of the Balkans, Academic Press, London/New York.
1978:
'Nature Reserves and National Parks in Bulgaria', *L'Espace Géographique*, 1, 69–72.
'The Geography of the Vinodol Region', *British-Croatian Review*, 14, 7–11.

1979:

'Prague and Sofia: An analysis of their Changing Internal City Structure', in R.A. French and F.E.I. Hamilton (eds), *The Socialist City: Spatial Structure and Urban Policy*, Wiley, Chichester/New York, 425–59.

'The Physical and Economic Geography of the Split Region', *British-Croatian Review*, 16, 7–11.

'The Maritime History of Split', *British-Croatian Review*, 16, 23–28.

'Aktuelle Entwicklungsprobleme in der Region Gross-London', *Geographische Berichte*, 24, 227–40.

1980:

'Ragusa: The City Republic of Dubrovnik', *History Today*, 30(1), 46–51.

'Public Transport in Eastern Europe: A Case Study of the Prague Conurbation', *Transport Policy and Decision Making*, 1, 209–29.

'Between East and West: Geography in Higher Education in Yugoslavia', *Journal of Geography in Higher Education*, 4(2), 43–53.

1981:

'Kafka's Prague', in J.P. Stern (ed), *The World of Franz Kafka*, Weidenfeld and Nicholson, London, 30–43.

'Greece', in H.D. Clout (ed), *Regional Development in Western Europe*, Wiley, Chichester/New York, 389–402.

'Conservation Problems of Historic Cities in Eastern Europe', *Department of Geography University College London Occasional Papers*, 39.

'Yugoslav Geography: A Case of Misguided Optimism', *Journal of Geography in Higher Education*, 5(2), 12–18.

1982:

'Historic Cities in Eastern Europe: Problems of Industrialisation Pollution and Conservation', *Mazingira: The International Journal for Environment and Development*, 6(3), 62–76.

1983:

'Praga e Sofia: Un Analisi Compariativa del Cambiamento della loro Struttura Interna', in R.A. French and F.E.I. Hamilton (eds), *La Citta Socialista: Structura Spaziale e Politica Urbana*, Franco Angeli Editore, Milan, 514–54.

'Cracow's Early Development', *Slavonic and East European Review*, 16, 197–225.

1984:

'Eleventh Century Poljica: An Analysis of its Historical Geography', in E.Pivcevic (ed), *The Cartulary of the Benedictine Abbey of St Peter of Guman (Croatia) 1008–1187*, David Arthur, Bristol, 23–37.

'Pollution in Prague: Environmental Control in a Centrally Planned Socialist Country', *Cities*, 1, 258–73.

'Medieval Trade in the Western Balkans', *Journal of the British-Yugoslav Society*, 3, 15–18; 4, 14–16.

1985:

'Cracow as Trade Mediator in Balkan-Polish Commerce 1590–1600', *School of Slavonic and East European Studies Occasional Papers*, 2, 45–76.

'Balkan Historic Cities: Pollution versus Conservation', in Anon (ed), *Proceedings of the Anglo-Bulgarian Modern Humanities Symposium London 1982*, School of Slavonic and East European Studies, London, II 75–97.

'Pollution in Post-War Czechoslovakia', *Transactions of the Institute of British Geographers*, 10, 17–44.

1986:

'Balkan Irrigation with Special Reference to Yugoslavia', *Journal of the British Yugoslavian Society*, 1, 9–15.

'Nuclear Power Production in Czechoslovakia', *Geography*, 71, 136–9.

'City Profile: Tirana', *Cities*, 3, 270–81.

1987:

'Bulgaria', in A.H. Dawson (ed), *Planning in Eastern Europe*, Croom Helm, London/Sydney, 67–101.

'Czechoslovakia', in A.H. Dawson (ed), *Planning in Eastern Europe*, Croom Helm, London/Sydney, 103–38.

'Greece', in H.D. Clout (ed), *Regional Development in Western Europe*, D. Fulton, London, 419–33 (Third Edition).

'Tirana', *Albanian Life: Journal of the Albanian Society*, 49(2), 14–29.

The City and the Environment in Socialist Countries, Forschungsschwer-punkt Umweltpolitik, Wissenschaftszentrum Berlin für Sozialforschung, Berlin.

'Cracow's Wine Trade Fourteenth to Eighteenth Centuries', *Slavonic and East European Review*, 65, 337–378.

1988:

'Space and History', *Area*, 20, 87–8.

'Czechoslovakia: Nuclear Power in a Socialist Society', *Environment and Planning C: Government and Policy*, 6, 269–87.

'Cracow's Transit Textile Trade 1390–1795: A Geographical Assessment', *Textile History*, 19, 23–60.

1989:

'Stadt und Umwelt im Sozialismus', in H. Schreiber (ed), *Umweltprobleme in Mitte und Ost Europa*, Campus Verlage, Frankfurt/New York, 60–90.

'Post-War Regional Economic Development: A Comparison between Bulgaria and Greece', in E.P. Dimitriadis and A.Yerolympos (eds), *Space and History: Urban Architectural and Regional Space*, Ververidis-Polychronides, Thessaloniki, 203–10.

'Air Pollution in Poland', *Geographica Polonica*, 56, 155–77.

'Czechoslovakia's Ecological Disaster' and 'Bulgaria's Dirty Smoke Stacks', *People*, 16(3), 8–11.

1990:

'Public Transport Development in Nineteenth Century Prague', in J.Komlos (ed), *Economic Development in the Habsburg Monarchy and the Successor States: Essays*, East European Monographs, Boulder/Columbia University Press, New York, 27–40.

'The Industrial Development of Prague 1800–1850', in J. Komlos (ed), *Economic*

Development in the Habsburg Monarchy and the Successor States: Essays, East European Monographs, Boulder/Columbia University Press, New York, 41–67.

'Housing Policy in Bulgaria', in J.A.A. Sillince (ed), *Housing Policies in Eastern Europe and the Soviet Union*, Routledge, London/New York, 170–227.

'Czechoslovakia: Geographical Prospects for Energy Environment and Economy', *Geography*, 75, 253–5.

'Bulgaria: Geographical Prognosis for a Political Eclipse', *Geography*, 75, 263–5.

'Development and the Environment: A Case Study of Hvar Island, Yugoslavia', *Mediterranean Social Sciences Network Newsletter*, 4, 20–39.

'Viticulture on Hvar Island, Yugoslavia', *Journal of Wine Research*, 1, 139–157.

'Agricultural Geography of Eastern Europe', *Chemistry and Industry*, 12, 386–90.

'Property Research in Eastern Europe: Beyond the Socialist City', in Anon (ed), *Property Research in Europe*, Royal Institute of Chartered Surveyors, London, 57–70.

1991:

'1992 and the Balkan Mediterranean Countries', in Y.Özkan (ed), *Europe and the Mediterranean Countries*, Center for Mediterranean Studies, Ankara, 39–55.

'Czechoslovakia', in D.R. Hall (ed), *Tourism and Economic Development in Eastern Europe and the Soviet Union*, Belhaven Press, London/New York, 154–72.

'Bulgaria', in D.R. Hall (ed), *Tourism and Economic Development in Eastern Europe and the Soviet Union*, Belhaven Press, London/New York, 220–35.

1992:

'Geographical Aspects of East-West Environmental Policy', in M. Jachtenfuchs and M. Strübel (eds), *Environmental Policy in Europe: Assessment Challenges and Perspectives*, Nomos Verlagsgesellschaft, Baden-Baden, 177–96.

'Ethnic Groups in Cracow', in M. Engman (ed), *Ethnic Identity in Urban Europe*, Dartmouth, New York University Press, 241–65.

'Ethnic Residential Patterns in the Cities', in M. Engman (ed), *Ethnic Identity in Urban Europe*, Dartmouth, New York University Press, 375–89.

'State of Environment Depends on State of Mind', *University of Tennessee Forum for Applied Research and Public Policy*, 7(1), 104–5.

'Dubrovnik v negoviya zenit', *Geografiya* (Sofia), 46(3), 5–8.

'Agriculture on Hvar during the Venetian Occupation: a Study in Historical Geography', *Geografski Glasnik*, 54, 45–62.

1993:

#*Environmental Problems in Eastern Europe*, Routledge, London/New York.

'Regional Planning in Post-War Bulgaria', in L. Collins (ed), *Proceedings of the Second Anglo-Bulgarian Symposium, Blagoevgrad 1985*, School of Slavonic and East European Studies, London, 112–52.

'Yugoslavia', in M. Bainbridge (ed), *The Turkic Peoples of the World*, Kegan Paul International, London/New York, 297–342.

'Ethnicity as a Cause of Migration in Eastern Europe', *GeoJournal*, 30, 241–8.

@'International Migration between East and West in Europe', *Ethnic and Racial Studies*, 16, 467–91.

'Pollution in Bulgaria: The Way Ahead?', *Newsletter: British-Bulgarian Friendship Society*, 2(12), 24–8.

1994:

Trade and Urban Development in Poland: An Economic Geography of Cracow from its Origins to 1795, Cambridge University Press, Cambridge.

'The Economic and Demographic Development of Albania since 1945', in L.di Comite and M.A.Valeri (eds), *Problemi Demo-Economici dell'Albania*, Argo, Bari, 15–63.

'Settlement and Population during Venetian Rule (1420–1797): Hvar Island, Croatia', *Journal of European Economic History*, 23, 1–37.

'Minorités Nationales et Groupes Ethniques en Bulgarie: Distribution Régionale et Liens Transfrontaliers', *Espace Populations Société*, 3, 299–309.

'The Political Geography of the Dubrovnik Republic', *Acta Geographica Croatica*, 29, 77–98.

'Saving Bulgaria's Coastal Heritage', *Newsletter: British-Bulgarian Friendship Society*, 3(15), 22–4.

1995:

#*Proces Przeksztalcen Spoleczno-Gosposarczych w Europie Srodkowej i Wschodniej po roku 1989*, Turpress, Toruń.

'Historical Evolution', in W.I. Stevenson (ed), *Interpreting the Balkans*, Royal Geographical Society/Institute of British Geographers, Geographical Intelligence Paper 2, London, 21–35.

'Insediamento e Popolazione a Hvar (Lesina) nel Periodo Veneziano 1420–1797', *Proposte e Ricerche: Economia e Societa nella Storia dell'Italia Centrale*, 18(34), 125–62.

'National Minorities/Ethnic Groups in Bulgaria: Regional Distribution and Cross Border Links', *Regions and Regionalism* (Łódź/Opole), 2, 61–85.

'Poland's Migration Problems: A Post-Communist Legacy', *Geografski Vestnik*, 67, 141–61.

'Közép-Európa: Valóság vagy Földrajzi Fikció?', *Foldrajzi Kozlemenyek*, 119, 232–50.

'Market Economics and Regional Diversity', *International Geopolitical Research Colloquium on Euro-Atlantic Security, Garmisch Partenkirchen 1995: Conference Report*, Marshall Center, Garmisch, 27–8.

'National Minorities and Ethnic Groups in Bulgaria', *Newsletter: British-Bulgarian Friendship Society*, 4(16), 7–11.

1996:

#*The Changing Shape of the Balkans*, UCL Press, London.

#*Central Europe after the Fall of the Iron Curtain: Geopolitical Perspectives Spatial Patterns and Trends*, P.Lang, Frankfurt am Main/Berlin.

#*Environmental Problems in Eastern Europe*, Routledge, London/New York (Paperback – Partially Updated – Edition).

'Recreational Planning in Bulgaria', in L. Collins and T. Henniger (eds), *The Third Anglo-Bulgarian Symposium, London 1988: Proceedings*, School of Slavonic and East European Studies, London, 35–65.

'"For Bread and Freedom": Geographical Implications of the Polish Diaspora', in G. Prévélakis (ed), *The Networks of Diasporas*, Cyprus Research Center, Nicosia, 335–51.

'The Black Sea Coastal Region: Geographical Problems and Perspectives', *Problemi na Geografijata* (Sofia), 1, 43–56.

'Ethnic Problems in Post-Communist Eastern Europe', *Revista Geografică* (Bucharest), 2–3, 154–60.

1997:

'Joseph Crabtree and the Polish Connection', in B. Bennett and N. Harte (eds), *The Crabtree Orations 1954–1994*, The Crabtree Foundation, London, 210–18.

'Czechoslovakia in Transition: Migration before and after the "Velvet Divorce"', *IMIS-Beitrage* (Institut für Migrationsforschung und Interkuturelle Studien, Universität Osnabrück), 6, 35–63.

1998:

Central Europe after the Fall of the Iron Curtain: Geopolitical Perspectives Spatial Patterns and Trends, P. Lang, Frankfurt am Main/Berlin (Second Edition).

'Bulgaria', in D. Turnock (ed), *Privatisation in Rural Eastern Europe: The Process of Restitution and Restructuring*, Edward Elgar, Cheltenham/Northampton Maine, 69–92.

'Political Transformations and the Environment (A Co-Review)', in H.van der Wusten (ed), *Proceedings of the Conference on Transformation Processes in Eastern Europe: Volume 4 Political Transformation of the Environment*, ESR, The Hague, 11–22.

'Central Dalmatia in Balkan Historical Geography: Evidence from a Frontier Region', in E.P. Dimitriadis, A.P. Lagolopoulos and G. Tsotsos (eds), *Roads and Crossroads of the Balkans from Antiquity to the European Union*, University Studio Press, Thessaloniki, 57–66.

'Geographical Problems in East Slovakia', *Regions and Regionalism* (Łódź/Opole), 3, 187–203.

'La Diversité Ethnique de la Pologne', *Géographie et Cultures*, 27, 79–98.

1999:

#*Shock-Shift in an Enlarged Europe: The Geography of Socio-Economic Change in East-Central Europe after 1989*, Ashgate, Aldershot.

#*The States of Eastern Europe: Volume 1 The Northern States; Volume 2 The Southern States*, Ashgate, Aldershot.

'Restructuring the Balkans and the Influence of the Adriatic Sea: A Geographical Appraisal', in L. di Comite and R. Pace (eds), *Integrazione Politica ed Integrazione Economica nel Bacino Mediterraneo*, Cacucci Editore, Bari, 87–115.

'The Role of Foreign Direct Investment in the Regional Development of Central and Southeast Europe', in Anon (ed), *Regional Prosperity and Sustainability: Proceedings of the Third Moravian Geographical Conference, Slavkov u Brna*, n.p., Brno, 16–23.

@'Rural Diversification in Bulgaria', *GeoJournal*, 46, 183–91.

'The Geography of Foreign Direct Investment in Central-East Europe during the 1990's', *Wirtschafts-Geograpische Studien*, 24–25, 40–70.

2000:

'The Czech Republic', in D. Hall and D. Danta (eds), *Europe goes East: Enlargement Diversity and Uncertainty*, The Stationery Office, London, 59–79.

'Slovakia', in D. Hall and D. Danta (eds), *Europe goes East: Enlargement Diversity and Uncertainty*, The Stationery Office, London, 168–91.

'The Role of Foreign Direct Investment in the Czech Republic during the 1990s', *Moravian Geographical Reports*, 8(1), 2–16.

@'Ethnicity in Eastern Europe: Historical Legacies and Prospects for Cohesion', *GeoJournal*, 47, 1–17.

'Bulgaria cannot be Marginalised', *Newsletter: British-Bulgarian Friendship Society*, 9(1), 1–2.

'Money should keep Flowing in: Foreign Direct Investment in Bulgaria', *Newsletter: British-Bulgarian Friendship Society*, 9(2), 1–2.

2001:

#*Environmental Problems in Eastern Europe*, Routledge, London/New York (Second Edition).

'The States System', in D. Turnock (ed), *East Central Europe and the Former Soviet Union: Environment and Society*, Arnold, London, 30–40.

2004

@'Foreign Direct Investment and City Restructuring', N. Milanowic (ed), *Transformation of Central and East European Cities: Globalisation Europeanisation or Cross-Border Regionalisation?*, United Nations University Publications, Tokyo, in Press.

'Foreign Direct Investment in Bulgaria', in D. Turnock (ed), *Foreign Direct Investment in East Central Europe and the Former Soviet Union*, Ashgate, Aldershot, 209–23.

PART ONE
THEMATIC STUDIES

Chapter 1

Foreign Direct Investment and Economic Transformation in Central and Eastern Europe

Michael J. Bradshaw

Introduction

'FDI has not assisted the early transition, but it has come as the proof of the success of reform rather than as a catalyst of growth. [Put another way] FDI per capita and growth are positively correlated, but rather than improving the investment climate, FDI requires that the investment climate is already good' (Åslund 2002 p.434). The purpose of this chapter is to examine the relationship between FDI and the processes of economic transformation in CEE (adopting the UN definition to cover the 19 transition economies listed in Table 1.1); thereby setting the scene for the more specialist and regionally oriented chapters that follow. To achieve this aim, the chapter is divided into six sections. Following this brief introduction, I turn back the clock to consider the role that East-West trade played in the Soviet period and the institutions that were created to manage foreign trade under central planning. The next section considers the theoretical relationship between FDI and economic transformation. It is shown that 'internationalisation' was seen as an essential component of 'economic transition' and that substantial inflows of FDI are therefore a measure of success, or as the EBRD terms it 'progress'. The following section turns to consider the actual pattern of FDI in CEE, the region's role in global FDI, the distribution of FDI within the region and its economic significance to the host countries. The final substantive section seeks to explain this pattern of FDI in terms of motivations and barriers. The chapter concludes by identifying key issues for further consideration in other thematic chapters and regional case studies.

East-West Trade, Technology Transfer and Socialist Economic Integration

When considering the processes of economic transformation it is always necessary to point out that when the Soviet era came to an end in 1989 in ECE and 1991 in the Soviet Union, it left behind a legacy upon which the new market democracies had

Table 1.1   FDI inflows into CEE, 1990–2001 ($ mln)

	1990–95 Average	1996–01 Average	1996	1997	1998	1999	2000	2001	Cumulative flow 1996–2001 (%)
Albania	42	91	90	48	45	41	143	181	0.4
Belarus	12	227	105	352	203	444	90	169	1.0
Bosnia & Hercegovina	–	84	2	1	55	149	131	164	0.4
Bulgaria	57	610	109	505	537	819	1002	689	2.7
Croatia	120	1048	516	551	1014	1635	1127	1442	4.7
Czech Republic	947	3779	1428	1300	3718	6324	4986	4916	16.9
Estonia	165	371	150	267	581	305	387	538	1.7
Hungary	1863	2081	2275	2173	2036	1944	1643	2414	9.3
Latvia	116	370	382	521	357	348	408	201	1.6
Lithuania	36	457	152	355	926	486	379	446	2.0
Macedonia	17	148	12	16	118	32	178	530	0.7
Moldova	31	84	24	79	74	37	138	150	0.4
Poland	1396	6869	4498	4908	6365	7270	9342	8830	30.7
Romania	162	1119	263	1215	2031	1041	1025	1137	5.0
Russia	1167	3128	2579	4865	2761	3309	2714	2540	14.0
Serbia & Montenegro	82	231	0	740	113	112	25	165	0.9
Slovakia	147	849	251	220	684	390	2075	1475	3.8
Slovenia	100	269	194	375	248	181	176	442	1.2
Ukraine	206	625	521	624	743	496	595	772	2.8
Total	6666*	22440	13551	19115	22609	25363	26564	27201	100

Source: UNCTAD (2002) pp.305-6.

Note * excludes Bosnia and Hercegovina

to be built. It goes without saying that this legacy represented a shaky foundation upon which to build a market economy, but it is clear, after more than nearly a decade, that those initial conditions matter when it comes to explaining the relative performance of the states of CEE and their constituent regional economies. Although the Soviet development strategy sought to maximise the level of economic self-sufficiency within the 'Soviet bloc' through the development of Comecon and the promotion of the 'Internationalist Socialist Division of Labour', these centrally planned economies (CPEs) were not totally isolated from the global economy, although intra-Comecon trade dominated their foreign economic relations. 'Socialist Economic Integration' (see Marer and Montias 1980), as developed within the framework of Comecon, was based on bilateral agreements that promoted plan coordination between member states. Certain countries were assigned specialisations to deliver goods to the Comecon market; for example, Ikarus (Hungary) specialised in the production of coaches, while various member states invested in the huge pulp and paper complex at Ust'-Ilimsk in Eastern Siberia. However, the reality, in terms of trade turnover, was that Comecon-orchestrated multilateral and bilateral integration was relatively unimportant compared to the supply of energy and raw materials from the Soviet Union and the return trade of manufactured good and agricultural products. In every case, the Soviet Union was the dominant trading partner, as trade relations were used by Moscow to keep the ECECs in line.

Nevertheless, as economic growth faltered in the 1970s, even the FSU saw the benefit in expanding economic relations with the industrially developed capitalist West and East-West trade started to flourish. For the FSU, increased trade with the West was financed by exports of energy and raw materials to Western Europe. The resource-poor economies of ECE lacked this option, and increased imports of Western technology and equipment had to be financed by loans, resulting in an ever-increasing debt burden. The expansion of trade relations with the capitalist West required the development of an elaborate bureaucracy that served to integrate foreign economic relations within the CPEs and also isolated the domestic economy from outside influence. Thus, an expansion of East-West trade did not mean an internationalisation of the domestic economy. The state exercised a monopoly over foreign economic relations and a system of foreign trade organisations (FTOs) was used to link the domestic economy to foreign companies. In the case of the FSU, the Ministry of Foreign Trade controlled the majority FTOs, which were organised on industry lines. The FTOs handled both import and export activities and often the domestic producer had no idea that their goods were being exported and certainly saw little benefit. Equally, Western equipment was often ordered by FTOs without consulting the enterprises to which it was finally delivered. The literature is full of anecdotes of how the equipment was often inappropriate and/or unexpected and often left out in the Siberian winter while the enterprise managers figured out what to do with it.

The expansion of East-West trade and technology transfer generated a substantial body of research in the West as analysts sought to determine the impact of such activity on the productivity of the economies of the Soviet Union and ECE; furthermore, a strategic embargo was maintained to insure that no militarily sensitive technologies were exported. This was often a source of conflict between Western states (including Japan). The consensus that emerged from this research seemed to be that while the actual level of technology transfer was modest, the fact that it was concentrated in specific sectors, such as the automotive and chemical industries, meant that it had a proportionally greater impact upon the productivity of those industries (OECD 1984; Parrott 1985). However, the scale of that impact, and the diffusion of benefits into the wider economy, was limited by the very nature of the planned economy. In other words, technology transfer could partially compensate for the technological backwardness of key sectors, but it could not bring about a transformation of the economy, only radical reform could do that. In the late 1980s, Mikhail Gorbachev attempted such a reform and the opening-up of the Soviet economy was an important component of Perestroika. Some ECECs, principally Hungary, had already introduced reforms to allow foreign companies to create joint ventures with domestic enterprises. In 1987 the Soviets allowed the creation of joint ventures. The hope was that increased foreign involvement in the domestic economy could promote increases in efficiency. Despite the euphoria of the time, this was a forlorn hope; the centrally planned economy had proved unsustainable and the collapse of Soviet authority prompted the revolutions across the ECE in 1989 and eventually the collapse of the Soviet Union itself in 1991.

For the purposes of the current discussion, it is important to note that there was trade between East and West during the 1970s and 1980s. For the most part, Western companies were kept at arm's length by the agencies of the central planning system, but companies such as the Italian car maker Fiat, played a major role on the development of the automotive industry in Poland and Russia. The supply of equipment and machinery also meant that Western companies had some knowledge of industrial enterprises in the East. As only the high priority enterprise would have been supplied with foreign technology, this also meant that Western companies were familiar with the most modern enterprises (the ones worth buying). However, the vast majority of domestic enterprises in CEE had been totally isolated from the international market and its competitive pressures: instead they had benefited from captive domestic markets in a shortage economy. For example, concepts such as marketing and customer care were totally alien. This made domestic producers a soft target for import competition and many eagerly sought a foreign partner to help them adjust to the new realities of an emergent market economy. The heady days of the early 1990s saw many a Western business executive sent east to look for rich pickings, and although most returned disappointed, some positioned themselves to reap the benefits of the sale of the century, the privatisation of the formerly state-owned economies.

Economic Transformation and Internationalisation: A Theory

Just as the Bolsheviks lacked a blueprint for converting the emergent market economy of Tsarist Russia into a socialist economy, so the new states of CEE lacked a blueprint for transforming their failing CPEs into sustainable market economies. The international community, principally in the guise of the USA and EU, the IMF and the World Bank, with its newly created European Bank for Reconstruction and Development (EBRD), came to the rescue with a package of policy measures that became known as the 'Washington Consensus' (Smith 2002; Williamson 1993). Not surprisingly, the aim of this policy prescription was to create a set of 'fully functioning market economies' as understood in Washington and, to a lesser degree, Brussels. The early 1990s saw heated debate among economists as to the best way to 'make a market economy', consultants travelled to the ECE capitals and Finance Ministers and officials travelled west. Seemingly vast sums of money were spent on so-called 'technical assistance' to aid in the process of transformation, or transition as it was called at the time. A sub-discipline of 'transition economics' soon emerged and with it the 'four pillars' of transition were identified, the policy prescription that would turn consensus into success transition. These four pillars were: privatisation, liberalisation, stabilisation and internationalisation. Within this paradigm there was considerable debate as to the correct sequencing of policy measures and the rate at which reform should be introduced, the latter usually seen as an alternative between shock therapy and gradualism. Many of the more technical aspects were simply overrun by events. In many instances reform progressed as quickly as politically desirable and/or possible. The international financial institutions monitored events and rewarded progress. This overly prescribed notion of post-socialist development was not without its critics (Stark 1992); but it soon became the dominant trope of transition and by the mid-1990s possible membership of the EU was linked to the creation of a 'fully-functioning market economy'.

There is neither the space nor the need to delve deep into the technicalities of transition economics (Åslund 2002; Lavigne 1999), for the purposes of the present discussion, the pillars of privatisation and internationalisation are most relevant. Under Soviet-style socialism the state owned the means of production whereas in the market economy the vast majority of firms are privately owned. Therefore, to create a market economy it is necessary to privatise the state-owned enterprises (SOEs). At the same time, one must create conditions to foster new private enterprise, usually in the form of small- and medium-sized enterprises (SMEs). The processes of privatisation and SME formation are closely related as SMEs were often spun-off SOEs in the run-up to privatisation (World Bank 2002 p.67). Across the region a variety of different models of privatisation were tried and the process is far from complete (Åslund 2002 pp.253-303). Small-scale privatisation proved relatively simple; while the privatisation of the large industrial SOEs has been most problematic. That said, privatisation has been crucial to the attraction of FDI, as it has provided an opportunity for foreign companies to obtain equity ownership, and often control, of SOEs. As we shall see, the timing and openness of the privatisation process is a crucial factor in explaining the dynamics and geography of FDI in CEE. However, the purchase of shares during privatisation is far from the only vehicle available for FDI in transition economies.

As we have noted, the centrally planned economy was closed to FDI, it was only late in the Soviet period that joint ventures were permitted. The policy prescriptions of the Washington Consensus required the transition economies to open up their economies to foreign trade and investment. This presented a problem. The poor state of domestic industry meant that local producers could not compete against imported goods. Increased import competition contributed to the 'transitional recession' that accompanied the collapse of the centrally planned economy. The orthodox economic explanation is that internationalisation is good as it opens the domestic economy to increased competition and this promotes competition. This is all well and good if domestic producers have the capacity to respond to that competition, in most CEE states they did not. Western interests demanded market access, but local industry wanted protection. In some instances, joint ventures enabled domestic producers to find a strategic partner who could provide access to the capital and know how needed to meet the competition. Many enterprises simply failed. The joint venture provided the means for initial forays into the new markets of CEE, but as the reform process gathered momentum, privatisation and legal reforms provided the opportunity for portfolio investment and the establishment of 100% foreign-owned subsidiaries. The latter, often linked to the construction of greenfield facilities, required that the legal framework be sufficiently robust to protect the interests of the investor. In simple terms, a foreign company would not build a new facility unless it was certain that its ownership of that investment was secure and that it could repatriate its profits. Thus, as we shall see, the stability of the domestic business environment soon became a major factor in determining the relative attractiveness of the various countries. In some states there was, and still is, a strong domestic lobby resisting foreign investment and ownership. In Russia, foreign companies were largely shut out of the privatisation process and powerful domestic lobbies sought to block the introduction of legislation that would favour the interests of foreign companies. By comparison, the majority of ECECs, and in particular the eight 'first-wave' EU accession states (Czech Republic, Estonia, Hungary, Latvia, Lithuania, Poland, Slovakia and Slovenia – scheduled to join on May 1, 2004) are anxious to attract as much as FDI as they possibly can, seeing it as an essential source of capital, technology and business know-how. So who has succeeded and who has failed?

Economic Transformation and Internationalisation: The Reality

"With the collapse of central planning, that is, once external commerce has finally become subordinated to economic considerations and state monopoly over foreign trade was abolished, the EU quickly emerged as the largest trading partner" (Kaminski 2001 p.4). However, unsurprisingly, prior to the collapse of the central planned economies, the CEECs were not major recipients of FDI. According to

UNCTAD (2002 p.7), between 1986 and 1990 they only accounted for 0.1% of global FDI flows; in fact until 1990 total FDI inflows into the region were less than $1bln. The collapse of the Soviet system and the subsequent transition prompted a relative investment boom (Table 1.1 and Figure 1.1). In 1995 FDI inflow exceeded $14bln and by 2001 had reached $21bln. Between 1995 and 2001 the region's total inward FDI stock increased four-fold from $40bln to $160bln (UN 2003 p.1). In relative terms, the region's share of global FDI flows increased from 2.2% in the 1991–92 period to 3.7% in 2001 (UNCTAD 2002 p.7). In fact, in 2001 FDI into CEE showed a growth of just 2% against a backdrop of a decline in global FDI inflows of 40% (Ibid p.68). Thus, it is clear that transition in ECE attracted the interest of global investors and that the region fast became a favoured location for FDI. Hunya (2002 p.4) quotes an UNCTAD press release in 2001 concerning a survey of 129 major TNCs. According to the survey, the area with the best prospects was CEE, with 60% of the respondents expecting increasing investment. However, as is clear from Table 1.1 and Figure 1.2, the majority of FDI inflow is concentrated in a relatively small number of countries.

The dynamics and geographical distribution of FDI are closely tied to the timing of the privatisation process and the 'progress' made by the various states in terms of transition to the market and the process of EU accession. Overall, four countries account for the majority of FDI: Czech Republic, Hungary, Poland and Russia. At the beginning of the decade Hungary was the lead recipient because it opened up to investors more quickly – having the fact that it had already experimented with joint ventures during the late Soviet period – and it started privatisation earlier. Thus, the opportunities for foreign companies were greatest there. As the other countries launched their privatisation programmes so they captured a greater share of FDI, with Poland leading the way. Most recently, their privatisation programmes have neared completion and new countries have emerged as front-runners: for example, in the late 1990s Croatia and Slovakia started to capture a larger share of FDI. As the process of EU enlargement gathered pace and the number of accession states was increased to eight, so the net has widened. Whereas between 1996 and 2001 the eight 'first wave' accession states captured 67.2% of the total FDI inflow, the level in 2000 was 73.0% and in 2001 70.8%. Outside of these states, Russia is the only significant recipient of FDI. However, Russia can be considered a special case. In relative terms, given the size of its economy and its resource wealth, Russia has badly under-performed when it comes to attracting FDI (Bradshaw and Swain 2003). A difficult business environment, the impact of the 1998 financial crisis and a hostile domestic business community all contribute to this poor performance; but there are signs that this is about to change. As noted earlier, foreign companies were effectively shut out of Russia's privatisation process, but the economy is now showing signs of recovery: under President Putin the business environment is improving and investor confidence is growing.

On February 11 2003, the oil company BP announced that it had reached agreement with Alfa Group and Access-Renova to combine their interests in Russia to create the country's third biggest oil company (BP 2003). According to the BP

Figure 1.1   Dynamics of FDI in CEE, 1990–2003

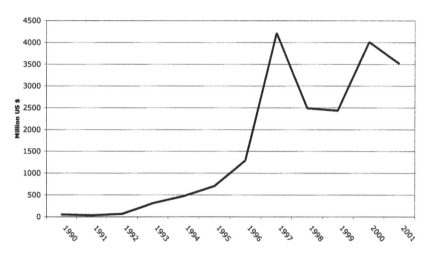

Source: http://unctad.org/fdi.

Figure 1.2   Top ten recipients in FDI in CEE, 2000 and 2001

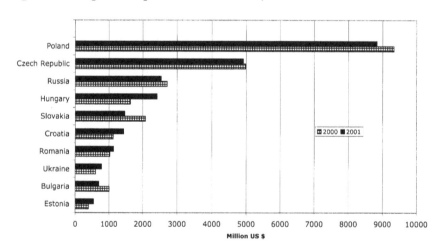

Source: UNCTAD (2002) pp.305–6.

announcement, the company will pay $6.75bln to AAR for its 50% stake in the company to be known as TNK-BP. Table 1.1 puts the scale of this deal in perspective. In recent years total FDI into Russia has averaged $2.0-2.5bln, but if this continues in 2003 the BP investment is added, the FDI into Russia could come close to $10bln. It is possible that this purchase will not actually be recorded as FDI since the company being purchased is actually an international affiliate registered in London, but the new company will be a Russian entity and the deal represents a major commitment to the Russian market. Thus the question remains, will the BP investment trigger a substantial inflow of FDI into Russia? Already two major Russian oil companies, Yukos and Sibneft, have merged to ward off possible acquisition by foreign oil majors. In an analysis of the BP deal, Sundstrom (2003) concluded that the Russian government's own estimate of an annual FDI inflow of $6-7bln over the next three or four years was realistic. Therefore, as the privatisation process concludes and EU enlargement proceeds, there may be a relative shift in investor interest further eastwards. As Sundstrom (2003 p.3) puts it: 'while there is no argument that FDI flows to Russia have been small, there is also the perspective that the quite phenomenal experience with FDI inflows in Central Europe [i.e. the northern parts of ECE] is rather exceptional for the region, and has a lot to do with the prospect of being candidates or EU membership and integration'.

The absolute value of FDI inflow does not convey the relative importance of FDI to a host economy. Table 1.2 shows the relative importance of FDI in terms of percentage of GDP. From these data it is clear that there are large variations in the importance of FDI to national economies, both in terms of the CEE average and the world average. In addition, in every instance the relative importance of FDI increased between 1995 and 2000. This suggests that economic activity related to FDI grew far more quickly than domestic activity. One would expect to find high levels of dependence in small economies such as the Baltic States, but the scale of dependence in Czech Republic and Hungary does seem exceptionally high suggesting that the attraction of FDI has been a cornerstone of domestic economic policy. Another way of assessing the relative importance of FDI is to measure its share of gross capital formation, in other words the share of foreign capital investment in total investment. Initially, only the Baltic States and Hungary showed high shares in the 20% range, but more recently Bulgaria, Czech Republic, Poland and Slovakia have exhibited a high share of FDI in gross capital formation (UN 2003 p.2). Hunya (2002 pp.9-15) provides additional evidence of the high levels of FDI penetration among certain CEECs. In 1999, in Hungary foreign investment enterprises (FIEs) accounted for 72.9% of equity capital, 46.5% of employment, 82.2% of investment, 73.0% of sales and 88.8% of export sales. In Poland FIEs accounted for 50.5% of equity capital, 29.4% of employment, 63.1% of investment, 49.0% of sales and 59.8% of export sales. Thus, the majority of FDI has been concentrated in a small number of countries for whom it has been a critical component of economic recovery. Of course, more modest levels of FDI are also of great significance to smaller states. For example, on a per capita basis FDI in

Table 1.2   Inward FDI stock as a percentage of GDP in CEE, 1995 and 2000

Country/region	1995	2000
World average	10.0	20.0
CEE average	5.4	18.9
Albania	8.7	15.4
Belarus	0.5	11.9
Bosnia and Hercegovina	1.1	8.1
Bulgaria	3.4	26.4
Croatia	2.5	27.1
Czech Republic	14.1	42.6
Estonia	14.1	53.2
Hungary	26.7	43.4
Latvia	12.5	29.1
Lithuania	5.8	20.6
Macedonia	0.7	10.9
Moldova	6.5	35.7
Poland	6.2	21.3
Romania	3.2	17.7
Russia	1.6	7.7
Serbia and Montenegro	2.7	15.6
Slovakia	4.4	24.2
Slovenia	9.4	15.5
Ukraine	2.5	12.1

Source: United Nations (2003) p.3.

Estonia matches that of Hungary. So who is investing in ECE and which sectors of the economy are the major recipients?

Most FDI comes from member states of the EU and the USA. The importance of particular EU states in a given ECEC is a function of proximity. EU accession states have been the focus of EU investment activity: the share of EU states is above 80% for Czech Republic, Estonia, Hungary, Poland and Slovenia and close to 80% in Slovakia (UN 2003 p.3). There are also some strong neighbourhood effects: in Estonia's FDI in 2001, Finland accounted for 25.4% and Sweden 39.5% (Ibid p.6). The USA is currently a significant investor in Croatia (23.6%) and Russia (34.0%). In 2000 the EU was a source of 27.9% of Russia's FDI. There are also marked differences between states in terms of the sectors of the economy are the major beneficiaries of FDI. Since most CEECs lack natural resource wealth, FDI has tended to be attracted to the manufacturing and service sectors. On the other hand, Russia and some of the other CIS states such as Azerbaijan and Kazakhstan have substantial mineral wealth that has attracted foreign investors (Heinrich et al. 2002; Meyer and Pind 1999). According to the UN (2003 pp.3-5), FDI in 2000 in the manufacturing sector represented more than one-third of inflows in Czech Republic

(38.1%), Hungary (36.8%), Poland (39.3%), Slovakia (53.2%) and Slovenia (40.6%). In most CEECs the privatisation of the banking sector, telecommunications and utilities has boosted the role of the service sector. Within the manufacturing sector there is a certain degree of specialisation. For example, the automotive sector is particularly significant in Czech Republic, Hungary and Poland where foreign penetration exceeds 80%. This has occurred both through the acquisition of domestic producers and greenfield investments (Bradshaw and Swain 2003; Pavlínek 2002). According to Hunya (2002 p.11), in Hungary in 1999 FEIs accounted for 73% of the country's manufacturing sales and employed 46% of the labour force in the manufacturing sector. In sum, it is clear that the collapse of the CPEs and the subsequent opening up of the transition economies has presented a major opportunity for international capital. However, the process of internationalisation has been highly selective and only certain countries and economic sectors have been the target for FDI. To understand this pattern it is necessary to consider the motivation of both the host governments and the foreign investors and potential barriers to foreign investment. This is the task of the next section.

Motivations and Barriers

The academic literature is replete with theories that seek to explain the process of FDI and why certain countries, regions and sectors are favoured over others. Rather surprisingly, there appears to be relatively little published work on these issues in the context of economic transition in CEE: see for example the reference list in UNCTAD (2002). There are various research papers from international institutions, such as the World Bank and the EBRD, that seek to link FDI to progress in transition. The academic literature, such that it is, seems divided between country studies that mention FDI and industry case studies that examine one sector in detail, such as Pavlínek (2002) on the automotive sector. A review of the wider transition economics literature and the more specialist FDI literature suggests reasons as to why a CEEC might want to attract FDI and motivations on the part of the investor. All transition economies faced the task of restructuring their economies to better meet the needs of the new market system and the demands of the market, both domestically and internationally. In theory at least, investment by foreign companies – be it via privatisation, post-privatisation acquisition or the construction of greenfield facilities – can assist in this process.

First, FDI provides a source of capital above and beyond what can be generated by the domestic economy. In the early stages of transition this was crucial as transitional recession left the state and the emergent private sector short of capital. Second, FDI can be a vehicle for technology transfer, both embodied technology in the form of machinery and equipment and disembodied technology in terms of production and management know-how. Third, the presence of foreign companies can help improve corporate governance and expose domestic companies to international standards e.g. through the subcontracting process. Fourth, FDI provides the state with income from the privatisation process and subsequently from taxation. However, there has been a tendency in CEE for the various governments to offer incentives (usually tax holidays) to encourage FDI and this may serve to reduce the level of income generated. Fifth, research suggests that the activities of FIEs increase the level of foreign trade in a host economy. Deichmann (2001) demonstrates that there is a positive relationship between the levels of international trade and the scale of FDI across the transition economies. Much of this trade is actually internalised within supply chain of the foreign company. Kaminski (2001 p.36) reports that: 'between 1993 and 1998 the total value of exported parts and components from CEECs to the EU grew almost fivefold (4.8 times) while that of manufactures almost threefold (2.8 times). Imports of parts and components grew four times, while those of manufactures ... increased less than three times (2.8 times) over the same period'. Sixth, all of these benefits are supposed to generate spin-offs and multipliers in the domestic economy beyond the FIE itself. These range form the increased consumer demand generated by the workforce through to pressures to improve the level of technology and managerial skills among indigenous enterprises, both through subcontracting and a positive demonstration effect. Finally, success in attracting substantial levels, for reasons discussed below, is seen as evidence of successful economic transition and the attainment of international standards of corporate governance.

There are very few studies that have critically appraised this long list of benefits, much less the assumption that free trade is good for everyone (for it tends to be assumed that FDI is a good thing). However, Smith and Pavlínek (2000 p.230), in their research on FDI in Czech Republic and Slovakia conclude that inward investment 'may not be the godsend that many of the proponents of FDI in transition suggest'. They identify five tendencies that suggest limited positive impacts: with the exception of Škoda, inward investment has had only limited effect on local employment; skills upgrading is limited; while the introduction of new management practices is a significant factor, it has not enhanced the democratisation of the workplace; inward investment has resulted in the emergence of substantial differences in intra-regional wage levels between firms with FDI and the local economy; and consequently FDI has had a limited impact on local supply chains. The country case studies presented later in this volume shed more light on the specifics of FDI and should enable a critical evaluation of the positive and negative impacts of FDI on the host economies.

Foreign multinationals are making investments in CEE to promote their own business interests. The literature suggests three possible motivations. First, gaining access to new sources of resource supply – not a prime motivation in ECE, but significant in Russia and elsewhere in the CIS. Second, gaining access to new markets for products, since the CEECs represent a large market for consumer goods that were in short-supply during the socialist period and for which there is now – as personal incomes recover – a large pent-up demand, especially for cars and electronic and white goods. Third, seeking efficiency savings through reduced production, principally labour costs; given that the CEE workforce is well educated and relatively cheap by EU standards and can be marshalled to produce goods for both the domestic market and export. The prime motivation of a given multinational will have a direct influence upon the decision as to where to locate, both in terms of which country and where in the chosen country. Proximity to the EU seems to be a key factor. Kaminski (2001 p.32) notes that 'a 60 mile line from the border from the border of the EU captures the whole territory of Slovenia, around 50% of the territory of the Czech Republic, one fourth of the territory of Slovakia and an area almost reaching Budapest: the capital and industrial centre of Hungary'. However, geography alone does not explain the pattern of FDI described earlier, for there are specific barriers which limit the level of FDI in particular countries.

Table 1.3 presents the average per capita level of FDI in a selection of CEECs, together with two measures of the investment climate: the World Economic Forum's index of global competitiveness, which is a ranking of countries, and the EBRD's transition scores (Lankes and Venables 1997 p.561). The table suggests that there is a close relationship between the transition scores and perceived investment risk. These measures are then used to produce an overall ranking. There would seem to be a positive relationship between the level of FDI and a positive investment environment. Put another way, successful transition fosters higher levels of FDI.

Table 1.3   Indicators of the general investment climate

	FDI $ per capita 1997–2001 average	World Economic Forum Index	EBRD* Transition Score 2002	Investment Climate Ranking @
Estonia	287.9	26	4	1
Hungary	202.8	29	4	2
Czech Republic	412.2	40	4	3
Slovakia	175.8	49	4	4
Lithuania	140.2	36	4–	5
Slovenia	128.6	28	4–	6
Poland	189.9	51	4–	7
Latvia	152.6	44	4–	8
Croatia	244.4	58	3+	9
Bulgaria	86.9	62	4–	10
Romania	57.5	66	3+	11
Russia	24.6	64	3	12
Ukraine	13.0	79	3	13

* Range from 0 (no progress) to 4+ (fully functioning market economy) and measures 'progress' in privatisation, liberalisation, enterprise performance and financial sector reform. @ A simple sum of the ranks on the three indicators.

Source: EBRD (2002); Fankhauser and Lavric (2003) p.21; World Economic Forum (n.d.).

However, the relationship between FDI and growth is not that straightforward. As Hunya (2002 p.2) notes 'growth can be generated by FDI through additional investment measures and the transfer of technology and capabilities, as well as through improved access to export markets. On the other hand, foreign investors react positively to the consolidation of market-economy rules and resumption of economic growth'. It has already been noted that Russia has failed to attract FDI relative to the size of its domestic market and resource wealth (Bradshaw 2002). Survey research (Ahrend 2000; Westin 1999) makes it clear that there are a host of problems that deter foreign investors: a high tax burden and an unfair tax system; a lack of international accounting standards; unclear and overly bureaucratic systems of regulation; crime and corruption; inadequate protection of property rights; and problems with customs and checkpoints. Table 1.4 presents an index that seeks to assess the level of corruption across the globe: the index is a score from between 10 (highly clean) and zero (highly corrupt). For the sake of comparison, the UK was ranked 10th with a score of 8.7 and the USA 16th with a score of 7.7. There is considerable variation among the CEECs with most of the major recipients of FDI ranked more highly. However, this survey does suggest that all the CEECs have considerable progress to make before their reach the levels of corruption risk found in the EU. That FDI has tended to concentrate in the more economically progressive and politically stable states is no surprise, but one should also remember that a number of them have been embroiled in ethnic conflicts and remain politically

Table 1.4 Transparency International's Corruption Perception Index, 2002

Country	Global Rank	CPI 2002 Score	Survey Used	Standard Deviation	High-low Range
Slovenia	27	6.0	9	1.4	4.7–8.9
Estonia	29	5.6	8	0.6	5.2–6.6
Hungary	33	4.9	11	0.5	4.0–5.6
Belarus	36	4.8	3	1.3	3.3–5.5
Lithuania	36	4.8	7	1.9	3.4–7.6
Bulgaria	45	4.0	7	0.9	3.3–5.7
Poland	45	4.0	11	1.1	2.6–5.5
Croatia	51	4	4	0.2	3.6–4.0
Czech Republic	52	3.7	10	0.8	2.6–5.5
Latvia	52	3.7	4	0.2	3.5–3.9
Slovakia	52	3.7	8	0.6	3.0–4.6
Russia	71	2.7	12	1.0	1.5–5.0
Romania	77	2.6	7	0.8	1.7–3.6
Albania	81	2.5	3	0.8	1.7–3.3
Ukraine	85	2.4	6	0.7	1.7–3.8
Moldova	93	2.1	4	0.6	1.7–3.0

Source: Transparency International (2002).

unstable to the extent that they are virtual no-go areas. Furthermore, this instability has clearly had a negative impact upon the attractiveness of neighbouring states. In a very real sense, the process of FDI has picked out the relative winners and has assisted them in their transformation to the market; and in doing so has probably helped those countries that are probably least in need. There are still large parts of post-socialist Europe that remain plagued by political instability and who remain impoverished and dependent upon international aid and assistance rather than foreign investment.

Conclusions

It is clear that the collapse of the Soviet Union and the processes of economic transformation that have followed have presented international capital a major opportunity to seek out new markets and to incorporate new locations into their global production networks. The pattern of FDI in CEE also makes it clear that the process of EU enlargement has imposed a particular spatial logic upon these processes of internationalisation and incorporation that favour certain states, and cities and regions within those states. This chapter has painted a picture with a very broad brush; it is for the subsequent chapters to examine particular issues, industries, countries and regions in more detail. The relationship between economic transformation and FDI is not straightforward. It is clear that the attraction of FDI is an important part of 'progress', but it is also clear that individual states, and their constituent regions, have first to create the right conditions to attract FDI. Just as elsewhere in the world, CEECs are in competition with one another to attract FDI. They have to offer incentives and provide facilities. The underlying assumption is that the costs paid to attract FDI are more than outweighed by the benefits that accrue to the host economy. But is that always so, for Smith and Pavlínek (2000) suggest otherwise. As we have seen, the process of internationalisation is highly selective, for it benefits some and not others; and in doing so it contributes to the process of uneven development so that existing inequalities may be amplified. The chapters that follow provide some of the further evidence required to reach a judgement on the impact of FDI on the regional development of region.

References

Ahrend, R. (2000), 'Foreign Direct Investment in Russia-Pain without Gain? A Survey of Foreign Direct Investors', *Russian Economic Trends*, 9(1), 26–33.

Åslund, A. (2002), *Building Capitalism: The Transformation of the Former Soviet Bloc*, Cambridge University Press, Cambridge.

BP (2003), *BP Creates Strategic Partnership in Russia*, BP Press release, February 11 <www..bp.com/centres/press/p_r_details_p.asp?id=963>.

Bradshaw, M.J. and Swain (2003), 'Foreign Investment and Regional Development', in

Bradshaw, M.J. and Stenning A.C. (eds), *The Post-Socialist States of ECE and the FSU*, Pearson, Harlow.

Deichmann, J.I. (2001), 'Distribution of FDI among Transition Economies in CEE', *Post-Soviet Geography and Economics*, 42, 142–52.

EBRD (2002), *Transition Report*, EBRD, London.

Frankhauser, S. and Lavric, L. (2003), *The Investment Climate: Joint Implementation in Transition Countries*, EBRD Working Paper 77, London.

Heinrich, A., Kusznir, J. and Pleines, H. (2002), 'Foreign Investment and National Interests in the Russian Oil and Gas Industry', *Post-Communist Economies*, 14, 495–507.

Hunya, G. (2002), *Recent Impacts of FDI on Growth and Restructuring in Central European Transition Countries*, WIIW Research Report 284, Vienna.

Kaminski, B. (2001), *How Accession to the EU has affected External Trade and FDI in Central European Economies*, World Bank Policy Research Working Paper 2578, Washington, D.C.

Lankes, H.P. and Venables, A.J. (1997), 'FDI in Eastern Europe and the FSU: Results from a Survey of Investors', in S. Zecchini (ed), *Lessons from the Economic Transition: CEE in the 1990s*, Kluwer and OECD, London, 555–61.

Lavigne, M. (1999), *The Economics of Transition: From Socialist Economy to Market Economy*, Macmillan, Basingstoke.

Marer, P. and Montias, J. M. (eds) (1980), *East European Integration and East-West Trade*, Indiana University Press, Bloomington Ind.

Mayer, K.E. and Pind, C. (1999), 'The Slow Growth of FDI in the Soviet Union Successor States', *Economics of Transition*, 7, 201–14.

OECD (1984), *East-West Technology Transfer*, OECD, Paris.

Parrott, B. (ed) (1985), *Trade, Technology and Soviet-American Relations*, Indiana University Press, Bloomington Ind.

Pavlínek, P. (2002), 'Restructuring the CEE Automobile Industry: Legacies Trends and Effects of FDI', *Post-Soviet Geography and Economics*, 43(1), 41–77.

Smith, A. (2002), 'Imagining Geographies of the "New Europe": Geo-Economic Power and the New European Architecture of Integration', *Political Geography*, 21, 647–70.

Smith, A. and Pavlínek, P. (2000), 'Inward Investment, Cohesion and the "Wealth of Regions"', in J. Bachtler, R. Downes and G. Gorzelak (eds), *Transition Cohesion and Regional Policy in CEE*, Ashgate, Aldershot, 227–42.

Stark, D. (1992), 'The Great Transformation? Social Change in Eastern Europe', *Contemporary Sociology*, 21, 299–304.

Sundstrom, N. (2003), *FDI and the BP-TNK Deal: the Broader Picture*, Citibank: Schroder Salomon Smith Barney, Economic and Market Analysis, London, 14th February. Transparency International (2002), *Transparency International Corruption Perceptions Index 2002*, Transparency International Secretariat, Berlin <www.transparency.org>.

UN (2003), *World Investment Directory Volume III: Central and Eastern Europe 2003*, UN, New York.

UNCTAD (2002), *World Investment Report 2002*, UN, New York.

Westin, P. (1999), 'FDI in Russia', *Regional Economic Trends*, 8(1), 36–43.

Williamson, J. (1993), 'Democracy and the "Washington Consensus"', *World Development*, 21, 1329–36.

World Bank (2002), *Transition: The First Ten Years*, The World Bank, Washington, D.C.

World Economic Forum (n.d.), World Economic Forum Index: a measure of global competitiveness <www.weforum.org>.

Chapter 2

Technologies of Transition: Foreign Investment and the (Re-)Articulation of East Central Europe into the Global Economy[1]

John Pickles and Adrian Smith

Introduction: Foreign Investment and Technologies of Transition

'Foreign direct investment (FDI) leads to economic growth, and economic growth leaves to poverty reduction: this line of reasoning forms the foundation of contemporary thinking about FDI' (CUTS 2003 p.1). These words help in understanding why attempts to (re-)integrate post-socialist societies into the global economy lie at the heart of the various projects of political and economic transition in ECE. We highlight this as a process of re-integration, for two main reasons. First, prior to the development of Soviet-style state socialism, many ECE societies were closely integrated with the global economy. Even in the 19th century the 'second serfdom' of the region was predicated on integration into a wider European economy through the export of largely agricultural products to the West (Berend and Ránki 1974; Dunford 1998). During the period of Soviet-style state socialism, these societies were always partially and unevenly integrated into the wider global economy. This was particularly the case in the former Soviet Union, where primary commodity and energy exports underpinned late-Soviet economic policy thinking, but is was also true in ECE as Western European firms, especially from Germany, sought lower cost production through outsourcing to SOEs in the region under outward processing trade arrangements (Begg et al. 2003).

Since 1989, this reintegration has taken many forms and has been part of the myriad ways in which political-economic policy has scripted a series of 'technologies of transition' (Smith 2002; Smith and Pickles 1998). Central to this political-economic globalisation has been the opening of these societies to FDI. This has occurred alongside other forms of reintegration, including legislative and political alignment for EU accession, production for new, notably EU export markets (Begg et al. 2003, Smith et al. 2003), convergence of community and cultural consumption, and expanding circuits of periodic and permanent migration.

FDI has played a very particular and important role in the globalisation of ECE economies. Indeed, it has been seen as a paradigmatic moment expressing the hopes and aspirations of the post-socialist countries as they sought to shift away from the relative autarky of Soviet-style economies. FDI was perceived as a way of injecting fresh capital, know-how, technology and economic practices into economies perceived as being 'exhausted' and redundant. FDI was upheld as a central element in the kinds of economic strategies that states in the region – both national and local – should pursue to ensure that their exhausted economies were returned to some form of strength. Of course, this refocusing on the importance of FDI occurred at the same time when economies the world over were also emphasising the importance of supply side initiatives such as the attraction of inward investment. The result was two-fold. The ECECs became integrated into an increasingly global discourse of neoliberal, supply side policy making, and they became part of a virtual zero-sum game in which competition to attract FDI operated within an emerging context of winners and losers. This chapter focuses on these political and cultural economies of FDI and the ways in which it is reconfiguring the regional economies of the region.

Political Economies of FDI

Thirteen years of post-socialist transformation in which FDI has played a key discursive – if not universally significant material – role have generated a large literature on the role of inward investment in regional reconstruction. It is not our purpose to review this literature in its fullness here. Rather, we wish to highlight some of the key and emergent concerns over the role that FDI has in transition. In other words, we wish to begin to problematise the discursive construction of the centrality of FDI to 'successful' transition. This, however, is not to suggest that FDI has been irrelevant and wholly problematical. Indeed, in many cases FDI has had a dramatically transformative role in the reconfiguration of political-economic life. We begin with these transformations, but then turn to a problematising of the claims made for them. We suggest that the universalising role that FDI has been given – as the all-powerful force for economic reconstruction – could never have borne fruit. In contrast we focus on the geographically uneven nature of these processes; on the islands of capitalist development and dynamism, alongside highly uneven and contested experiences of FDI-led transformation.

We begin by briefly evaluating the particular position of the economies of ECE within this context to shed some light on the relative position of the region in global flows of capital. Indeed, in 2002, ECECs experienced faster rates of annual growth (3.7%) than the OECD countries (global growth rates were 3.1%), and FDI into the region has continued to increase throughout the 1990s and early years of the 21st century (EBRD 2003). FDI inflows into the 19 ECECs (most coming from the EU) have continued to increase in total amounts year over year, from $25.1bln in 2001 to $27.9bln in 2002 (with expected net FDI into the region in 2003 of $28.1bln)

(EBRD 2003). Inflows are, of course, unevenly distributed, with declines between 2001–2002 in ten countries and increases in nine (UNCTAD 2003). In 2002, 60% of total FDI inflows measured in terms of the value of privatisation related sales or assets ($20bln) went to just five ECECs: Czech Republic, Hungary, Poland, Slovakia, and Slovenia, with Czech Republic receiving more than $9bln of this, and Slovakia and Poland about $4bln each. Hungary received less than $1bln for the same period (WIIW 2003), while since 1989 more than 80% of all FDI has gone to Czech Republic, Hungary and Poland (Gradev 2001 p.3).

The type of investment is also important, with the bulk of new FDI being related to privatisation of former SOEs (UNCTAD 2001 p.1). Such differences in national and sectoral patterns of FDI reflect broader geographies of uneven development, economic marginality in Europe, and the central role of the 'global triad' of Europe, North America and Japan/East Asia in FDI flows (Dicken 2003). In particular, they illustrate the strategic importance of specific regions and sectors in ECE as primary recipients and beneficiaries of FDI in addition to the specific role that FDI in ECE has played in reworking divisions of labour and production networks in the wider European economy. Where privatisation has been largely completed (such as in Hungary) FDI is focused predominantly on new greenfield investments and crossborder mergers and acquisitions. According to UNCTAD (2001) this pattern is likely to dominate FDI inflows into other countries (such as Czech Republic and Poland) in the next few years, while privatisation purchases are likely to continue to predominate in those countries where privatisation has only partially occurred (such as the Russian Federation). FDI also has unevenly distributed consequences. In part, these emerge from the different goals that key actors have in pursuing FDI. In particular, FDI has been perceived as playing different roles in economic transformation by investors and recipients.

Investors

The motives and consequences of FDI in ECE (particularly from the EU) are geographically and sectorally complex. Many foreign companies invest to take advantage of economic opportunities in privatisation offerings, wage gaps, potential markets, and/or differential labour and environmental regulatory protections. Through their investments and subsequent firm restructuring these companies may improve enterprise efficiency, introduce new forms of corporate governance, create stable jobs, and increase incomes for workers and managers alike, although the balance among these depends in large part on the kinds of strategy they adopt. Based on extensive surveys, Gradev (2001) has suggested three primary strategies for FDI in ECE: market-seeking investment, efficiency oriented investment, and low-wage seeking.[2] First, market-seeking strategies linked to ECE's new and growing markets have been pursued primarily by SMEs from EU member states, along with large MNEs such as Danone, Interbrew, Nestlé and Shell; leading supermarket chains such as Carrefour and Tesco; and pharmaceutical, agricultural input, and mobile telephone firms (Gradev 2000 p.9).

Second, efficiency-oriented investments have sought factor-cost advantages, with a direct relation to low labour costs. Two primary kinds of investments seem to have emerged: (i) large MNEs (in automobiles, electronics, chemicals, plastics and rubber, and pharmaceuticals) whose locational strategies seem determined primarily by their position in the value chain, with labour intensive industries emerging across the region, and capital-intensive industries exhibiting a clear preference for the larger urban markets; and (ii) SMEs (in clothing, shoes, and leather) 'operating on the basis of outward processing and integrated in the supply and marketing chains of large European department stores and major garment and sportswear producers' (Gradev 2001 p.11; see also Begg et al. 2000, 2003; Pickles 2000; Smith et al. 2002). Third, low-wage production has generally been seen to be the primary reason for much FDI in ECE (Pickles 2000), particularly in view of the enormous wage gaps between Western Europe and ECE, and even among countries within the region (e.g. Bulgarian and Romanian wages are now three times lower than those of Hungary). Nonetheless, wage differentials should not be over-emphasised. Gradev's (2001 p.13) research indicates that investors adopt 'a much more complex process of strategic decision making and investment motivations' than a low wage strategy would suggest, including political climate and connections with ruling elites, opportunities for special concessions in the privatisation process, previous business links, subsidies for energy and other inputs, existing levels of technical and managerial know-how, geographical location and proximity to transport hubs and raw materials, and broader infrastructural concerns (such as telecommunications and their reliability). Gradev (2000) found similar patterns to those we are also finding in the clothing industry (Begg et al. 2000, 2003; Pickles 2002; Smith 2003). While cheap labour is an important – even determining – factor, labour quality, educational levels, and existing infrastructure are all important. Many firms also avoid the lowest cost labour pools because they are found furthest from larger urban areas, transport routes, and may have deeper problems associated with high unemployment (Gradev 2001 p.13).

Recipient States

For many of the political and economic elites in ECE, FDI was seen as providing a panacea for national and regional economic revival. As Dunning (1993 p.20), for example, has argued MNEs 'are uniquely able to supply many of the necessary ingredients for economic growth, a reshaping of attitudes to work and wealth creation, the redesigning of the business and legal framework, especially with respect to property rights and contractual relationships'. He goes on to suggest that economic transition strategies centred on FDI may 'be reasonably expected [to] markedly improve the economic lot of [ECE] citizens' (Dunning 1993 p.25). From the perspective of investors and recipients alike, 'the lesson seems clear: with time and diligent reform, foreign capital may start to come into its own' (Gradev 2001 p.1). From this perspective FDI was discursively constructed as a necessary element for an economic transition to capitalism in three key, interrelated respects. First, FDI was widely seen as key to establishing dynamic national and regional economies. To take one typical example of this kind of policy discourse, the Slovak Ministry of the Economy estimated in the early 1990s that at least SKK12.5bln per year (approximately \$350mln) of FDI was required to generate 'a low level of positive economic growth', ideally reaching between SKK25–40bln/yr (\$0.7–1.1bln) (Balaž et al. 1995 p.470). Of the total industrial investment needed to fully modernise the industrial base, it was argued that 60–75% should be derived from FDI largely because of the capital scarcity operating in transitional economies such as Slovakia (Hajko 1994; Sojka 1994). Later in the 1990s these estimates of the levels of investment required to 'revitalise' the industrial base were increased to SKK216.9bln (Privarova 2001). In Slovakia and elsewhere in the world where FDI has been relied upon as a national and regional strategy, FDI was constructed as a key element in the rejuvenation of the exhausted economies of the region. A key element of such discourses concerned how the integration into global production networks (Dicken et al. 2001) and value chains (Smith et al. 2002) would provide the potential basis for an economic and technological upgrading of production throughout ECE.

Second, FDI was seen as a catalyst for regional learning. The transference of both 'soft' and 'hard' technologies and knowledge of business practices was seen to be a strength of a strategy focused on FDI. Soft technologies, such as 'Western' management techniques and business organisation practices were seen as lacking in ECE, and so FDI was perceived as being the best way to ensure the learning of the business of western capitalist economies: see Thrift (1998) on the role of management technologies in what he calls 'soft capitalism'. FDI was also seen as providing hard technologies to stimulate regional and national revival. Through capital investment in new plant, technology and management systems, the perceived technological backwardness of the ECECs was to be overcome. Third, FDI was seen as a basis for social regional upgrading. In restructuring the economy from plan to market, FDI was seen as an instrument for generating backward and forward linkages among private firms through which institutional and technical learning would be possible and a new social contract would emerge. Indeed, the 2003 World Investment Report has recently suggested that: 'there was also a tendency of firms (including foreign affiliates) in several CEECs, particularly those slated for accession to the EU, to shed activities based on unskilled labour and to expand into higher value-added activities, taking advantage of the educational level of the local labour force' (UNCTAD 2003 p.22). The consequences were assumed to be an emerging virtuous circle of higher-value production and work at the national level, embedded in and articulated with an emerging trans-European regional world.

The Political Economies of Actually Occurring FDI

For most people and places in ECE such virtuous circles and their associated regional worlds of production have failed to materialise, although in some places and in some sectors there is increasing evidence of new industrial complexes emerging. These cases further enhance the image of FDI as a mechanism by which peripheral parts of the region are increasingly forced to position themselves within flows of global capital in order to restructure the perceived outmoded, inefficient industries that represent the legacy of state socialism (Dunning 1993). The result of this type of global engagement has invariably been forms of 'defensive restructuring' (Lipietz 1992, Pavlínek and Smith 1998) – a kind of 'low road' response based on low cost labour, east-west wage differentials and relative isolation of new production plants from local economic interactions. This is what Grabher (1994) has characterised as the ECE version of the 'cathedrals in the desert' scenario. As Pavlínek and Smith (1998 p.634) have argued, these cathedrals 'are important sites of global capitalist integration, but result in thoroughly "disembedded" regional economies' (see also Smith and Swain 1998), in which old state socialist era forms of local autarky in which enterprisers had few local economic linkages and limited supply chains (Smith 1998) are reproduced by new institutional and enterprise structures that are similarly unintegrated into local economic space, but are located at the international scale in global circuits of capital. This results, Pavlínek and Smith (1998) argue, in a form of vertical integration into multinational production networks located outside the region (Pavlínek 1998a), usually situated within pan-European and global value chains. For Grabher (1992) this results in a truncating of foreign investment plants as 'islands' of often advanced technological production in a sea of unconnected and underdeveloped local linkages, in which those linkages that did exist under state socialism have become broken, if not, eradicated.

If the early experience of industrial restructuring led by FDI largely created landscapes of cathedrals in the desert, it is also clear that a decade later the picture has become more complex. Reflecting ways in which, in the words of Amin and Thrift (1994), the global can be 'held down' in local economies and the competitiveness of less favoured regions can be enhanced through an engagement with global capital, there is emerging evidence to suggest that in some, albeit relatively few, manufacturing sectors such as automobiles the development of integrated and often localised supply chains have begun to transform regional economies in quite dramatic – and sometimes problematic – ways (see Grabher 1992, 1994; Havas 1997; Pavlínek 1998, 2002; Sadler and Swain 1994; Sharp and Marz 1997; Smith and Ferenčíková 1998). Driven in part by the discovery of Porteresque cluster strategies (Porter 2003 – see also Martin and Sunley (2003) and Smith (2003) for critiques) national and regional states in ECE are beginning to recognise *some* of the dramatic forms of regional embeddedness that can be achieved in *some* industrial sectors. Thus, in recent years the 'cathedrals in the desert' arising from large automobile buy-outs have begun to generate a wide diversity of local subcontracting relationships as well as deepening integration into the global structure of car production and consumption. For some scholars trying to understand these emerging patterns notions of regional industrial clusters have proved useful in understanding FDI in broader conceptual terms (Buček 2003; also Smith, 2003 for more critical consideration).

Outside of the manufacturing sector, FDI is seen to have played a different role. As in manufacturing, there was an early emphasis on establishing joint ventures or buy-outs of existing service sector infrastructures, particularly in retailing. Most notable in that regard was the purchase in the early 1990s of the Prior department store network in the former Czechoslovakia by KMart, followed by its subsequent sale to Tesco (Wrigley 2000). However, many such brownfield acquisitions were used in the 1990s to gain a variety of knowledge for large service multinationals. For example, Tesco, like other EU-based food retailers such as Carrefour, utilised the knowledge it gained through the Czech and Slovak department store network to then begin to engage in superstore and hypermarket development which has transformed the retail and urban landscapes of larger towns and cities, particularly in Central Europe.

The State and the Discursive Construction of FDI

In addition to the particular strategies of Western MNEs, inward investment patterns and strategy have been mediated by the action of national and regional states in ECE. States throughout the region embraced FDI (albeit to varying degrees) as a key element in their transition and restructuring strategies. Indeed, most states developed inward investment agencies to promote themselves and their available investment opportunities, literally advertising wholesale sectors and branches of industry for sale. Initially drawing upon EU-funded assistance programmes – such as PHARE – these agencies drew upon and developed Western models of investment promotion and place selling. Such place selling focused particularly on comparative regional advantages in tourism, food processing, and information services. But it has also focused on a much deeper recomposition of social forces, including the introduction of new work practices, forms of global (and European) integration and new class structures and power relations. In their briefing paper, CUTS (2003) illustrate this growing role of the state in attracting FDI and the degree to which the state willingly adjusts itself to the demands of potential investors. They identify three 'generations' of effective state investment policies: liberalisation; facilitation and promotion; and changing the entire society.

The first step governments make when trying to create a more positive FDI regime is to liberalise FDI-specific investment laws including restrictions on entry, size, ownership form, and repatriation of profits. From 2002 the WTO Trade Related Investment Measures Agreement also eliminated specific restrictions on export requirements, import restrictions, and local content requirements. In the second generation, liberalisation policies are no longer seen to be adequate to effective recruiting of FDI. It is now vital that a state seeking FDI should also (i) reduce burdensome regulations, (ii) create investment promotion agencies to market the country and its investment opportunities, and (iii) provide investment incentives to prospective investors (CUTS 2003 p.3). In the third generation, countries have to change fundamental institutional and economic characteristics in order to change the risk-reward ratio that investors use when considering FDI. The agenda is vast, and will take time to legislate, let alone to implement. To change the risk side of the equation, countries seeking to attract FDI must create the conditions for political stability, transparency, eliminate corruption, change property rights, and foster macroeconomic stability. They must also change the return side of the equation by enhancing education and skill levels, adopting a competition policy that regulates the entry and behaviour of powerful multinationals into the economy in ways that supports competition, provides support for key service sectors of the economy (such as financial organisations), and implement policies to reduce tariff protections and barriers to trade (CUTS 2003 p.4).

Investment promotion, in this view, is part of such broader processes of state reconfiguration. The neoliberalisation of post-socialist states has meant that the state has become an enabler for the development of market and commodity relations and has been actively engaged in reordering the balance of power between capital, labour, and the state (Smith, 2002; but see also Tickell and Peck, 2003 for a wider discussion). The creation of an attractive business climate for FDI thus involves parallel processes of class transformation. Because post-socialist ECE lacked any existing capitalist social classes, these processes of social transformation have resulted in the emergence of small scale, undercapitalised petty capitalists, legally and illegally newly enriched social groups of one sort or another (including former apparatchiks and criminals), and technocratic elites with special access to foreign contacts and capital, alongside an increasingly large proportion of people living in difficult economic circumstances. In these processes the state and its inherited managerial apparatus has had to 'be able and willing to behave as if [these groups] constituted a "real" capitalist class' (Radice n.d. p.7). But, as Radice suggests, such groups are not always willing to accept the role of junior partners to foreign capital and have, as a result, oscillated between 'obedient neoliberalism and more-or-less belligerent nationalism' (Ibid), and have at times frustrated any expectations by the state or new investors that acquisitions would be easy and their restructuring straightforward.

FDI has thus been an important element and tool of broader social transformations, creating deep systemic changes in the fabric of post-socialist lives and geographies. That such a high proportion of FDI has been concentrated so heavily in Czech Republic, Hungary, Poland, Slovakia and Slovenia in the first decade of transformation itself indicates the consequences for regional restructuring of certain types of risk averse strategies and calculations of political stability. The consequences also reach deep into the social fabric of these regions. Gradev (2001 p.8) characterises this process as one of social and cultural engineering that has 'provided a rare opportunity not only for enterprise profit maximization in a turbulent environment, but also for moulding the economic and social context of change and pointing it in the desired direction'. The geopolitical uses and consequences of FDI were seen clearly by the US ambassador to Hungary. In response to his rhetorical question: 'why isn't there a new Marshall Plan to help CEE?', the ambassador responded: 'well, there is – it is here – and it is called private foreign investment' (quoted in Gowan 1995 p.10). The consequences for the state are far from clear. Enjoined to dismantle the bureaucratic planning structures and to replace them with a liberal state, states throughout ECE have had to enact an unprecedented number of new laws (particularly in those contexts where EU accession is looming) at the very time that structures of civil society and citizen participation are to be expanded. Not surprisingly, the state is all too frequently caught between citizens distrustful of its continued exercise of power and social interests pressing it to act to resolve social problems and meet social needs: 'the state is importuned on all sides, and blamed for all disappointments. The outcome is a tendency towards both instability and paralysis, reinforced by voter passivity' (Radice n.d. p.4).

Labour and Working with/against FDI

The state, in conjunction with business and financial institutions, has played a major role in the design and implementation of restructuring policies, and foreign corporations have argued vigorously for the kinds of policies and regulations that will facilitate a stable and open business climate. In this framing, the social dialogue between the state, capital, and labour is one in which organized labour is expected to withdraw from any active participation in management while newly privatised enterprises jettison their social welfare functions. While foreign capital promises to restructure production vigorously at the enterprise level, it has often to make compromises in the interests of retaining workforce skills and local networks of suppliers and customers. Nevertheless, a process of differentiation has clearly set in, between workers who benefit from the higher wages and security of employment provided by foreign-owned firms, and the restructured unemployed, who include disproportionately the unskilled, women, and ethnic minorities (Radice n.d. pp.2–3; but see also Begg and Pickles, 1998 and Pickles and Begg, 2000).

In reporting the results of a European Trade Union Institute survey of investors, state officials, and labour organizers across ECE, Gradev (2001 pp.14–19) suggests that there have been several distinct effects of FDI on labour markets across the region. All countries in the region have faced pressing problems with unemployment and, in this regard, FDI has made an important contribution to labour market change throughout the region. In part, the rise in unemployment has been a function of the very restructuring and labour shedding wrought by foreign companies entering the region. 'In the case of a "late transformer" such as Bulgaria, the "optimisation" of the workforce in foreign companies has tended to release between 25% and half the workforce' (Gradev 2001 p.14). But FDI has also created new jobs, stabilised those jobs that have survived restructuring, and increased productivity, value added, and overall efficiency of work through the introduction of new management practices, new product mixes, and new technologies. In Hungary, for example, the majority of employees now work in foreign capital enterprises (67% of total employment and 55% of manufacturing workers) (Gradev 2001 p.14).

One consequence of these changing patterns of employment and unemployment has been the differentiation of access to work and highly uneven levels of wages for those in work. On average, wages are higher in foreign-owned firms, with wage differentials between foreign-owned and domestic firms ranging from 10–13% in Slovenia, 30% in Czech Republic, 50% in Poland, and 20–100% in Romania (Gradev 2001 p.15). For the emerging elite class of skilled workers, wages, benefits, and their broader integration into the international structure of the firm are occurring quickly. By contrast, lower-skilled workers in SMEs (clothing, leather and footwear) are often squeezed by managers, who are themselves operating under tight price pressure from international buyers and contractors. Workers in these industries are more likely to be paid in cash, more likely to be required to work overtime (sometimes unpaid), and more likely to receive fewer social benefits. In the next section we contrast two industries that typify these high and low road models of FDI and illustrate the regional consequences of each.

Regional Transformations and the Combined and Uneven Development of FDI

In this section we turn to the differentiated nature of investment strategies and the ways in which they both shape and are shaped by the geographies of ECE labour markets. We do this through two case studies to illustrate two aspects of regional transformations engendered by an engagement with inward investment. The first concerns the reconfiguration of industrial supply chains through replacement of old supply networks with new just-in-time (JIT) systems centred on the role of FDI in the automobile sector. The second example examines the politics of production associated with inward investment, global contracting and low wage labour in the apparel industry.

Automobiles

One of the key areas of activity engaged with significant flows of Western capital has been that associated with the automobile sector. Initially focused on a number of strategic investments in both joint ventures and greenfield sites, the automobile sector has become one of the largest recipients of inward investment, and as such has been positioned as a central pillar in the promotion strategies of several ECE states (Pavlínek 2002; Privarova 2001; Sadler and Swain 1994). The Czech and Hungarian examples have often been cited to highlight the significance of the automobile sector (Sadler and Swain, 1994, Pavlínek and Smith, 1998). Similar trends and trajectories can also be found in Slovakia, which had a relatively small motor vehicle manufacturing sector in the late 1980s, but which has witnessed significant growth over the last 13 years (Buček 2002; Pavlínek and Smith 1998; Privarova 2001). For example, motor vehicles was only one of three sectors that witnessed any increase in employment in the second half of the 1990s as most manufacturing sectors experienced downsizing of quite dramatic proportions. The growth of employment in motor vehicles is clearly related to the strategic investment of VW in Bratislava and the development of its associated supply chain. But this growth in motor vehicles employment – comprising in 1999 only 3.8% of total industrial employment – has done little reduce the overall dramatic decline in industrial employment in Slovakia over this period. However, it is also apparent that FDI has had a dramatic effect on labour productivity. The motor vehicles sector saw the most rapid improvement in labour productivity in the late 1990s by far, as a whole series of capital investments and new managerial techniques gave rise to increases in productivity. For example, motor vehicles saw the highest sectoral level of investment in 1999, double that of the next most significant level of investment in food processing (calculated from ŠÚSR 2000 p.123).

The strategic significance of the automobile sector has also been underlined by its key role in Slovak exports, which saw a rapid growth from only two percent of total manufacturing exports in 1993 to 10.3% in 1997. It is clear that the dynamism in the motor vehicles sector, most of which is seen in the greater Bratislava region, is closely linked to the position of Volkswagen (VW) Bratislava as a key export platform. Indeed, it has dramatically transformed the Slovak automobile industry from relative marginality under Czechoslovak state socialism to a key element in the corporate and production networks of VW in Europe. While for much of the 1990s Bratislava existed as an extended workbench of VW Wolfsburg, more recently the competitiveness of, and high levels of productivity in, the Bratislava plant led to the decision to locate production lines for the new Colorado SUV and the development of a host of associated component plants to establish a quasi-JIT system in the Bratislava regional economy and beyond. This resulted in VW Bratislava becoming one of the key strategic enterprises in Slovakia employing around 7,000 workers, seeing a tripling in turnover between 1997 and 1999 and now accounting for 16% of Slovakia's total exports. VW Bratislava has thus now become the fifth largest manufacturer of cars in ECE, behind Lada in Russia, Škoda in Czech Republic, and Fiat and Daewoo in Poland. This has also led to follow-on supply chain investments (Privarova 2001) and the emergence of other EU car manufacturers in Slovakia, such as the recently announced Peugeot investment. Consequently, the automobile sector has shifted from being characterised in the mid-1990s as a cathedral in the desert centred on VW Bratislava (Smith and Ferenčíková 1998), to one in which there is an emerging and quite complex set of localised supply chains. In 2000 VW decided to invest in the construction of an industrial park near Malacky, about 30km from the plant, for between 12 and 15 of VW's key European suppliers, working on a JIT basis for the Bratislava plant. It is estimated that these suppliers will employ between 1,200 and 1,500 directly <http://www.slovensko.com/investor/german.htm>.

However, as we shall see in the apparel sector, while the VW Bratislava investment increased employment in the plant to around 2,000 workers in the late 1990s, the plant is operating in a constrained labour market. The recruitment of qualified and skilled personnel has been a problem for the plant and as a result in the mid- to late-1990s VW Bratislava launched a series of high-profile job vacancy advertising campaigns. The retention of employees has also been difficult because of the relatively low wages paid to workers to maintain cost competitiveness (albeit higher than average industrial wages), because of the stresses of a high level of long distance commuting that many workers undertake and because the labour intensity of production is high. However, the strategic significance of investment in the automobile sector in Slovakia and elsewhere in ECE cannot be put in doubt. Notwithstanding the significant levels of supply chain development, the automobile does remain one of the very few sectors in which significant growth has occurred. One result is that FDI flows remain quite regionally concentrated as localised JIT systems develop. Consequently, the extent to which the experience and lessons of the automobile sector can be generalised to other areas of the economy and to other regions in which car manufacturing does not play a role appear to be limited.

Apparel and Textiles

The process of dramatic rearticulation of economic activity with the world economy has not only been experienced in the automobile sector. Since the 1980s, and in particular throughout the 1990s, the apparel and textile industry in ECE has seen a dramatic engagement with the global – or more correctly – the EU economy. This engagement has to a very large extent been based on the expansion of outward processing contract production from the EU to firms in ECE (Begg et al. 2000, 2003; Pickles 2002; Smith 2003). Firms across the region have been variously involved in pan-European production systems in which EU retailers, contractors and manufacturers have utilised preferential EU trade regulations governing outward processing (within a context of a wider liberalisation of EU and world apparel trade) to establish quite dense networks of contracting. Often this has involved the use of producers in ECE in forms of cut-make-trim operations in which the intellectual capital embodied in design, marketing, retailing and control over the whole apparel production process remains centred in the main markets (generally the EU and the USA) (Smith 2003). Often this outward processing system involves contractual relations between EU and USA-based firms and domestically-owned production units in ECE. In this sense, then, FDI has been limited in the apparel sector, with a strong focus instead on forms of non-equity collaboration and strategic alliance.

However, since the mid-1990s FDI in the apparel sector has grown in both the apparel and the textiles production sectors. Central to this increased capitalisation has been the recognition that upstream investment in textiles has enabled western contractors to benefit from improved quality fabrics, which in turn has enabled an overall increase in value-added of apparel products assembled in ECE. Italian textile manufacturers, for example, faced major recapitalisation decisions later than their north European counterparts, and this is reflected in the emerging geographies of the increasing levels of Italian textile FDI in ECE. While German manufacturers established close sourcing relationships with established large textile producers in Czech Republic, Poland and Slovakia, Italian manufacturers are currently investing in greenfield factories in Bulgaria and Romania. The Italian case illustrates the historical and regional dynamics of FDI in apparel and textiles. Most of the growth in apparel production in the 1990s has been on an outward processing trade basis. This is a system of outsourcing for assembly of materials produced in EU countries to make use of large wage differentials in ECE labour markets. Initially the system was supported as a customs practice that protected the export of inputs (cloth, threads and buttons) from the EU for assembly in ECE from tariffs and quotas on re-entry into the EU. The system continues as an outsourcing assembly system, but now the customs benefits have largely ended as EU markets have increasingly been opened for products from ECECs as part of the wider EU-accession process (Begg et al. 2003). Alongside this rapid and widespread growth in outsourcing of assembly production based on contracting has emerged a growing amount of FDI in the industry. There are many reasons why FDI might be attractive to apparel manufacturers, including the relatively low-cost of initial investment needed. But in recent interviews it has become much clearer that quite large-scale investments (for example into Bulgaria from Greece and Turkey) are also now occurring.

The picture that emerges is one of increasing integration of ECE apparel producers in supply chains tied particularly to the EU, but also to the USA, in a highly competitive global industry. Price pressure on contracting has resulted in diverse firm and regional strategies to maintain competitiveness, and these include turning to local markets and subcontracting production to other firms to maintain employment and production during low production periods, upgrading local production and design capacities, joint ventures with EU firms, the outsourcing from Western Europe of logistical, planning, and design functions to ECE hubs and factories, and new greenfield investments on the part of EU-based or Turkish firms in preparation for EU accession in 2004 or 2007. At the same time, the intensity of competition and the pressure on prices in the industry is also rapidly producing parallel tendencies towards further relocation as buyers and contractors begin to plan for subcontracting relations in regions with even cheaper labour pools and similar histories of industrial production. In Slovakia, for example, wages have slowly begun to increase partly as a result of economic integration and political accession with the EU. As a result, Slovak firms and foreign buyers and manufacturers with investments in Slovakia are now using subcontracting links with Ukrainian producers to out-source production, while upgrading logistical facilities and downgrading production facilities and employment in Slovak factories (see also Kalantardis et al. 2003). The regional impacts of FDI in the apparel industry thus remain ambiguous, tied as they are to ever shorter investment cycles and the rapidly growing challenge to producers of competition from Chinese (and other) sources, particularly as trade liberalisation in clothing is completed with the ending of all quantitative quotas on imports to all WTO members countries on January 1, 2005.

Conclusion

In the past 15 years, FDI has been an important tool in the restructuring of Soviet-style economies and for opening them to the influences of the global economy. Such forms of reintegration have occurred in particular with economic actors and production networks in the EU, and in this regard FDI has been an important policy as well as economic tool of a wider European integration process. In its own terms, FDI has also had very distinct regional, sectoral and organisational consequences for countries in ECE. Industries have been restructured, employment has been lost, and new processes of class formation and differentiation have emerged quickly. In certain sectors, such as automobiles, investments have not only resulted in firm restructuring and integration of production relations with broader global corporate cultures, but regional economies have experienced substantial transformations as investments have spawned new subcontracting relations across a wide area. In other industries, such as apparel, tight contracting and buying relations on a global scale have produced much more fragmented responses to economic integration, and a much less clear role for FDI. The search for cheap labour markets certainly drives both strategies of FDI, but the opportunities of exploiting extremely low wage, yet skilled labour pools in textile and apparel producing regions throughout ECE has generated a region-wide resurgence of the industry.

FDI has also served as a discursive strategy for reconfiguring social relations within the state. The state itself has adopted an active role in selling itself and its industries, opening them to the interests of global investors. In some cases this has produced an open, neoliberal state (as in the case of Bulgaria after 1997) with few formal barriers to entry. In other cases, concern about foreign ownership of national resources generated state policies that sought to privatise large sectors of industry into the hands of managers (such as in post-independence Slovakia under the nationalist policies of the Mečiar government until 1998). In both cases the need for markets and investment capital opened the economy to FDI and subcontracting, with important consequences for the political role of organized labour in each. Finally, FDI has been and continues to influence conceptual frameworks within the academy. Much of this reconfiguration and reorientation has been productive, and has certainly stimulated new thinking on the regional dynamics and economic geographies of transformation. But FDI has also served as an important tool in the social engineering of the transition, and it remains vital that regional economic consequences of FDI be assessed in parallel with the ways in which these projects are simultaneously reordering the social and political lives of people in the region.

Notes

[1] This chapter results in part from research funded by the US National Science Foundation (Award BCS/SBE 0225088) for a project entitled 'Reconfiguring Economies, Communities and Regions in Post-Socialist Europe: A Study of the Apparel Industry' in which the authors

are working with Bob Begg, Milan Buček and Poli Roukova. It also draws upon research conducted by Adrian Smith (working with Mick Dunford, Jane Hardy, Ray Hudson, Al Rainnie and David Sadler) on an Economic & Social Research Council project entitled 'Regional Economic Performance, Governance and Cohesion in an Enlarged Europe' (L213 25 2038) which was part of the 'One Europe or Several?' Programme. We are grateful to the ESRC and NSF for supporting this research. The normal disclaimers apply.
2 Here we draw on Gradev's recent (2001) research which is continuing at ETUI by focusing on issues of convergence, economic integration and social cohesion http://www.etuc.org/etui/projects/default.cfm.

References

Amin, A. and Thrift, N. (1994), 'Living in the Global', in A. Amin and N. Thrift (eds), *Globalization Institutions and Regional Development in Europe*, Oxford University Press, Oxford, 1–22.

Balaž, P., Ferenčíková, S., Filip, J. et al. (1995), *Medzinárodné podnikanie*, Sprint, Bratislava.

Begg, R. and Pickles, J. (1998), 'Institutions Social Networks and Ethnicity in the Cultures of Transition: Industrial Change Mass Unemployment and Regional Transformation in Bulgaria' in J. Pickles and A. Smith (eds), *Theorising Transition: The Political Economy of Post-Communist Transformations*, Routledge, London, 115–46.

Begg, R., Pickles, J. and Roukova, P. (2000), 'A New Participant in the Global Apparel Industry: The Case of Southern Bulgaria', *Problemi na Geografiata*, 3–4, 121–52.

Begg, R., Pickles, J. and Smith, A. (2003), 'Cutting it: European Integration, Trade Regimes and the Reconfiguration of East-Central European Apparel Production', *Environment and Planning A*, 35, 2191–2207.

Berend, I. and Ránki, G. (1974), *Economic Development in East–Central Europe in the 19th and 20th Centuries*, Columbia University Press, New York.

Buček, M. (2003), *The Automobile Industry Cluster in Slovakia*, European Urban and Regional Studies Conference, Barcelona.

CUTS (Center for Competition Investment and Economic Regulation) (2003), *Investment Policies that Really Attract FDI*, CUTS Briefing Paper 3/2003.

Dicken, P. (2003), *Global Shift*, Sage, London.

Dicken, P., Kelly, P., Olds, K. and Yeung, H. (2001), 'Chains and Networks Territories and Scales: Towards a Relational Framework for Analysing the Global Economy', *Global Networks*, 1(2), 89–112.

Dunford, M. (1998), 'Economies in Space and Time: Economic Geographies of Development and Underdevelopment and Historical Geographies of Modernization', in B.Graham (ed), *Europe: Place Culture Identity*, Arnold, London, 53–88.

Dunning, J. (1993), 'The Prospects for FDI in Eastern Europe', in P. Artisien, M. Rojec and N. Svetlicic (eds), *Foreign Investment in CEE*, Macmillan, London, 16–33.

EBRD (2003), 'EBRD Region Continues to Outperform Global Economy: But Sustaining Growth is becoming more Difficult to Manage', *EBRD Press Release*, 22 April.

Gowan, P. (1995), 'Neo-Liberal Theory and Practices for Eastern Europe', *New Left Europe*, 213, 3–60.

Grabher, G. (1992), 'Eastern Conquista: the Truncated Industrialisation of East European Regions by large West European Corporations', in H. Ernste and V. Meier (eds), *Regional Development and Contemporary Industrial Response*, Belhaven Press, London, 219–33.

Grabher, G. (1994), 'The Disembedded Regional Economy: the Transformation of East German Industrial Complexes into Western Enclaves', in A. Amin and N. Thrift (eds), *Globalization Institutions and Regional Development in Europe*, Oxford University Press, Oxford, 177–95.

Gradev, G. (2001), 'EU Companies in Eastern Europe: Strategic Choices and Labour Effects', in G. Gradev (ed), *CEE Countries in EU: Companies' Strategies of Industrial Restructuring and Relocation*, European Trade Union Insititute, Brussels, 1–19.

Hajko, J. (1994), 'Bez zahraničných investícií sa hospodárstvo nerozhýbe', *Trend*, 26 January, 1 and 4–5.

Havas, A. (1997), 'FDI and Intra-Industry Trade: the case of the Automotive Industry in Central Europe', in D. Dyker (ed), *The Technology of Transition*, Central European University Press, Budapest, 211–40.

Kalantardis, C., Slava, S. and Sochka, K. (2003), 'Globalization Processes in the Clothing Industry of Transcarpathia, Western Ukraine', *Regional Studies*, 37, 173–86.

Lipietz, A. (1992), 'The Regulation Approach and Capitalist Crisis: An Alternative Compromise for the 1990s', in M. Dunford and G. Kafkalas (eds), *Cities and Regions in the New Europe*, Belhaven, London, 309–34.

Martin, R.L. and Sunley, P. (2003), 'Deconstructing Clusters: Chaotic Concept or Policy Panacea?', *Journal of Economic Geography*, 3, 5–35.

Pavlínek, P. (1998), 'Foreign Investment in the Czech Republic', *Professional Geographer*, 50, 71–85.

Pavlínek, P. (2002), 'Transformation of the CEE Passenger Car Industry: Selective Peripheral Integration through FDI', *Environment and Planning A*, 34, 1685–709.

Pavlínek, P. and Smith, A. (1998), 'Internationalization and Embeddedness in ECE Transition: the Contrasting Geographies of Inward Investment in the Czech and Slovak Republics', *Regional Studies*, 32, 619–38.

Pickles, J. (2002), 'Gulag Europe? Mass Unemployment New Firm Creation and Tight Labor Markets in the Bulgarian Apparel Industry', in A. Smith, A. Swain and A. Rainnie (eds), *Work Employment and Transition*, Routledge, London, 246–73.

Pickles, J. and Begg, R. (2000), 'Ethnicity State Violence and Neo-Liberal Transitions in Post-Communist Bulgaria', *Growth and Change*, 31, 179–210.

Porter, M. (2003), 'The Economic Performance of Regions', *Regional Studies*, 37, 549–78.

Privarova, M. (2001), 'Slovakia: FDI Industrial Relations and Development of the Economy', in G. Gradev (ed), *CEECs in the EU Companies' Strategies of Industrial Restructuring and Relocation*, European Trade Union Institute, Brussels, 173–95.

Radice, H. (nd), *The Role of Foreign Capital in Eastern Europe: Implications for a Socialist Strategy* <http://eszemelet.tripod.com/ango12/radiceeang2.html>.

Sadler, D. and Swain, A. (1994), 'State and Market in Eastern Europe: Regional Development and Workplace Implications of FDI in the Automobile Industry in Hungary', *Transactions of the Institute of British Geographers*, 19, 387–403.

Sharp, M. and Marz, M. (1997), 'Multinational Companies and the Transfer and Diffusion of

New Technological Capabilities in CEE and the Former Soviet Union', in D. Dyker (ed), *The Technology of Transition*, Central European University Press, Budapest, 95–125.

Smith, A. (1998), *Reconstructing the Regional Economy: Industrial Transformation and Regional Development in Slovakia*, Cheltenham, Edward Elgar.

Smith, A. (2002), 'Imagining Geographies of the "New Europe": Geo-Economic Power and the New European Architecture of Integration', *Political Geography*, 21, 647–70.

Smith, A. (2003), 'Power Relations Industrial Clusters and Regional Transformations: Pan-European Integration and Outward Processing in the Slovak Clothing Industry', *Economic Geography*, 79: 17–40.

Smith, A. and Ferenčíková, S. (1998) 'Inward Investment Regional Transformations and Uneven Development in ECE: Case Studies from Slovakia', *European Urban and Regional Studies*, 5, 155–73.

Smith, A. and Pickles, J. (1998), 'Introduction: Theorising Transition and the Political Economy of Transformation', in J. Pickles and A. Smith (eds), *Theorising Transition: The Political Economy of Post-Communist Transformation*, Routledge, London, 1–22.

Smith, A. and Swain, A. (1998), 'Regulating and Institutionalising Capitalisms: the Micro-Foundations of Transformation in CEE', in J. Pickles and A. Smith (eds), *Theorising Transition: The Political Economy of Post-Communist Transformation*, Routledge, London, 25–53.

Smith, A., Rainnie, A., Dunford, M. et al. (2002), 'Networks of Value Commodities and Regions: Reworking Divisions of Labour in Macro-Regional Economies', *Progress in Human Geography*, 26, 41–63.

Smith, A., Buček, M., Pickles, J. and Begg, B. (2003), 'Global Trade European Integration and the Restructuring of Slovak Clothing Exports', *Ekonomický časopis*, 51, 731–48.

Sojka, J. (1994), 'Dôsledky nedotiahnutej reprodukcie fixného kapitálu ponesú budúce generácie', *Trend*, 19, January 7.

ŠÚSR (2000), *Ročenka priemyslu 2000*, ŠÚSR, Bratislava.

Thrift, N. (1998), 'The Rise of Soft Capitalism', in A. Herod, G. O'Tuathail and S. Roberts (eds), *Unruly World? Globalization Governance and Geography*, Routledge, London, 25–71.

Tickell, A. and Peck, J. (2003), 'Making Global Rules: Globalization or Neoliberalization?' in J. Peck and H. Wai-chung Yeung (eds), *Remaking the Global Economy*, Sage, London, 163–81.

UNCTAD (2001), 'Central Eastern European Transnationals set to become "Prominent Players" in World Investment', *UNCTAD Press Release* TAD/INF/PR26, September 18.

UNCTAD (2003), *World Investment Report 2003: FDI Policies for Development: National and International Perspectives* <www.unctad.org/en/docs//wir2003overview_en.pdf>.

WIIW (2003), *WIIW Database on FDI in CEECs and the Former Soviet Union with Special Attention to Austrian FDI Activities*, WIIW, Vienna.

Wrigley, N. (2000), 'The Globalization of Retail Capital: Themes for Economic Geography', in G. Clark, M. Gertler and M. Feldmann (eds), *The Oxford Handbook of Economic Geography*, Oxford University Press, Oxford, 215–32.

Chapter 3

Foreign Direct Investment: A Sectoral and Spatial Review

Andrew Dawson

Introduction

FDI has become a much-used phrase in recent years. But the phenomenon is not new. The movement of money from one country to another in the pursuit of a return is as old as civilisation. Indeed, it can be argued that economic development, and the civilities that are associated with it, have depended in part – and sometimes in large part – upon such movements. Conversely, barriers to investment reduce opportunities to exploit the fact that the earth's surface is a varied place, that some parts are better suited to particular types of economic activity than others, and that increasing prosperity depends not only upon improvements in the productivity of labour, but also in making more productive use of the globe's natural resources. At some times in the past, economic growth has been achieved largely by moving labour to places in which it can be used more effectively, but at others – such as the present, when many countries have adopted stringent controls over immigration – the role of foreign investment becomes more important. Places that are cut off from foreign money become or remain poor; those that attract foreign capital prosper.

That said, foreign investment can be a slippery subject. Its definition is both complex and subject to debate. According to the United Nations Conference on Trade and Development (UNCTAD 2001 pp.275–6) FDI is an 'investment involving a long-term relationship and reflecting a lasting interest and control of a resident entity in one economy in an enterprise resident in an economy other than that of the foreign direct investor', with an implied influence on the management of the enterprise. Three components of FDI are mentioned: equity capital, reinvested earnings and intra-company loans, with 'FDI stock' reckoned as the value of the share of capital and reserves attributable to the parent enterprise. But foreign investors may also gain a voice in management through non-equity forms of FDI such as subcontracting, management contracts, franchising and product sharing. FDI is also difficult to measure. According to the Russian government <http://www.gks.ru> investment received in 2000 amounted to $10.98bln – 40% in direct investments, 59% in other investments (mainly trade and other credits) and the remaining 1% in portfolio investments. But the UN figure of $2.7bln quoted in

Table 3.1 suggests that many of the credits in the government's list may well have been no more than promises at that time. It is for this reason that comparisons of the relative success of countries in attracting investment in this chapter will be based on the United Nations' more conservatively-defined data.

Table 3.1   FDI inflows for ECE and the FSU, 1995–2000

	1995	**1996**	**1997**	**1998**	**1999**	**2000**
World	331 068	384 910	477 918	692 544	1 075 049	1 270 764
Developed Countries	203 462	219 688	271 378	483 165	829 818	1 005 178
ECE/FSU*	16 268	15 591	24 006	25 364	27 701	29 519
Albania	70	90	48	45	41	92
Armenia	25	18	52	232	130	133
Azerbaijan	330	627	1 115	1 023	510	883
Belarus	15	105	352	203	444	90
BiH	**	**	1	10	90	117
Bulgaria	90	109	505	537	819	1 002
Croatia	114	511	540	935	1 474	899
Czech Rep.	2 562	1 428	1 300	3 718	6 324	4 595
Estonia	202	151	267	581	305	398
Georgia	5	45	243	265	82	197
Hungary	4 453	2 275	2 173	2 036	1 944	1 957
Kazakhstan	964	1 137	1 321	1 152	1 587	1 249
Kyrgyzstan	96	47	83	109	35	19
Latvia	180	382	521	357	348	407
Lithuania	73	152	355	926	486	379
Macedonia	10	12	16	118	32	170
Moldova	67	24	79	74	39	128
Poland	3 659	4 498	4 908	6 365	7 270	10 000
Romania	420	265	1 215	2 031	1 041	998
Russia	2 016	2 479	6 638	2 761	3 309	2 704
Serbia & Montenegro	45	102	740	113	124	29
Slovakia	195	251	206	631	356	2 075
Slovenia	176	186	321	165	181	181
Tajikistan	15	16	4	30	21	24
Turkmenistan	100	108	108	64	80	100
Ukraine	267	521	624	743	496	595
Uzbekistan	120	55	285	140	121	100

* aggregates rounded; ** value below 1
Source: UNCTAD (2001) pp.291–5.

Changing Opportunities for FDI in CEE (including all FSU states)

These thoughts have particular resonance with regard to the CEECs which were characterised by CPEs until 1989–91, but which have since been in a process of economic transition to a greater or lesser extent. The CPE, as pursued in the FSU after 1917 and in ECE after 1945, was characterised by a high degree of economic isolation. Not only was trade with non-communist countries severely limited, but it was highly constrained even within the Soviet bloc whose members' individual economic plans were linked through the Comecon. In a situation in which international trade was a state monopoly and in which Soviet leaders determined to a large extent the pattern of economic specialisation within the bloc, international trade was divorced from movements of capital. Neither could state enterprises – let alone individual holders of money – acquire land or property or invest in what might have appeared to be promising economic activities in other countries in the group, even if they had been minded to do so. Mechanisms for such action did not exist. Nor – given currency inconvertibility – was there any way in which they could invest in other parts of the world. Moreover, non-communist investors were given no encouragement to expand into the bloc: capitalist money was anathema. By the 1970s, the rest of the world had begun to invest – on a very small scale – in a limited number of ECECs. In particular, Hungary, Poland and Yugoslavia – noticing the economic gaps that were opening between the communist bloc and other parts of the world – had begun to seek out Western firms with whom to establish what were called 'joint ventures': the provision of factories and hotels which were paid for in part by Western money or which made Western products under licence. However, even in these countries, most of the economy remained untouched by such initiatives.

And so it was, when the communist governments of the region collapsed, taking with them their characteristic style of economic management, that CEE was revealed to be an area making many old-fashioned and unwanted products, at levels of productivity far below those in many other parts of the world. As a result, the first stage of economic transition was characterised by an inflow of foreign goods, the collapse of many domestic enterprises, the retreat of population from many of the remoter places in which economic activities had been established, and substantial rises in unemployment. It was also characterised by the privatisation of many enterprises, the establishment of free markets with regard to much of the economy, the abandonment of the public monopoly on foreign trade, and a movement towards currency convertibility. The door was thus opened for foreign investors.

Moreover, it was opened at a time in the world's economic history when institutional change was favouring FDI to a greater extent than it had for some 75 years. Ever since the dislocation caused to international economic interchange by the First World War, the inter-war abandonment of the gold standard, the great depression and the subsequent flight into protectionism, the opportunities not only for international trade, but also the international movement of both labour and capital, had been severely constrained. Governments, anxious to protect voters from unemployment and falling standards of living, had raised barriers against foreigners, thus exacerbating the problem. And, even after the Bretton Woods settlement of 1944 – designed to put an end to such 'beggar-my-neighbour' policies – fixed exchange rates and draconian controls over currency movements had inhibited the working of international markets. The retreat of not only CEE, but also China and India, into the isolation which accompanied central planning and the public ownership of the means of production, merely reduced still further the opportunity for the 'invisible hand' to guide investment to those places in which it would be employed most productively. But, the system proved to be too rigid. The appearance in the 1960s of such placeless currencies as the Euro-dollar undermined government control of the international money markets; American overspending forced the dollar off the gold standard in 1971; and, before long, many Western governments had abandoned fixed exchange rates. The privatisation of many publicly-owned industries in the so-called 'mixed' economies of Western Europe, Japan, Latin America and Australasia in the 1980s encouraged investors to look more widely across the world; and the gradual opening of China at the same time also helped to encourage a sharp upturn in international investment. It has been suggested that, by the early 1990s, the ability to invest abroad had returned to about the level that had existed at the start of the century (Anon 2002a).

In these circumstances, the attraction of CEE might appear to have been considerable. Not only were there potentially large markets in the more populous countries – Russia, Poland, Romania and Ukraine – but there was also the possibility of using low-wage – though generally well educated – labour to produce goods for export to more developed, and in some cases neighbouring, economies. However, in reality, the door was only half open. In every country in the region, irrespective of the level of enthusiasm to become more open to the rest of the world, governments retained control of substantial elements of the economy. Education, health, rail and bus transport and the generation of electricity – not to mention defence, police and justice – remained largely or entirely in public ownership. In some countries, such as Belarus, Serbia and Montenegro and Ukraine, little privatisation occurred; and in many cases there was a strong reluctance to allow foreigners to buy 'strategic' industries – a category that has proved to be broadly interpreted by the more nationalist of local politicians (Anon 2002b) – and especially land. Moreover, there was – as usual after any revolution – a degree of confusion and some lack of order. In the absence of a well-established body of law and precedent in favour of a property-owning democracy, titles to ownership in some countries, and the means to enforce them, proved to be weak. Governments changed rules arbitrarily, without warning, and many banks were dodgy. Much foreign currency – in some cases given by Western aid agencies and in others amassed by crooked entrepreneurs – was stolen, returned to some of the wealthiest countries in Western Europe, and laundered through the purchase of expensive villas on the Mediterranean or apartments in London and Paris. Civil order also became precarious in some countries, and violence against both property and the person – including foreign business-people – increased markedly. In some places, full-scale civil war broke out. Not surprisingly, some countries proved to be more attractive to foreign investors, but others much less so (Artisien-Maksimenko 2000; Collis et al. 1999; Deichmann 2001; Du Pont 2000; Meyer and Pind 1999; Stare 2001; Turnock 2001; Woodruff 1999).

Nor, as has already been noted, have the CEECs been the only focus of investor attention. By far the largest flows of capital – accounting for about 80% of foreign investment – occur between the world's most developed economies: in North America and Western Europe; and, even among the transition and developing countries, China and Brazil have proved to be three or four times more attractive on aggregate than even the most successful country in ECE (Tables 3.1 and 3.2). Even by 2000, after a decade of transition and with the establishment of more stable conditions in Russia by the incoming President Putin, the region, with between 6% and 7% of the world's population, was only attracting about 2% of FDI. Competition for investment has been fierce. Nevertheless, some sectors of the transition economies have done well, and some countries – or at least some regions within them – have attracted substantial investment. Moreover, the contribution of FDI to fixed capital formation has become increasingly important: by the end of the 1990s it was rising rapidly and had reached 18% for the region as a whole, with much higher values for some countries (Table 3.3). At a time when local capital markets have been insufficiently developed to provide the funds needed for the widespread consolidation of enterprises and large-scale modernisation, foreign sources have played a crucial role. Using an Inward Investment Index, based on GDP, employment and exports of countries, the UN has shown that by the late 1990s several ECECs were attracting more than they might have expected (UN 2001, pp.10–11).

It is of some interest to note the origin of this capital. By 2001, Poland, the most successful, had attracted 906 foreign firms, 70% of which were from the EU and 19% from North America. Germany was the largest single source, with 22% of the total. However, France was the major source, by value, of direct investment, with 19% of the total, followed by USA with 15% and Germany with 13%, the Netherlands and Italy with 7.9%, and South Korea with 5%, while 4% came from Russia <http:// www.polamb.nl/weh>. This pattern contrasts with that of Russia, which attracted about $14bln of foreign investment in 2001. No less than 16% of this substantial sum apparently originated in Cyprus, a country with a tiny economy and one at some distance from any part of the Russian Federation, but which has been a lightly-regulated haven through which much money has passed in recent years, with a government naturally sympathetic to a fellow member of the Orthodox-Cyrillic group of nations. More expectedly, both the USA and UK contributed 11% <http://www.gks.ru> with about 8% each from France, Germany, The Netherlands and Switzerland.

It should also be noted that, ten years after the collapse of central planning, the first legitimate reverse flow of capital – in which CEE enterprises have begun to acquire Western businesses – has started to appear (Stare 2002, Table 5). In other words, the capital markets of at least a few of the CEECs have begun to mature. Much of this was to other countries in Europe. In 2001, the largest single contributor

Table 3.2　Inward and outward FDI stock for ECE and the FSU, 1990–2000

| | INWARD | | | OUTWARD | | |
	1990	1995	2000	1990	1995	2000
World	1 888 672	2 937 539	6 314 271	1 717 444	2 879 380	5 976 204
Developed Countries	1 397 983	2 051 739	4 210 294	1 637 265	2 621 165	5 248 522
ECE/FSU*	3 652	42 701	150 257	616	6 573	19 833
Albania	**	201	517	**	48	82
Armenia	**	34	574	**	**	33
Azerbaijan	**	352	4 510	**	**	652
Belarus	**	50	1 243	**	8	16
BiH	**	66	282	**	13	40
Bulgaria	108	445	3 404	**	105	88
Croatia	**	477	4 927	**	703	1 052
Czech Rep.	1 363	7 350	21 095	**	345	784
Estonia	**	674	2 840	**	68	429
Georgia	**	32	489	**	**	**
Hungary	569	10 007	19 863	197	383	2 012
Kazakhstan	**	2 895	9 341	**	**	18
Kyrgyzstan	**	144	438	**	**	1
Latvia	**	616	2 081	**	231	241
Lithuania	**	352	2 334	**	1	29
Macedonia	**	33	380	**	**	5
Moldova	**	93	444	**	18	19
Poland	109	7 843	36 475	95	539	1 491
Romania	766	1 150	6 439	66	121	122
Russia	**	5 465	19 245	**	3 015	11 637
Slovakia	81	1 268	4 892	**	374	320
Slovenia	666	1 763	2 865	258	504	655
Tajikistan	**	25	120	**	**	**
Turkmenistan	**	200	660	**	**	**
Ukraine	**	910	3 843	**	97	106
Uzbekistan	**	255	956	**	**	**

* aggregates rounded; Serbia & Montenegro figures are below 1; ** value below 1
Source: UNCTAD (2001) pp.301–11.

to such outflows – the Russian firm Gazprom – held equity investments in no less than 13 transition states and nine West European countries. Nevertheless, not too much should be made of this: most of Gazprom's business was still conducted inside Russia, none of the world's hundred largest MNEs in 2000 originated in CEE (UNCTAD 2001 pp.90–2,114–6) and the total stock of FDI held by firms from the region was small (Tables 3.2 and 3.4).

Table 3.3   Inward FDI flows as percentages of gross fixed capital formation in ECE and the FSU, 1995–1999

	1995	1996	1997	1998	1999
World	5.3	5.9	7.5	10.9	16.3
Developed Countries	4.4	4.8	6.1	10.6	17.0
ECE and FSU	9.0	7.0	10.0	13.0	18.0
Armenia	12.2	6.2	19.5	75.7	42.9
Azerbaijan	73.1	67.9	78.0	64.2	38.8
Belarus	**	2.6	9.9	3.5	9.9
Bulgaria	4.5	8.1	44.0	37.9	41.4
Croatia	3.9	12.6	11.0	18.1	31.3
Czech Rep.	15.4	7.7	8.0	23.6	44.5
Estonia	21.8	12.9	20.6	38.3	23.6
Georgia	3.9	16.4	64.6	66.1	23.5
Hungary	49.7	23.5	21.4	18.3	18.8
Kazakhstan	25.1	31.4	36.7	30.4	43.3
Kyrgyzstan	31.2	11.3	37.2	50.3	17.7
Latvia	26.7	41.0	49.3	21.5	21.3
Lithuania	5.2	8.4	15.2	35.4	20.3
Macedonia	1.3	1.6	2.4	18.9	5.2
Moldova	29.1	7.1	20.5	19.7	17.8
Poland	15.5	15.1	14.5	15.9	17.8
Romania	5.5	3.3	16.3	25.3	16.6
Russia	2.8	2.8	8.0	5.7	11.0
Slovakia	4.0	3.7	2.8	7.8	5.9
Slovenia	1.8	2.0	3.7	1.8	2.0
Tajikistan	5.1	3.3	0.9	7.5	4.8
Ukraine	3.1	5.6	6.2	9.0	8.1
Uzbekistan	2.0	0.8	5.3	3.6	2.3

Albania, BiH, Turkmenistan and Serbia & Montenegro figures are below 0.1; ** value below 0.1
Source: UNCTAD (2001) pp.312–24.

The Sectoral Split

As already intimated, not all economic sectors have proved to be equally attractive to investors from outside CEE. Some activities have not been opened to foreigners, while others have simply proved to be unappealing. Agriculture is one such. Although it remains a major sector in several countries – employing more than a fifth of the labour force in Poland and Romania – and although some of the enormous state and collective farms (together with their centralised processing plants) have survived the collapse of central planning, it has attracted

Table 3.4   FDI flows from ECE and the FSU, 1995–2000

	1995	1996	1997	1998	1999	2000
World	355 284	391 554	466 030	711 914	1 005 782	1 149 903
Developed Countries	305 847	332 921	396 868	672 027	945 687	1 046 335
ECE/FSU*	785	1 103	3 718	2 576	2 509	4 379
Albania	12	10	10	1	7	6
Armenia	**	**	**	12	13	8
Azerbaijan	175	36	64	137	336	179
Belarus	8	3	2	2	**	**
BiH	8	29	**	**	**	**
Bulgaria	**	**	**	**	17	**
Croatia	6	30	186	97	35	28
Czech Rep.	37	153	25	127	90	118
Estonia	2	40	137	6	83	157
Georgia	2	**	7	44	11	17
Hungary	43	**	431	481	249	532
Kazakhstan	**	**	1	8	4	4
Latvia	**	3	6	54	17	8
Lithuania	1	**	27	4	9	13
Macedonia	**	**	1	1	1	1
Poland	42	53	45	316	31	126
Romania	3	2	**	**	16	**
Russia	358	771	2 597	1 011	1 963	3 050
Slovakia	8	52	95	146	**	23
Slovenia	6	8	26	11	38	48
Tajikistan	**	**	**	**	17	6
Turkmenistan	**	**	**	68	**	8
Ukraine	10	**	42	**	7	1
Uzbekistan	139	**	118	60	**	57

* aggregates rounded; Kyrgyzstan, Moldova and Serbia & Montenegro figures are below 1;
** value below 1
Source: UNCTAD (2001) pp.296–300.

almost no FDI. This has been in part because of the unwillingness of many governments to allow foreigners to buy land. It has also been for structural reasons: much of the farm land in Poland and the former Yugoslavia – in which collectivisation made little progress – is in small and fragmented holdings; while in other countries the breakup of collective farms recreated a mass of small peasant farms. In both cases, Western investors found it difficult to put together land deals that would allow them to create a suitable basis for large-scale agriculture. It is also the case that much farming in the region is blighted by low standards of cultivation and plant and animal hygiene. Some Western investors in pastoral activities were obliged to start by slaughtering the stock they acquired.

There have also been substantial barriers to foreign investment in the fuel and energy industries. Once perceived to be amongst the most important elements of the CPEs, the demand for coal, brown coal and lignite fell sharply in the 1990s when more realistic prices were introduced and some of their principal industrial customers went into decline. However, governments, struggling to rationalise industries in which workers enjoyed some of the highest wages in the CPEs, faced truculent labour forces and found privatisation politically difficult. Meanwhile, foreign investors noted the growing presumption in the international community against the burning of coals of low calorific value and high sulphur content, which have formed a large part of those mined in the region. They also noted the low technical standards of many mines in the region, and the declining reserves and appalling accident records of those in Russia and Ukraine in particular. Similarly, there has been no wish, either on the part of governments in the region or foreign investors, to privatise nuclear power stations. Faced with the abysmal safety standards revealed by the 1986 Chernobyl disaster, Western governments have felt obliged to provide the money to upgrade or close these stations. However, it has been clear that the future of many of them is not only short, but will continue to be expensive, and that returns from the sale of electricity will not cover their continuing costs of improvement and ultimate decommissioning.

More hopeful have been the oil and gas industries, where much foreign interest has been generated. In particular, the shortage of such fuels in much of Western Europe, together with the uncertainties associated with Middle Eastern supplies and the relatively high cost of further developments in the northeast Atlantic, have all encouraged Western firms to invest in the FSU and especially the countries of what was Soviet Central Asia. Notwithstanding the 'Wild East' reputation of Azerbaijan, much Western investment has been attracted to its Caspian Sea fields and to the construction of new pipelines which will avoid Russia, and especially its turbulent north Caucasus region, and deliver oil and gas to the Mediterranean coast. Nevertheless, the stakes are also high in Russia itself, where the production and pipeline infrastructure created in the 1970s and 1980s in such remote areas as the Volga Basin and western Siberia is now in desperate need of replacement, but where the process of privatisation has created one of the world's largest companies – Gazprom – which has not been anxious to allow Western rivals access to its reserves.

Conversely, manufacturing has provided many opportunities for investors. The sale of many state-owned enterprises and the relatively low barriers to the import of many manufactured goods into Western European and other developed countries have encouraged a wide range of firms to move into ECE in particular. From Volkswagen's takeover of Škoda, in the Czech Republic (Pavlínek 2002), to food-processing, chemicals, engineering, pharmaceuticals, clothing and many others, Western companies have taken the chance to buy into and rationalise existing industry, and also in some cases to establish new branch plants. The two largest foreign investments in Poland have been in the automotive industry – by Fiat and Daewoo (OECD 2000).

Similarly, the service sector has offered foreign investors a wide range of opportunities. Under central planning, it was less developed than would have been the case in a market economy, and some types of service – especially those connected with the private ownership of property and the conduct of privately-owned business – were almost entirely absent. Economic transition has thus engendered considerable openings for development. That said, it has not been entirely easy for foreign investors to enter these markets. The provision of services depends in large part upon person-to-person communication, where lack of fluency in the local language is a serious barrier. Furthermore, many personal services are offered by small, local enterprises. It is unlikely that foreign investors would have much interest in, let alone access to, such investment opportunities. Moreover, it has been in this sector, above all, that state ownership has continued. In most transition states there has been no question of privatising defence, education, health, police and many forms of transport services – all important sources of employment – or even allowing any foreign contribution towards many of them. However, several countries have succeeded in attracting substantial FDI to their financial, telecommunications, wholesale and retail, and tourist industries. Some indication of the relative attractiveness of the various sectors is given by inflows of foreign investment into Poland in 2001. About 28% went to manufacturing – with a third going to food processing (Walkenhorst 2001) – though most went into the service sector, with no less than 27% of the total going to financial services and 28% to trade and repairs <http://www.polamb.nl/weh>.

The Spatial Pattern

The map of investment has also been varied. As indicated above, some countries have proved to be far more attractive to FDI than others, and there have also been marked contrasts within states. During the 1990s, the clear winners, in terms of investment per capita of the population, were Czech Republic and Hungary (Table 3.2). Substantial inflows of foreign capital – much provided by the private sector, but some from the EU – allowed these countries to run foreign-account deficits. The rapid privatisation of manufacturing industries led to many Western takeovers, followed by considerable restructuring, capital investment and notable increases in productivity. Western firms have also opened wholly-owned branch plants in these countries and ploughed money into their financial and telecommunications industries. Poland has also succeeded in attracting substantial investment into a range of manufacturing and service industries, while Azerbaijan and Kazakhstan have attracted oil and gas investment. Some of the smallest countries – Croatia, Estonia, Latvia, Slovakia and Slovenia – have also proved to be attractive to foreign investors. In contrast, those countries which have been slow or simply unwilling to relinquish the public ownership of production have attracted next to no investment.

Prominent among these have been Belarus, for ideological reasons; Serbia and Montenegro, because of the turmoil caused in the region by the former Serb leader Slobodan Milošovič; and Ukraine, because of its inability to establish smoothly-working democratic institutions capable of making and carrying through clear decisions. Russia – by far the region's largest country, but another in which institutional change has been slow – has also attracted far less investment than its size might suggest. The contrast between its stock of about $20bln worth in 2000 and China's $347bln is marked. Similar, but not identical, contrasts are also to be found in the contribution that FDI has made to capital formation in these countries (Table 3.3). Thus, Azerbaijan and Kazakhstan have relied on foreign investment for more than a third of their capital investment in recent years. Countries, such as Poland, Russia and Ukraine, have received a much smaller proportion, raising most from within the country. However, there is also a group of countries which, although they have received only small amounts of FDI, have been heavily reliant on it for capital formation. These include Armenia, Bulgaria and Georgia. It is clear that these countries, whose economies are weak, would be in a far worse position in the absence of even the small amounts that they have received.

There is also a striking contrast within countries. In general, large, and especially capital, cities have been far more successful in attracting investment than smaller settlements and rural areas. Even under central planning, in many of the countries, the capital city was by far the largest, most prosperous and most economically developed part of the country, as well as being the place with the best international transport links, not to mention being the seat of government. It is hardly surprising that many Western firms should have opened their first branches in such places (Anon 2002c; Pavlínek 1998). However, there have been other places that have also attracted investment. Thus, non-capital, but large, historic cities, such as Kraków and St Petersburg, have attracted the attention of hotel chains and other tourist-related industries. Enterprising local mayors and governors, offering the prospect of well-regulated societies, have enticed foreign investors to such places as Nizhny Novgorod and Yekaterinburg (Anon 1999; Ruble and Popson 1998; Turnock 2001; Zimine and Bradshaw 1999); and some border areas have benefited, particularly those adjoining the EU in Hungary and Poland, from their relatively low labour costs (Kiss 2001). Some of Russia's Pacific coast has also faired relatively well, though the absolute level of foreign investment has not been high (Kontorovich 2001). Rural and remote areas, in contrast, have attracted almost no foreign money, except where it has been allowed to participate in the oil and gas industries. It is ironic that the admission of several ECECs to the EU seems to have created something of a shadow over those parts of them that will lie close to what will become the bloc's new eastern frontier: investors have shied away from what will become heavily-policed border zones (Anon 2001), and the UN indicate that there were almost no foreign affiliates in eastern Poland in 1999 (UN 2001 p.65).

A Tale of Two Cities

These contrasts are clearly illustrated by the fortunes of the two largest, and neighbouring, cities in Poland: Łódź and Warsaw. The case of Łódź is of especial interest, for its authorities – in their search for FDI – are now attempting to reproduce conditions similar to those that led to its foundation and much of its early growth. Poland disappeared from the map of Europe in the late 18th century, partitioned by its neighbours. However, a vestigial remnant within Russia, the Congress Kingdom, was recognised by the Congress of Vienna in 1815; and the Russian government accorded it a somewhat greater degree of local autonomy with regard to internal affairs than other parts of the country, so creating a potential which a handful of Polish leaders exploited. Situated between the craftsmen of Saxony, to the west, and the huge potential demand for goods, and especially textiles, in the Russian Empire, to the east, the Kingdom's Minister of Finance, Xavery Drucki-Lubecki, and Director of its Department of Industry and Crafts, Stanisław Staszic, encouraged Western capitalists and skilled workers to settle in the Kingdom, while Rajmund Rembieliński, the Prefect of Mazovia, oversaw the laying out of a modest settlement in an otherwise entirely rural area, eighty miles to the southwest of Warsaw (Ginsbert 1961).

Little development took place, however, not least because, after 1830, the Congress Kingdom was cut off from markets in the rest of the empire and the town had no local supplies of fuel. Western textile producers wishing to gain access to the Russian market moved instead to Białystok, the nearest site to Western Europe within Russia proper, or to Moscow. However, the removal of the barrier between the Congress Kingdom and the rest of the empire in the 1850s, the arrival of the railway in the 1860s – providing a link with the coalfields of Upper Silesia and with St Petersburg and Moscow – and the imposition of a swingeing tariff on Western European imports – the so-called gold tariff – of 1873, provided powerful incentives for FDI. Manufacturers from Belgium, Britain and France, but especially Germany, flocked to the nearest available settlement within this protected market which provided appropriate transport links, but also an existing textile industry with its local infrastructure and labour force, much of which was already owned and run by expatriate Germans. Within 20 years Łódź had become 'the Polish Manchester', and developed not only as Poland's principal centre of manufacturing of cotton textiles, but also of woollens, worsteds and linen. In 1886, out of the total value of woollen textile production in the Congress Kingdom of Rb29mln, 12mln was from plants owned by Germans and 13mln from those in the hands of people who had settled in Poland within the previous 25 years. In 1914, 400 of the 540 plants were owned by foreigners, including all except one of those employing more than 500 workers (Sroka 1914). Only a handful of these mills were located other than in Łódź and its environs. In the same year, three-quarters of the region's jobs depended on textiles, and most of its products were exported to other parts of the Russian Empire. Together with its satellites of Pabiance and Zgierz, Łódź had a population of more than half a million, in a townscape dominated by huge mills and factory chimneys, and covered in soot. Thanks to FDI, it had become the second largest Polish city after Warsaw.

The contrast between that 19th century development and what occurred in the twentieth was stark. Following the re-establishment of the Polish state at the end of the First World War, relations with its neighbours were fraught. War with Russia, a territorial dispute with Lithuania and problematic relationships with Germany and Czechoslovakia, meant that Poland was at odds with all of them. The self-imposed isolation of the Soviet Union severed Polish links with its principal pre-1914 market, while rampant domestic inflation in the 1920s and world recession in the 1930s further reduced the scope for international trade. There was almost no further foreign investment, and many mills closed. Unemployment in Łódź, as elsewhere in the country, was high. Nor was the period between 1945 and 1989 much better. Obliged to pursue economic policies dictated by Soviet leaders, and with only minimal contact with the non-communist world, Poland created an economic infrastructure which proved to be but poorly-suited to the needs of its citizens (Slay 2000). At a time when the textile industry in northwest Europe was facing severe competition from southern Europe and developing countries, production in Łódź – now entirely in mills seized by the state – revived, in part to supply markets in other communist countries in the Third World under protective agreements worked out by Comecon, and in part because the Polish market was closed to all but state-owned Polish producers. Many older factories were closed, but new mills were built in the 1970s on the edge of the city. The burning of coal was phased out as electricity replaced steam power. However, much of the town centre continued to bear earlier grime. More people moved in from the countryside, in part to work in other industries which were established in the city, and the population of the region rose to one million. Much new housing was constructed, but it was in drab, high-rise blocks on peripheral estates that lacked shops and places of entertainment. Moreover, the economy, like that of many another isolated industrial city in central and eastern Europe, was still dominated by a single industry, and one that, by 1989, had been cut off from the stimulus of competition for 50 years.

Not surprisingly, as Poland opened up to international trade, there was a sharp fall in the demand for its products. High levels of unemployment re-appeared – by 2000, the city was suffering more than twice the rate in other large Polish cities and six times that in Warsaw – and weeds grew from the roofs of abandoned mills. Despite the establishment of a Special Economic Zone (SEZ), where foreign companies are offered a reduction in local taxes, there has been only a little foreign investment: Coats Viyella in textiles and a small number of other multinational firms in other sectors of the city's economy. City administrators have spoken about a new 21st century wave of FDI that might flow to the city, once the Polish motorway network, with its junction of west-east and north-south routes outside the city, is eventually complete <http://www.uml.lodz.pl> but neither the intrinsic characteristics of Łódź and its environs nor its location do it many favours. Not only has little money been available to do more than pedestrianise the main street – thus leaving much of the city in its unattractive, communist form – but the cosmopolitan population of 19th and early 20th century Łódź, which proved to be so attractive to potential investors in Western Europe, has long since vanished. The region around the city is almost devoid of historic or scenic interest. Furthermore, a connection with the European motorway network is unlikely for many years and the city also lacks an airport with more than a handful of regular links, even within Poland. Set in the middle of the country, Łódź is distant from both the ports to the north and the Germany frontier to the west: in short, from places that are already benefiting from their close links with other parts of the EU and beyond. Entry into the EU in 2004 will probably mean the abandonment of the SEZ. During the 1990s, there was a substantial movement of young people into further and higher education (indeed the city has one of the highest take-up rates in the country – which may serve as some attraction for FDI) and the city authority claims that both the cost of living and of labour are cheap. However, there has also been a substantial movement of population from the city, in search of better prospects elsewhere. The authority's aspiration that Łódź 'will become one of the more important centres of industry, trade and services in Europe' is unlikely to be fulfilled in the foreseeable future.

Not so with Warsaw. Like Łódź, it lies far from mineral resources and fuels, but has relied on its status as an administrative centre and the focus of Polish life to attract investment. In the 19th century, much of this came from within the Congress Kingdom or the wider Russian Empire, rather than from abroad, though the city also attracted investors from Western Europe, who established a wide range of manufacturing and service industries, catering to the population of the empire as a whole. However, the economy of Warsaw has always been more broadly based than that of Łódź, and its centrality to Polish life has been unchallenged. Moreover, as the largest city and the one with the best developed international transport links, it was in the prime position to attract foreign investment after 1989. The effects have been marked. The city is by far the most important location of foreign affiliates in the country (UN 2001 p.65). By the end of 1999, 13,200 foreign-owned firms – about one third of all the foreign-owned outlets in Poland – were operating there <www.warszawa.um.gov.pl>: 44% in retailing, 23% in banking and finance and 12% in manufacturing. Foreign investors have refurbished some of the city's principal hotels, and many specialist Western retailers and fast-food outlets have one or more branches within the city, especially in the CBD. Several foreign-owned chains of hypermarkets are operating. 27 foreign banks, including all the major ones, have established branches. In 2002, no less than 83 of the 100 largest firms in the country employed Western accountants as their auditors, and 82 of these accounts were with firms based in Warsaw (Rzecz Pospolita, 18 Czerwca 2002). Coca-Cola and Daewoo are amongst the foreign firms that are involved in manufacturing. The skyline, dominated since the early 1950s by the Soviet skyscraper Palace of Culture and Science, now includes a series of tall buildings, many of which have been constructed by or for foreign companies, in part for offices and in part for luxury apartments for foreigners working in Warsaw. Unlike much of the rest of the country, the city enjoys almost full employment. On the eve of the EU's formal invitation to Poland to become a member, *The Times* (December

12 2002) noted that much of the [Warsaw] region's economic mini-boom has been financed by foreign investment.

Conclusion

The effects of FDI have not been merely superficial or local. Foreign investors, putting money into activities for which not only Warsaw – but also Poland and CEE as a whole – enjoy comparative advantages, are not only restructuring the economies of these places, but embedding them increasingly into the European and world economy (Weresa 2001), in ways that later chapters in this volume will examine in more detail. It is likely that some places will never attract substantial foreign investment, and that their economies will continue to decline. But others have already done so and will continue to do so. The era of CEE's economic isolation is over.

References

Anon (1999), 'Yekaterinburg', *Business Central Europe*, February p.62.

Anon (2001), 'Lublin', *Business Central Europe*, July/August p.58.

Anon (2002a), 'A Survey of International Finance', *The Economist*, May 18, pp.26–7.

Anon (2002b), 'Privatisation in Eastern Europe', *The Economist*, November 16, p.82.

Anon (2002c), 'The Balts and the EU', *The Economist*, December 14, pp.37–8.

Artisien-Maksimenko, P. (ed) (2000), *Multinationals in Eastern Europe*, Macmillan, Basingstoke.

Collis, C., Berkeley, N. and Noon, D. (1999), 'Attracting FDI to East European Economies: Can Lessons be Learned from the UK Experience?', in N.A. Phelps and J. Alden (eds), *FDI and the Global Economy: Corporate and Institutional Dynamics of Global-Localisation*, The Stationery Office, London, 139–156.

Deichmann, J.I. (2001), 'Distribution of FDI among Transition Economies in CEE', *Post-Soviet Geography and Economics*, 42, 142–52.

Du Pont, M. (2000), *FDI in Transitional Economies: a case study of China and Poland*, Macmillan, Basingstoke.

Ginsbert, A. (1961) *Łódź: Studium Monograficzne*, Wydawnictwo Łódźkie, Łódź.

Kiss, J.P. (2001), 'Industrial Mass Production and Regional Differentiation in Hungary', *European Urban and Regional Studies*, 8, 321–8.

Kontorovich, V. (2001), 'Economic Crisis in the Russian Far East: Overdevelopment or Colonial Exploitation?' *Post-Soviet Geography and Economics*, 42, 391–415.

Meyer, K.E. and Pind, C. (1999), 'The Slow Growth of FDI in the Soviet Union Successor States', *Economics of Transition*, 7, 201–14.

Pavlínek, P. (1998), 'FDI in the Czech Republic', *The Professional Geographer*, 50, 71–85.

Pavlínek, P. (2002), 'Restructuring the CEE Automobile Industry', *Post-Soviet Geography and Economics*, 43, 141–77.

Ruble, B.A. and Popson, N. (1998), 'The Westernization of a Russian Province: The Case of Novgorod', *Post-Soviet Geography and Economics*, 39, 433–66.

Slay, C. (2000), 'The Polish Economic Transition: Outcome and Lessons', *Communist and Post-Communist Studies*, 33, 49–70.

Sroka, A.R.S. (1914), *Przemyśl I Handel Kró lestwa Polskiego 1914r*, n.p., Warsaw.

Stare, M. (2001), 'Advancing the Development of Producer Services in Slovenia with FDI', *Service Industries Journal*, 21, 19–34.

Stare, M. (2002), 'The Pattern of Internationalisation of Services in Central European Countries', *Service Industries Journal*, 22, 77–91.

Turnock, D. (2001), 'Location trends for FDI in ECE', *Environment and Planning C*, 19, 849–80.

UNCTAD (2001), *World Investment Report 2001*, UN, New York.

Walkenhorst, P. (2001), 'The Geography of FDI in Poland's Food Industry', *Journal of Agricultural Economics*, 52, 71–86.

Weresa, M. (2001), 'The impact of FDI on Poland's Trade with the EU', *Post-Communist Economies*, 13, 71–83.

Woodruff, D.M. (1999), *Money Unmade: Barter and the Fate of Russian Capitalism*, Cornell University Press, Ithaca, N.Y.

Zimine, D.A. and Bradshaw, M.J. (1999), 'Regional Adaptation to Economic Crisis: The Case of Novgorod Oblast', *Post-Soviet Geography and Economics*, 40, 335–53.

Chapter 4

Foreign Direct Investment and International Trade

Alan Smith

This chapter will examine the contribution of FDI to the modernisation and upgrading of the manufacturing sector of the former CPEs of ECE and the FSU as reflected through the structure of exports. The first section will examine the structures of production, investment and trade in the Soviet era and the demand this has created for modernisation through FDI. The second section will examine the size and distribution of FDI flows to the region and the third section will examine the relationship between FDI and changes in the structure of exports from the transition economies to the EU. The final sections will ask whether the patterns outlined in the analysis indicate the presence of 'agglomeration' effects that may result in a long-lasting, regional divergence between some transition economies and the rest of Europe.

Structures of Production, Investment and Trade in the Soviet Era

These structures were heavily influenced by the Stalinist policy of industrialisation – adopted in the Soviet Union in the 1930s – which gave a high priority to investment in, and production of, heavy industrial goods and a relatively low priority to the production of consumer goods and agricultural produce. This resulted in the large-scale production of energy-intensive metallurgical and civil engineering products (with a relatively low emphasis on high-technology products outside the armaments and aviation industries) and the mass production of basic staple household goods, with a very low emphasis on luxury goods. Levels of production and trade flows between the USSR's constituent republics during the entire Soviet era were determined by central planners in Moscow. Domestic industrial priorities were reinforced by a highly centralised state monopoly of foreign trade administered by the Ministry of Foreign Trade which limited trade and production links with the outside world, in an attempt to contain economic dependence on the West and ensure that imports were restricted to commodities consistent with government priorities. As a side effect of this policy, Soviet enterprises were protected from foreign competition and guaranteed a secure market for their products, regardless of quality standards.

The Stalinist model of industrialisation was extended to the ECECs that fell under Soviet hegemony after the Second World War with a consequent effect on the structure of trade. Comecon[1] administered a regional trade preference zone which limited contacts with the outside world, forcing the ECECs to redirect trade away from their inter-war partners in Western Europe (and Germany in particular) towards the Soviet Union. Trade patterns between the communist economies resembled the spokes of a wheel radiating from Moscow as the Soviet Union became the largest trade partner for each of the ECECs: exporting energy, raw materials and iron and steel goods which reflected her favourable resource endowments, while importing machinery and equipment, consumer goods and foodstuffs, reflecting the needs of the Stalinist model of industrialisation. From the early 1960s onwards, ECE industrial exports to the FSU embodied both low quality and technological obsolescence relative to the demands of Western markets. As late as the end of the 1980s, civil engineering products still predominated in exports of machinery and equipment: metallurgical machine-tools, boilers and turbines, hoisting equipment, agricultural equipment and machine tools to produce relatively obsolete products. Only Bulgaria and the GDR exported computing equipment to the FSU. Exports of consumer goods predominantly consisted of clothing, footwear, furniture, household utensils and unsophisticated medicines intended to meet the demands of a large, but relatively low-to-middle income market. Exports of consumer electrical goods and appliances, higher-quality cars and cosmetic goods were virtually non-existent. Similarly ECE exports of agricultural products largely consisted of products with a relatively low degree of processing, including grain and animal feedstocks, fruit and vegetables and tinned and canned items for household consumption (Smith 2000, Chapter 4).

FDI in the form of investment by a MNE to gain a controlling interest over a production facility in another country was effectively outlawed under communism. Investment flows between communist states that resulted in the acquisition of assets or titles of ownership in one foreign country by a legal person in another country were, in the main, not permitted. Although Comecon stimulated a number of joint investment projects in the 1970s and 1980s these were supranational in nature rather than multinational: largely joint construction projects in which a group of member states provided contributions in kind and were repaid in the form of outputs from the completed project while the venture remained the property of the host economy. The most successful joint ventures involved ECE contributions to the exploitation of Soviet energy sources for joint consumption (e.g. pipelines in the FSU paid for by deliveries of natural gas).

The slowdown in economic growth in the region in the mid-1960s contributed to the realisation that extreme protection from foreign competition and isolation from Western production techniques and foreign investment were contributing to technological backwardness. Attempts were then made to develop cooperation ventures with Western MNEs. During the 1970s and 1980s, all of the communist economies (except the GDR) enacted legislation permitting the operation of joint ventures involving Western MNEs, largely in an attempt to acquire access to

Western technology. However, these provided the MNEs with very restricted rights of ownership and in many cases amounted to little more than product-sharing agreements with capital inputs exchanged for output from the plant. The Comecon economies resorted to purchases of licences and imports of machinery and equipment for industrial modernisation on credit, largely as a substitute for multinational investment in the 1970s. The intention was that modernised plants would generate exports to the West to repay external debt. However, penetration of Western markets was limited by the patterns of production and trade already established between the Comecon states. ECE exports of manufactured goods to the EU in the late 1980s largely consisted of labour-intensive consumer goods (including clothing, footwear and furniture) and resource-intensive goods (iron and steel products, fertilisers and unsophisticated chemicals) and relatively simple engineering products (Smith 2000, Chapter 4). The inability to penetrate Western markets on a large scale with these products resulted in growing problems of external indebtedness in the late 1979s and 1980s and contributed to the collapse of the system.

The Role of FDI in the Reintegration of the Transition Economies into the World Economy

The disintegration of the Soviet economy between 1989–91 contributed to the collapse of the Comecon trading system and posed major problems for the CPEs in ECE, including the Baltic States. Firstly they were faced with the need to find new markets for manufactured products to preserve output and employment. Secondly, the redirection of trade to traditional interwar markets in Western Europe became a major political goal for most states, with the ultimate objective of EU membership. However, ECE exports to the EU in the late 1980s were concentrated on items with a relatively low income-elasticity of demand for which excess capacity already existed in Western markets and which also faced strong competition from producers in low wage economies. Furthermore, the existing capital stock was not suited to producing the sophisticated goods that were demanded by Western consumers. Consequently a major expansion of exports to the EU required extensive industrial restructuring, which in turn needed major imports of capital goods that could not be met by export revenues, resulting in current account deficits attracting substantial capital inflows. FDI was expected to play a critical role in this process, as a supplier of both financial capital and of technology, know-how and patent rights. Furthermore, Western MNEs benefited from economies of scale in marketing products in EU markets, including extensive supply and retailing outlets and high levels of advertising and brand-name recognition that could not easily be appropriated by newcomers. This, in turn, increased the dependence of enterprises in former CPEs on contacts with MNEs.

FDI Flows into Transition Economies since the Collapse of Communism

Table 4.1 provides a summary of the major aggregates of FDI into the transition economies. The first three columns provide estimates of the stock of FDI at the end of 2001 by total value, value per capita and percentage of GDP. Columns 3–5 measure net inflows of FDI taken from balance of payments data reduced to a per capita basis to give an indication of the progression of FDI inflows over the period, while the final column indicates FDI inflows as a percentage of GDP over the period

Table 4.1   FDI flows to ECE and the FSU, 1991–2001

	FDI Stock end 2001			FDI flows: $pc/per annum			FDI Flows as % GDP
	Total $ bln	$ pc	Stock as % GDP	1991–4	1995–8	1999–2001	1999–2001
Central East European States							
Czech Rep.	28.4	2,778	50.1	75	210	517	9.9
Hungary	23.8	2,356	45.9	156	244	161	3.3
Poland	46.4	1,201	26.3	10	77	181	4.3
Slovakia	5.9	1,092	28.7	25	39	260	7.2
Slovenia	2.0	1,018	10.8	40	120	99	1.0
Baltic States							
Estonia	2.9	2,118	52.2	106	181	211	5.6
Latvia	2.7	1,125	35.1	69	150	119	4.3
Lithuania	2.9	835	24.3	8	102	125	3.8
South East European States							
Albania	0.8	249	24.9	14	21	41	3.4
BiH	0.6	150	13.4	na	25	123	3.3
Bulgaria	4.0	504	29.5	7	39	99	6.5
Croatia	6.4	1,439	31.5	17	99	285	6.6
Macedonia	0.9	431	25.4	3	27	108	6.1
Romania	7.7	355	19.4	6	46	48	3.0
Serbia & Montenegro	1.6	149	14.6	na	213	101	0.6
Former Soviet Republics							
Belarus	1.4	143	11.7	na	16	20	1.4
Moldova	0.5	128	37.1	na	15	17	5.2
Russia	24.2	168	7.8	na	11	1	0.1
Ukraine	4.2	86	11.2	na	10	12	1.8

Source: Columns 1–3 from EIU (December 2002); Columns 3–6 estimated from EBRD Transition Reports.

from 1999–2001.[2] The stock of FDI at the end of 2001 in the former Comecon region and former Yugoslavia has been estimated at \$186.4mln, equivalent to \$456pc or 21.7% of the region's GDP (EIU 2002). However, this stock is distributed relatively unequally across the region on a per-capita basis, with the major share accumulating to Czech Republic, Hungary, Poland and Slovakia, along with the Baltic States and the two former Yugoslav republics of Croatia and Slovenia. The SEECs (excluding Croatia), as well as the former Soviet republics in Europe, have been substantially less successful.

Czech Republic, Estonia and Hungary have been the most successful in attracting FDI with a per capita stock of more than \$2,000, equivalent to more than 50% of GDP in Czech Republic and Estonia and 45.9% in Hungary. Meanwhile, Croatia, Latvia, Poland, Slovakia and Slovenia have attracted an FDI stock of over \$1,000 per capita. Inflows into Hungary have declined since the peak of 1995 but have accelerated into Czech Republic, Poland and Slovakia since the late 1990s. Lithuania's low stock, in relation to the Baltic states as a whole, can largely – but not entirely – be explained by low inflows in the early stages of the transition. Slovenia is an anomalous case, with a stock that is high in per capita terns (\$1,108) yet one of the lowest in the region in relation to GDP. Per capita FDI inflows into the SEECs have been substantially below the levels experienced elsewhere in ECE, although their relatively low income levels increases the size of the FDI stock expressed as a percentage of GDP.[3] Growing inflows into Bulgaria have taken the FDI stock to \$504pc, equivalent to 29.5% of GDP. Also, Macedonia has benefited from an FDI inflow equivalent to 12.7% of GDP in 2001 which has taken the FDI stock to 25.4% of GDP. However, Bosnia and Hercegovina, Romania and Serbia and Montenegro had FDI stocks at the end of 2001 that were below 20% of GDP. Finally, FDI flows into the former Soviet republics in Europe have been relatively sluggish. The stock in Russia was measured at only 7.8% of GDP at the end of 2001 and at just over 11% in Ukraine and Belarus. The high FDI stock in Moldova as a percentage of FDI largely reflects the very low level of per-capita GDP.

The Impact of FDI on Restructuring and Trade Flows

FDI can stimulate trade by the creation of production for export in the host country, or may replace trade flows by establishing production bases in the host economy which replace imports, or, in many cases, may involve a combination of the two. It is not always possible to measure the real contribution of FDI to trade by measuring the volume of exports and imports that are channelled through subsidiaries of multinational organisations that have invested in the host economy. For example, exports from an existing (brownfield) enterprise that has been taken over by a MNE will be classified as a growth in trade that results from FDI, even if the new owner does nothing to change production techniques or the volume of exports from the enterprise. Similarly, FDI may create positive externalities that improve the competitiveness of other enterprises in the economy which improves their export capability. Consequently, the remainder of this chapter will examine the relationship between FDI at the aggregate level and changes in the structure of exports of manufactured goods from the transition economies to the West.

Conventional neoclassical economic theory, derived from the Hecksher-Ohlin theory of factor proportions, predicts that transition economies that are labour-abundant and capital-scarce should specialise in the production of goods that are made by labour-intensive processes and export these, and should import goods that are made by capital-intensive processes from capital-abundant economies. Once FDI is introduced into the analysis, it would be expected that financial and physical capital would flow from capital-intensive economies to labour-intensive economies, attracted by the relatively high returns that can be obtained from combining scarce capital with abundant, and relatively-cheap, labour and that MNEs would develop production bases for export in low wage economies. Critically, the creation of 'production for export bases' may simply involve delegating basic labour-intensive functions to low wage economies without creating significant improvements in labour productivity in the host economy or may involve substantial investment in restructuring and upgrading production facilities.

Trade flows involving MNEs and FDI between countries that are each capital-abundant do not, in the main, involve the exchange of products produced by labour-intensive methods for products produced by capital-intensive methods but generate the simultaneous exchange of products produced by capital-intensive methods. Given that many of the ECECs possessed labour forces with a relatively high degree of education and training in engineering skills compared with low wage economies – but had an obsolete capital stock when the communist system collapsed – it might be expected that MNEs would seek to upgrade production in the host economies to generate a greater level of capital-intensive processes. Furthermore, MNE investment may create external economies of scale and other positive externalities (e.g. demonstration effects) which motivate entrepreneurs within the host economy to upgrade production techniques and/or stimulate investment in education and training. The rest of this section will examine the factor-intensity of exports from the transition economies, and changes in the factor intensity of exports, to see if there is any evidence of modernisation and upgrading of export structures in ECE manufacturing in the process of transition. An attempt will then be made to link this to aggregate levels of FDI.

Methodology

In order to ensure comparability of data between different exporting countries, the analysis will use trade partners' import data rather than the export data of the transition economies themselves. The most detailed and comprehensive set of trade data is provided on disk by the EU in the form of the COMEXT database (EC nd). The EU forms the largest market for all the transition economies and can be taken as representative of the general export structure of the individual economies. This data has been used in Tables 4.2 and 4.3. The factor intensity of the transition

Table 4.2 Exports of manufactured goods from transition economies to the EU, 1998

	Total €mln	Factor Intensity €mln			Factor Intensity % of total		
		Human Capital	Labour	Resource	Human Capital	Labour	Resource
Central East European States							
Czech Rep.	13,108	6,438	4,711	1,960	49.1	35.9	15.0
Hungary	12,902	8,440	3,357	1,104	65.4	26.0	8.6
Poland	13,311	4,590	5,631	3,089	34.5	42.3	23.2
Slovakia	5,009	2,579	1,447	983	51.5	28.9	19.6
Slovenia	4,978	2,357	1,748	873	47.4	35.1	17.5
Baltic States							
Estonia	1,105	1,105	468	158	43.3	42.3	14.3
Latvia	555	56	282	216	10.2	50.9	38.9
Lithuania	997	209	535	253	20.9	53.7	25.4
South East European States							
Albania	179	9	151	19	5.1	84.2	10.7
BiH	193	20	138	35	10.3	71.3	18.3
Bulgaria	1,848	320	797	731	17.3	43.1	39.6
Croatia	1,542	472	875	195	30.6	56.8	12.6
Macedonia	516	48	266	202	9.4	51.5	39.1
Romania	4,791	797	3,121	872	16.6	65.2	18.2
Serbia & Montenegro	852	155	324	372	183	38.1	43.7
Former Soviet Republics							
Belarus	380	61	180	139	16.1	47.5	36.3
Moldova	59	4	49	6	6.3	83.4	10.2
Russia	7,370	1,459	1,155	4,756	19.8	15.7	64.5
Ukraine	1,390	397	371	622	28.5	26.7	44.8
Intra-EU	–	–	–	–	61.8	23.3	14.8

Source: estimated from COMEXT database.
Categories derived from Wolfmayr-Schnitzer (1998).

economies exports has been derived from a highly simplified version of a more sophisticated set of categories devised by Wolfmayr-Schnitzer (1998). The version used here divides exports of manufactured goods into three main categories: (a) products involving processes that are intensive in human capital and which command above-average wage rates: largely items of machinery and equipment, including road transport equipment, domestic electrical appliances, chemical

Table 4.3 Change in exports of manufactured goods from transition economies to the EU, 1995–1998

	Total €mln	Factor Intensity €mln			Factor Intensity % of total		
		Human Capital	Labour	Resource	Human Capital	Labour	Resource
Central Eastern Europe							
Czech Rep.	5,613	3,540	1,859	215	63.1	33.1	3.8
Hungary	6.838	5,422	1,236	160	79.3	18.1	2.3
Poland	3,784	1,817	1,711	256	48.0	45.2	6.8
Slovakia	2,225	1,726	483	16	77.6	21.7	0.7
Slovenia	950	678	245	28	71.4	25.8	2.9
Baltic States							
Estonia	540	327	199	14	60.6	36.9	2.6
Latvia	198	17	142	39	8.5	71.7	19.7
Lithuania	466	96	303	67	20.6	65.0	14.3
South East European States							
Albania	66	3	64	–2	5.2	98.0	–3.2
BiH	174	17	124	33	9.8	71.2	19.0
Bulgaria	395	55	344	–4	13.9	87.1	–1.0
Croatia	20	30	38	–48	151.3	191.3	–242.4
Macedonia	21	–98	20	99	–456.4	95.2	461.7
Romania	1,681	366	1,333	–18	21.8	79.3	–1.1
Serbia & Montenegro	823	132	323	368	16.0	39.2	44.7
Former Soviet Republics							
Belarus	–50	–17	42	–76	na	na	na
Moldova	19	–2	25	–4	–10.3	132.3	–22.0
Russia	–782	–111	386	–1,057	na	na	na
Ukraine	378	104	162	112	27.4	42.8	29.8

Source: as for Table 4.2.

products including pharmaceuticals and cosmetics and photographic apparatus and optical equipment; (b) products involving labour-intensive processes: largely clothing, furniture, footwear, household equipment, tractors and less-sophisticated transport equipment and printed matter; and (c) products that are resource-intensive in production: largely iron and steel products, non-ferrous metals, glassware, pottery, paper and wood products (but not including exports of energy products, including oil, gas and coal and unprocessed raw materials).

Results

Table 4.2 provides data on the factor intensity of exports of the former CPEs to the EU-15 in 1998. It can be seen that commodities that are made by human capital-intensive processes accounted for 61.8% of trade between EU members, goods made by labour-intensive processes for 23.3% and goods made by resource-intensive processes for 14.8%. For simplicity these will subsequently be referred to as human capital-intensive goods, labour-intensive goods and resource-intensive goods. There is a significant degree of variation between the factor intensity structure of the exports of manufactures from the individual transition economies to the EU-15, ranging from Hungary at one extreme to Albania and Moldova at the other. Human capital-intensive goods accounted for 65.4% of Hungarian exports to the EU in 1998, a proportion that is higher than that prevailing in intra-EU trade. The next tier of states – Slovakia (51.5%), Czech Republic (49.1%), Slovenia (47.4%) and Estonia (43.3%) – also recorded high levels of exports of human capital-intensive goods, with Slovenia recording €1,179 per capita, Hungary €832, Estonia €762, Czech Republic €643 and Slovakia €475 per capita. Czech Republic, Estonia and Hungary had attracted large inflows of FDI by that time, while Slovakia and Slovenia had attracted only intermediate levels. Croatia and Poland occupied intermediate places with human capital-intensive goods accounting for 34.5% and 30.6% of exports of manufactured goods respectively. The two remaining Baltic States – Latvia (10.2%) and Lithuania (20.9%) – had relatively low proportions of human capital-intensive exports, while the SEECs – including Albania, Bulgaria and Romania, as well as the war-affected former Yugoslav republics – all recorded proportions below 20% (having attracted only relatively low inflows of FDI by 1998). Former Soviet republics in Europe also had low proportions of exports of human capital-intensive goods, although this partly reflected high proportions of exports of resource-intensive manufactured goods in the case of Russia and Ukraine. Nevertheless, Russia (€10 per capita) and Ukraine (€7) recorded very low levels of exports of human capital-intensive products. Exports of labour-intensive goods accounted for more than 50% of exports of manufactured good for eight countries: Albania (84.2%), Moldova (83.4%), Bosnia and Hercegovina (71.3%), Romania (65.2%), Croatia 56.8%, Lithuania (53.7%), Macedonia (51.5%) and Latvia (50.9%). Significantly, all had failed to attract significant inflows of FDI by the mid-1990s.

Table 4.3 provides a picture of changes in the structure of exports of manufactured goods from the ECECs to the EU in 1995–98 in order to examine if there has been any significant shift in factor intensity from labour-intensive towards human capital-intensive exports. The first four columns provide data on the growth in the value of exports of manufactured goods, broken down by factor intensity category and the remaining three columns indicate the share of the growth of manufactured exports that can be attributed to each category. Trade in human capital-intensive goods within the EU grew more rapidly than other categories and accounted for 75% of the growth of intra-EU trade in manufactured goods between

1995 and 1998, while trade in labour-intensive products accounted for only 14.4% and resource-intensive products for only 10.6%. As a result, the proportion of human capital-intensive goods in intra-EU trade in manufactures rose from 58.4% in 1995 to 61.9 % in 1998. The declining proportion of trade in resource-intensive products during 1995–8 is partly accounted for by falls in relative prices of raw materials following the Asian crisis of 1997. This has been reflected in absolute falls in exports of resource-intensive goods in seven transition economies over the period which has resulted in absolute falls in total exports of manufactured goods in the case of Russia and Belarus (Table 4.3, columns 1 and 4).[4] Hungary (79.3%) and Slovakia (77.6%) have shares of growth attributable to human capital-intensive exports that are higher than the intra-EU average. Column 5 in Table 4.3 shows that Slovenia (71.4%), Czech Republic (63.1%) and Estonia (60.6%) all have relatively high shares of growth attributed to human capital-intensive products and significantly increased the share of human capital-intensive exports in exports of manufactured goods during 1995–8. These economies also display a relatively high growth in the volume of exports to the EU over the period ranging from €674 per capita in Hungary to €372 in Estonia. Exports of human capital-intensive goods accounted for 48% of the growth of exports from Poland. Elsewhere, the growth of exports of human capital-intensive goods was relatively disappointing over the period. Russia, Belarus, Moldova and Macedonia all experienced falls in the value of exports of human capital-intensive goods, while exports in this category accounted for less than 22% of the growth of exports from Albania (5.2%), Latvia (8.5%), Bosnia and Hercegovina (9.8%), Bulgaria (13.9%), Serbia and Montenegro (16%), Lithuania (20.6%) and Romania (21.8%). These economies all displayed a high dependence on labour-intensive products in the growth of their exports over the period.

Changes in the product structure of exports from transition economies to the EU (Tables 4.4 and 4.5) provide a further analysis of the product structure of export growth from the transition economies to the EU during 1995–8. Table 4.4 shows that exports of machinery and equipment (including transport vehicles) accounted for over half of the growth of exports of manufactured goods from Slovakia (84.3%), Hungary (83.6%), Slovenia (75.7%), Czech Republic (69.5%) and Poland (54.2%). By 1998 exports of road vehicles constituted a significant proportion of total exports of manufactured goods from Czech Republic to the EU (38.2%), Slovakia (60.2%) and Slovenia (46.3%). Road vehicles accounted for 57.7% of the growth of exports from Slovakia to the EU and passenger cars for 51.3%. The corresponding figures for Czech Republic were 32.6% and 20.5%, and for Slovenia, 44.7% and 25.2% respectively. Hungary's exports of machinery and equipment were more diversified with exports of office equipment and telecommunications equipment of €2.6bn accounting for 20% of exports of manufactured goods, power-generating equipment for 20.6% and electrical equipment for a further 13.2% in 1998. Export growth from Hungary during 1995–98 was also more diversified with office equipment accounting for 17%, telecommunications equipment for 16%, power generating equipment for 24% and road vehicles for only 8.8%.

Table 4.4 Product structure of growth of exports of manufactured goods from Central Eastern Europe and Estonia to the EU, 1995–1998

	Czech Republic	Estonia	Hungary	Poland	Slovakia	Slovenia
Total	5,613	540	6,838	3,784	540	3,784
Machinery	3,902	261	5,716	2,052	261	2,052
% of growth of which	69.5	48.3	83.6	54.2	48.3	54.2
Office Equipment	110	–2	1,139	19	–2	19
Telecomms	55	184	1,088	437	184	437
Electrical	798	33	928	583	33	583
Power Generation	202	6	1,635	79	6	79
General Machinery	500	10	293	176	10	176
Road Vehicles	1,829	9	600	710	9	710
of which: Cars	1,154	0	293	239	0	239
Light Consumer Goods	**291**	**123**	**445**	**705**	**123**	**705**
% of growth	5.2	22.8	6.5	18.6	22.8	18.6
Ferrous and Non-Ferrous Metals (%)	**460**	**10**	**141**	**129**	**10**	**129**

Source: estimated from COMEXT database.

Telecommunications equipment accounted for 18.1% of Estonia's exports of manufactured goods and for 34% of the growth of Estonian exports over the period. Poland had the most diversified export structure (partly explained by geographical factors, including country size) with road vehicles accounting for 11.5% of exports of manufactured goods, light consumer goods for 25% and ferrous and non-ferrous metals for 16.7%. This diversity was reflected in the growth of Polish exports over the period although there are indications of increased specialisation in engineering products at the expense of clothing, textiles and furniture, with telecommunications and electrical equipment accounting for 27%, of export growth, road vehicles for 18.8% and light consumer goods for 18.6% and metal goods for only 3.4%.

Table 4.5 concentrates on the growth of exports of light consumer goods (clothing, furniture and footwear) from the SEECs and the former Soviet republics (including Latvia and Lithuania) to the EU. Total exports of machinery and equipment, including all forms of transport vehicles, were relatively insignificant in 1998, amounting to only €2.3bln, which was less than Slovakia (€2.6bln) and only marginally higher than Slovenia (€2.2bln) which has a population of only two million. Exports of machinery and equipment accounted for 17.5% of exports from Lithuania, 16.9% from Croatia, 15.2% from Ukraine, 13.2% from Romania and

Table 4.5 Growth of exports of light manufactured goods to the EU from SEE and the FSU, 1995–1998

	Value in 1998		Growth from 1995	
	€ mln	% of exports of manufactures	€ mln	% of growth of exports of manufactures
Albania	139	77.3	61	93.5
Belarus	131	34.5	42	–
BiH	129	66.6	116	68.5
Bulgaria	637	34.5	306	77.5
Croatia	693	45.0	43	204.0
Latvia	203	36.5	100	50.5
Lithuania	409	41.0	231	49.6
Macedonia	237	45.8	34	157.4
Moldova	35	58.9	16	84.6
Romania	2,769	57.8	1,181	70.3
Russia	175	2.4	4	–
Serbia & Montenegro	230	27.0	229	27.8
Ukraine	294	21.2	134	35.3
Total	6,081	47.2	2,497	73.2

Source: estimated from COMEXT database.

only 5.2% from Russia (where chemical products accounted for the major share of exports of human capital-intensive goods). Exports of machinery and equipment from these economies to the EU only grew by €494m during 1995–98, partly as a result in falls in exports from Russia and Macedonia. Dependence on exports of light consumer goods intensified for each of the SEECs and Moldova during 1995–98, as the share of light consumer goods in the growth of exports of manufactures exceeded the share of these items in the volume of exports and accounted for 73.2% of the growth of exports of manufactured products from the entire region. Albania (77.3%), Bosnia and Hercegovina (66.6%), Moldova (58.7%) and Romania (57.8%) had the highest dependency on exports of light consumer goods by the end of 1998.

Patterns of Specialisation

A pattern of export specialisation had emerged amongst the former CPEs by the late-1990s. Four ECECs (Czech Republic, Hungary, Slovakia and Slovenia) and the Baltic state of Estonia became increasingly specialised in the production of human capital-intensive goods. Three of these (Czech Republic, Hungary and Estonia) were the most successful in attracting FDI in the early and mid-1990s, which seems to have contributed to industrial restructuring with upgrading towards human capital-intensive production. These economies are converging on EU patterns of production which should increase their prospects for integration, following EU enlargement which is scheduled for 2004. Czech Republic and Estonia continued to attract increasing levels of FDI during 1999–2001, whilst inflows to Hungary fell from their peak in the mid-1990s. Slovakia attracted significant inflows of FDI in the late 1990s and the early 2000s following the replacement of the more nationalist-oriented, Mečiar government. It is also noticeable in the case of Slovakia that passenger cars accounted for 26.6% of exports of manufactured goods to the EU in 1998 and for 51.3% of the growth of exports of manufactured goods to the EU during 1995–98; while car components constituted a further 4.5%. of exports of manufactures. The growth of these exports accelerated in 1998 largely due to plants that had been acquired and reconstructed by Volkswagen, indicating that despite the relatively low quantity of FDI, it nevertheless played a significant role in upgrading production facilities and improving the structure of exports. Slovenia has shown less reliance on FDI as a proportion of GDP, seeking to limit FDI into the economy whilst pursuing active links with foreign MNEs, including subcontracting arrangements (Dyker et al. 2003). Renault has also been a major investor in the Slovene road vehicle industry which accounted for 20% of exports of manufactured goods to the EU in 1998. Nevertheless, Slovene exports have taken a declining share of the EU car market since the early 1990s which may reflect the possible disadvantages of more limited relations with Western MNEs that stop short of full FDI.

Poland and Croatia had a dual export structure, with intermediate levels of exports of human capital-intensive goods and a relatively high dependence on labour-intensive exports. In the case of Poland this partly reflects the size of the country and its regional diversity which is reflected in continued high levels of production of labour-intensive goods. FDI into Croatia in the early and mid-1990s was affected by problems of civil war and the much lower probability of early EU membership. However, both of these economies experienced a more than two-fold increase in FDI inflows during 1999–2001 and may be progressing towards upgraded production structures by the mid-2000s. The remaining SEECs and the former Soviet republics (including Lithuania and Latvia which are candidates for EU membership in 2004) have been less successful in both attracting FDI and penetrating EU markets for human capital-intensive goods – being more dependent on exports of labour-intensive goods – and also resource-intensive goods which are more vulnerable to changes in world market prices than other manufactured goods.[5] Light consumer goods (clothing, furniture and footwear) – a significant proportion of which are produced under outward-processing agreements – accounted for more than half the exports of manufactured good to the EU for Albania, Bosnia and Hercegovina, Moldova and Romania and for more than 40% for Croatia, Lithuania and Macedonia and accounted for a growing proportion of exports of the SEECs and former Soviet republics during 1995–98. The second group of economies appear to be diverging from intra-EU patterns of trade since their trade with the EU largely consists of trade in commodities with different factor endowments involving the exchange of labour-intensive and resource-intensive products for human capital-intensive goods.

Conclusion: Specialisation and Agglomeration Effects

What are the implications of dependence on labour-intensive exports for economic development, if the patterns outlined above persist in the long term? The Hecksher-Ohlin theorem predicts that increased specialisation and trade will increase demand for labour in labour-intensive economies and should bid up wage rates in labour-abundant economies (the Stolper-Samuelson theorem). Alternative schools of thought, broadly encompassed by 'the new international economics' attempt to explain intra-industry trade (which is effectively a specialised subset of intra-factor trade) between countries with similar factor endowments and levels of development by reference to oligopoly theory and the presence of internal and external economies of scale. When economies of scale are external to existing producers and are linked to locational factors, this creates the possibility that investment and manufacturing will cluster in a given location which provides investors with an initial advantage through agglomeration effects (Krugman 1991). The latter could reflect a specific geographical advantage, the presence of a skilled labour force and a social and political climate that is attractive to foreign investment. When external economies and agglomeration effects are significant, the possibility arises that investment will become concentrated in regions that generate an initial favourable climate to foreign investors which in turn generates external economies that cannot be acquired by 'latecomer' regions.

Which of these scenarios is more probable in the case of the transition economies? One of the problems facing the political economist is that it is difficult to predict outcomes with any certainty because each situation will be affected by specific historic, geographic, institutional, political and economic circumstances. Furthermore 'learning curves' may mean that governments may implement policies to avoid the repetition of unfavourable outcomes. However, the initial indications are that the northern ECEs ('Central East European states' in the tables) and Estonia have successfully attracted FDI, have upgraded production structures and have penetrated EU markets for human capital-intensive goods. These economies together with Latvia and Lithuania have been rewarded with the prospect of EU entry in 2004, which could strengthen agglomeration effects. Croatia could also join this group by the end of the decade. But the other SEECs and the former Soviet republics have been less successful and there are indications that FDI in the region is being largely attracted to labour-intensive and resource-intensive industries (Hunya 2002). Bulgaria and Romania have been offered the possibility of EU membership in 2007, provided that they can meet relatively strict economic conditions. This may help to counteract agglomeration effects and encourage increased FDI inflows. When agglomeration effects are significant, economies that fail to attract investment in the early stages of the transition may continue to face disadvantages in attracting investment in the later stages and may be forced to compete on grounds of lower real wages, either indefinitely, or for a long period. Lower wages may, in turn, imply that investment that is attracted to these economies will be concentrated in sectors where low wages are of major importance and exports will continue to embody a higher proportion of labour-intensive products. Under these circumstances it would be unwise to underestimate the possibility of growing divergence between transition economies, rather than convergence.

Notes

[1] Comecon was established in 1949. In the 1980s its membership consisted of Bulgaria, Cuba, Czechoslovakia, GDR, Hungary, Mongolia, Poland, Romania, USSR and Vietnam. Albania's membership lapsed in 1961 and Yugoslavia was only an associate member.

[2] The size of the FDI inflows per capita can be put into an international context by a comparison with inflows into the relatively poorer members of the EU. FDI inflows per annum during 1986–96 averaged $234pc into Ireland, $216 into Spain, $135 into Portugal and $88 into Greece.

[3] The ratio of FDI (measured in dollars) to GDP and other magnitudes such as gross fixed capital formation (measured in domestic currencies) is sensitive to exchange rate policies. Relatively less developed economies with exchange rates that are more depreciated relative to purchasing power parity rates have a corresponding upward bias in the ratio.

[4] This does provide some indication of the vulnerability to exogenous shocks for economies that are highly dependent on resource-intensive exports that they should not peg their currency too closely to the Euro.

[5] Bevan and Estrin (2000) demonstrate that FDI inflows to the region have been critically affected by 'announcement effects' about potential membership of the EU, with positive announcements about potential membership stimulating inflows of FDI.

References

Bevan, A.A. and Estrin, S. (2000), *The Determinants of FDI in Transition Economies*, Economic Policy Research Discussion Paper 2638 and London Business School Centre for New and Emerging Market Markets Discussion Paper, London.

Dyker, D., Nagy, A. and Spilek, H. (2003), "East" – "West" Networks and their Alignment: Industrial Networks in Hungary and Slovenia', *Technovation*, 23, 603–16.

EBRD (2002), *Transition Report*, EBDR, London.

EC (nd), *Comext Database* (CD format), Office for Official Publications of the European Communities, Luxembourg.

EIU (2002), *Economies in Transition*, EIU, London.

Hunya, G. (2002), *Restructuring through FDI in Romanian Manufacturing*, WIIW Discussion Paper 26, Vienna.

Krugman, P. (1991), *Geography and Trade*, Leuven University Press and MIT University Press, Leuven and Cambridge, Mass.

Smith, A.H. (2000), *The Return to Europe: The Reintegration of Eastern Europe into the European Economy*, Macmillan, Basingstoke.

Wolfmayr-Schnitzer, Y. (1998), 'Trade Performance of CEECs According to Technology Classes', in OECD (ed), *The Competitiveness of Transition Economies*, OECD, Paris, 41–70.

Chapter 5

Transformation of the Central and East European Passenger Car Industy: Selective Peripheral Integration through Foreign Direct Investment

Petr Pavlínek[1]

Introduction

At the end of the period of state socialism (1945–89), car (i.e. passenger car) manufacturing in CEE was not significant globally, only accounting for about six percent of world production. By Western standards the industry was unproductive and obsolete because it had been slow to adopt the changes that had swept through the industry in the 1980s in the capitalist West. With the exception of Russia and Serbia, the collapse of state socialism and the 'transition to capitalism' led to a transformation of the industry in the 1990s driven by FDI. Traditionally, the motor industry has been considered the key representative of modern manufacturing industry because of its size, its linkages with many other industries and its effects on overall economic development. Having recognised the importance of the car industry for the entire national economy, most CEE governments actively promoted its transformation through FDI. The industry played an important role in the reintegration of certain CEECs into the European economy. Some consider the restructuring of the car industry a 'symbol of sweeping changes' in the region and an indication of the success of post-socialist transformations (Havas 2000a p.95). However, although the transformation of the industry is generally regarded as being exceptionally successful by governments and the general public alike in countries such as Czech Republic, Hungary and Slovakia, the process of change has been geographically uneven and often contradictory in many respects.

Most studies of the industry's transformation in CEE have been related to analyses of inward investment inflows and questions about their effects on the integration of the region in the transnational production and distribution networks, national economic performance, regional development and enterprise restructuring (e.g. Grabher 1994, 1997; Havas 1997, 2000a, 2000b; Meyer 2000; Pavlínek 1998; Pavlínek and Smith 1998; Sadler et al. 1993; Sadler and Swain 1994; and Swain

1998). Because of this focus on FDI, most research has concentrated on the car industry as it has been a major recipient of inward investment. At the same time, the restructuring of other automotive sectors, such as the commercial vehicle industry, component sector and domestically-owned firms has been rather neglected (see, however, Havas 1995, 2000a; Pavlínek 2000, 2002a, 2002b, 2002c). Sadler et al. (1993 p.347) and Sadler and Swain (1994 p.400) view FDI in the region's automobile production as the latest 'spatial fix' for the global auto industry in its search for new markets and new low-cost production sites. Geographical analyses of particular investments, related transformations at individual plants and supplier networks, and their effects on local economic performance have shown that in exploiting this spatial fix foreign investors pursue very different strategies – not only in different industry sectors, but also within the car industry in order to capitalize on the particular local and regional comparative advantages (Grabher 1994, 1997; Pavlínek and Smith, 1998).

In this chapter I wish to further extend such analysis by proposing a classification of FDI in the CEE car industry in the 1990s, a classification which is based on the combination of path dependency and embeddedness of car manufacturers in local and regional economies. The concept of path dependency highlights the fact that the post-communist transformations do not take place in a vacuum or on a *tabula rasa* but in societies with social and economic ties from the past, with institutions that either survived state socialism or were built upon the state socialist institutions, and with attitudes, understandings, and behavioural patterns strongly shaped by the previous state socialist system. In other words, the course of post-communist transformation and its outcomes are strongly influenced by past developments. The behaviour patterns of the institutions and individual economic and social actors were associated with the previous system of state socialism. These patterns of behaviour that had developed and existed for more than 40 years did not disappear overnight because of price liberalisation and market introduction, as was assumed in the early 1990s in neoliberal economic approaches. Instead, they changed only gradually, and initially as mainly passive adjustments to new market conditions. In addition, the way in which the behaviour of institutions and individual actors changed was also affected by their previous experiences, habits and behaviour inherited from the period of state socialism. Previous economic development in a particular country, region or locality also affected the course of the transformation and the trajectories of possible future developments. Path dependent approaches thus stress the importance of institutional legacies and continuities in shaping the outcomes of CEE transformations (for example Grabher and Stark 1997; Hausner et al. 1995; Pickles and Smith 1998; Poznański 1993, 1996; Smith 1998; Stark 1992, 1996). However, it is important not to confuse 'path dependency' with past dependency. In this article I use the concept of 'path dependency' as being associated with strategic choices or path shaping strategies, which are often highly contingent, which actively influence present and future actions. Although it sets limits on the possible future trajectories, path dependency does not determine the strategies that are implemented. In contrast, past dependency is a form of historical determinism related to the idea that the past condemns present and future developments (Nielsen et al. 1995 p.4; Stark and Bruszt 1998 pp.6–7). In this paper, the post-communist transformations generally, and the car industry transformations in particular, are understood as a combination of path dependent evolutionary social and economic changes and active path-shaping changes based upon key decisions of social institutions and decisive actors (such as the decisions about privatisation strategies of domestic car manufacturers by the CEE governments and decisions by foreign car makers to invest in the region).

The path dependent nature of post-communist transformations is strongly affected by the fact that the social and economic actors do not function in isolation but are embedded in an intricate web of formal and informal social and economic relations (for example, Grabher 1993a; Grabher and Stark 1997). In Grabher's words (1993b p.4), 'embeddedness refers to the fact that economic action and outcomes, are affected by actors' dyadic relations and by the structure of the overall network of relations'. A number of classifications of networks in which these actors are embedded have been proposed. For example, Håkansson and Johanson (1993 p.35) distinguish between social and industrial networks. Thus, whereas social networks are dominated by actors' social exchange relations, industrial networks, defined as sets of connected exchange relations among actors engaged in industrial activities, are typified by interdependencies among actors, activities and resources. Jessop (1997 p.102) recognises three levels of embeddedness: the social embeddedness of interpersonal relations (Granovetter 1985 pp.490–3); the institutional embeddedness in systems and networks of inter-organisational relations; and the 'societal embeddedness of functionally differentiated institutional order'. In this paper, I focus on the embeddedness of foreign-owned car manufacturers in local and regional economies of CEECs. The degree of embeddedness of foreign investors in local and regional economic and social networks largely determines the extent of their overall impact on local and regional economies. Embeddedness is affected by the type of operation involved, that is its role within particular MNEs' production chain and its position within external business networks, and by the kinds of functions it performs (Dicken et al. 1994 pp.34–5). In the context of the region's car industry transformations, the degree of embeddedness of foreign car manufacturers in local and regional supplier networks has also been affected by the ability (or inability) of pre-existing networks of suppliers to transform itself and learn new forms of action to supply effectively car makers under new conditions and thus to remain part of active supplier networks.

The work on embeddedness of foreign companies in local and regional economies tends to focus on supply linkages which is part of the second type of embeddedness identified by Jessop (for example, Dicken et al. 1994). I follow the similar strategy in this paper by focusing on the embeddedness of foreign car manufacturers in industrial networks in terms of their supply linkages. However, foreign investors are typically also embedded socio-culturally at the local level – by providing a number of services to host communities. The degree of this socio-cultural embeddedness varies considerably, but the evaluation of socio-cultural embeddedness of foreign car makers in CEE is beyond the scope of this chapter. Here I investigate FDI-driven transformations in car manufacturing in the 1990s. Although I acknowledge the role of state socialist legacies in these changes, an historical analysis of the state socialist passenger car industry is beyond the scope of this chapter and has been adequately reported elsewhere (for example Havas 2000b; Pavlínek 2002a; Sadler and Swain, 1994; Sadler et al. 1993). Also I focus solely on the car industry and on the effects of FDI in its transformation. Commercial vehicle manufacturing and the component supply industry are also beyond the scope of this chapter as both require their own in depth analysis (Havas 1995; Pavlínek 2000, 2002a, 2002b, 2002c).

I will draw both on primary and on secondary data. The primary information presented is based upon in depth interviews which I conducted at the Czech Ministry of Industry and Trade and Škoda Auto in 1996, at Škoda Auto, Tatra, Daewoo-Avia and Karosa vehicle manufacturers in 1999, and with directors or top managers of twenty Czech automotive component firms in 2000 and 2001. I have used several sources of data to compile a production database for the region. Production figures can be found in statistical yearbooks of most CEECs. In general, the pre-1990 data are more consistent than the data after 1990 as some countries such as Czech Republic and Slovakia gradually stopped publishing car industry production data. In these cases, I have used the data provided by national automobile associations of the respective countries and also by the International Organisation of Motor Vehicle Manufacturers for 1997–2000 (OICA 2001). However, the various sources are not necessarily compatible. I begin with a brief characterisation of the region's car industry at the end of the period of state socialism and a discussion of the effects of trade and market liberalisation in the early 1990s. Second, I consider the consequences of FDI inflows for the industry including the immediate effects of foreign ownership on privatised companies and how it influenced production trends in different countries in the 1990s. Third, I present a typology of FDI-led car industry transformations in the 1990s. Fourth, the region's role in the European car production system is assessed. I summarise the main findings in the conclusion.

Effects of Market and Price Liberalisation on the Industry in the early 1990s

More than 40 years of state socialist development in ECE and 70 years in the FSU left behind a globally insignificant car industry with only 5.8% of the world production in 1990 (excluding former GDR). It was unable to satisfy domestic demand, and by Western standards was unproductive and obsolete as it had been slow to adopt the changes that swept through the industry in the 1970s and 1980s in the capitalist West (for example, Dicken 1998; Hudson and Schamp 1995; Law 1991; Wells and Rawlinson 1994). Many cars that were produced based on Western licence agreements were originally designed in the early 1960s and – typically – with the aid of technology that originated in the late 1960s and early 1970s. After substantial growth, especially between 1965 and 1980, the 1980s were years of stagnation and production decline. Car production declined in the 1980s in the former Soviet Union and Poland, the two largest CEE producers, while it stagnated in the rest of the region (see Pavlínek 2002a for details) (Figure 5.1). However, although Western views of the sector assume its uniform backwardness, important differences existed between the CEECs – not only in the various types of their car industries (see Havas 2000b; and Pavlínek 2002a for details), but also in the levels of technological advancement, which differed between individual producers and within their factories. Backwardness in the automotive industry was most pronounced in terms of management and the organisation of production rather than in terms of the hard technology employed (1996 and 1999 interviews – which refer to those conducted by the author at the Czech Ministry of Industry and Trade and Škoda Auto in 1996 and at the Škoda Auto, Tatra, Daewoo-Avia and Karosa vehicle manufacturers in 1999; Hardy 1998 pp.646–7). In some cases, such as Czechoslovakia's Škoda enterprise or Poland's Fabryka Samochodów Osobowych (FSO), the production facilities were modernised in the 1980s, largely using Western technology. So after Škoda was taken over by Volkswagen (VW) in 1991, there was no immediate need for investment, technology transfer or plant modernisation.

Three crucial developments affected the industry immediately after the collapse of state socialism (Pavlínek 2000 p.265). First, the established trade flows within the region broke down as Comecon (the Council for Mutual Economic Assistance) disintegrated and trade was reoriented towards Western Europe. The collapse of Comecon export markets especially affected car producers in the smaller CEECs dependent on exports to other members. They could not easily compensate for these losses by switching to Western markets since their vehicles were not competitive enough and their sale and distribution networks were poorly developed compared with those of Western makers. Second, the local car producers' domestic markets also collapsed because, after trade liberalisation, they were flooded with cheap, used, Western cars. The domestic demand for new cars also sharply declined because of the economic crises and high inflation associated with shock therapy which, in turn, undermined purchasing power and household savings. At the same time, prices for new domestic cars increased, making them less affordable for potential buyers. Third, the CEE markets, now open to Western trade and investment, were rapidly penetrated by Western car producers with their higher quality, better equipped, vehicles which appealed to the newly affluent classes across the region. As a result, sales of domestic and other CEE car producers collapsed in the early 1990s (see Pavlínek 2002a for details).

This situation plunged CEE car producers into a deep crisis as they lacked the capital to sustain pre-1989 production levels, let alone to finance the overall restructuring necessary for their long-term survival under new market conditions. Companies such as the Škoda and Bratislavské automobilové závody (BAZ) of the then Czechoslovakia and Poland's Fabryka Samochodów Małolitrażowych (FSM) quickly amassed large debts and were increasingly unable to finance their day-to-

Figure 5.1   Total car production (above) and percentage share of world production (below), 1950–1990

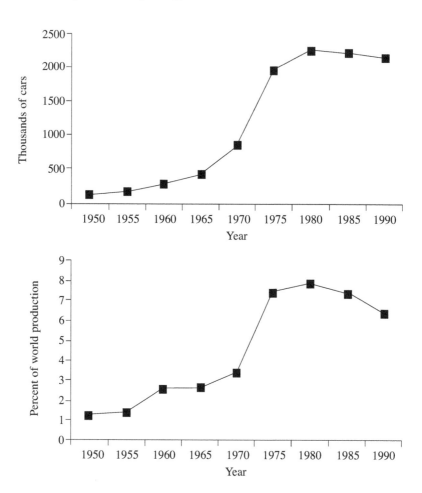

Source: National statistical yearbooks.

day operations (Anon 1991 p.9; Pavlínek and Smith 1998 p.626; Ratajczyk 1996 p.9). Given European and global production trends and the ever increasing competition in domestic and foreign markets, it became obvious that small CEE manufacturers with small domestic markets, such as former Czechoslovakia, Romania, Yugoslavia – and even Poland with its much greater market potential – could not survive without increased production and the substantial expansion of their exports. Two survival strategies were pursued by car makers and their respective governments. The most common strategy was a restructuring of the existing companies with their integration into European and, in some cases, global production networks, directed and financed by foreign capital. This strategy was successfully pursued in ECE (Czech Republic, Hungary, Poland, Slovakia and Slovenia) in the early 1990s and less successfully in Romania and Ukraine in the second half of the 1990s. The second strategy involved a continued reliance on domestic markets and price-based competition in which domestic makers concentrated on low-priced cars that were much cheaper than the cheapest foreign models. This approach was usually necessitated by a failure to find a foreign investor willing to invest under the conditions acceptable to CEE governments quickly enough (Dacia, Autozaz and Zastava); however, it was actively pursued as a viable alternative, as in the case of Russian car makers.

CEE Car Industry Transformation through FDI

It is important to view the industry's transformations in the context of the restructuring of the West European car industry over the past 15 years, as it profoundly affected the nature of change in the region during the 1990s. New production concepts, new work-organisation methods, and new types of inter-firm relations were quickly introduced in parts of CEE following FDI by Western companies, in many cases to exploit and build upon past traditions and experience of car making. West European car industry restructuring has been extensively analysed and documented by geographers (for example, Bordenave and Lung 1996; Hudson 1994, 1995, 1997a, 1997b; Hudson and Schamp 1995; Lagendijk 1997; Lung 1992; Sadler 1998, 1999). This was triggered by the opening of Japanese 'transplants' which produced superior quality vehicles and were twice as productive as West European manufacturers. Thus, the Japanese presence further intensified the fierce competition caused by overproduction and overcapacity in the West European market. To remain competitive, the West European manufacturers were forced to reorganise the industry, largely by introducing imitations of various forms of Japanese (mainly Toyota) car production strategies. The introduction of 'lean and flexible' production, JIT delivery strategies, attempts to achieve 'mass customisation' and a constant need to adopt new cost-cutting strategies involved new ways of work organisation and labour recruitment, new forms of capital-labour relations within plants, and new relationships between assemblers and suppliers. Such restructuring also led to the spatial reorganisation of the car industry in Europe. Similar strategies were selectively introduced in CEE in the 1990s, resulting in a profound transformation of the region's car industry.

   The CEE industry has seen a major inflow of foreign capital since the early 1990s, which has largely rebuilt its existing production capacity (Table 5.1, Figures 5.2 and 5.3, the latter based on data in Table 5.1). By 1996, the car industries of Czech Republic, Hungary and Poland were experiencing the highest degree of FDI penetration within their manufacturing sectors. Foreign-owned enterprises and joint ventures accounted for 85% of the production of the motor industry (including the

Table 5.1   Selected foreign investments in CEE car production, 1989–2003

Firm	Location/Original Local Partner	Investment $ mln	Year
BMW	Leipzig, East Germany	453	2001
BMW	Kaliningrad, Russia/Avtotor	50*	1999
Daewoo	Warsaw, Poland/FSO	800*	1995
Daewoo	Craiova, Romania/Automobile Craiova	900	1994
Daewoo	Zaporizhya, Ukraine/Avtozaz	1300	1997
Fiat	Bielsko Biała and Tychy, Poland/FSM	1800*	1992
Fiat	Nizhny Novgorod, Russia/GAZ	850	2000
Ford	Székesfehérvár, Hungary (components only)	182*	1990
Ford	Vsevolozhsk, Russia/Ford-Vsevolozhsk Ltd.	150*	1999
Ford	Minsk, Belarus (car assembly closed in 2000)	10*	1997
GM	Gliwice, Poland, Opel Polska	549*	1995
GM	Szentgotthárd, Hungary/Rába	951*	1991
GM	Eisenach, East Germany/Automobilwerke Eisenach	1000*	1990
GM	Elabuga,, Russia/ELAZ"	270	1996
GM	Togliatti, Russia/Avtovaz	100	2001
Isuzu	Tychy, Poland (engines only)	364	1997
Porsche	Leipzig, East Germany	45	2000
PSA Peugeot Citroën	Trnava, Slovakia	700	2003
Renault	Novo Mesto, Slovenia/Motor Vehicle Industry	54*	1993
Renault	Piteşti-Colibaşi, Romania/S.C. Automobile Dacia	100	1999
Renault	Moscow, Russia/Avtoframos	300	1999
Rover	Varna, Bulgaria/Daru+	120*	1993
Suzuki	Esztergom, Hungary/Autókonszern	406*	1990
Toyota/ Peugeot	Kolín, Czech Republic	1500	2001
VW	Mosel, East Germany	1360*	1990
VW	Dresden, East Germany	190*	1999
VW	Mladá Boleslav#, Czech Republic/Škoda	2000*	1991
VW	Györ, Hungary/Audi Hungária,	923*	1994
VW	Bratislava, Slovakia/ BAZ	680*	1991
VW	Poznań, Poland/Tarpan	241*	1993

*Denotes actually realized investment since the formation of a joint venture or the start of greenfield production. In other cases the figures are anticipated investment programs rather than actually realized investments. Year refers to the starting year of an investment program in the actually realized investments or the year when the investment commitment was made. Investment figures are only approximate.
"JV production halted in 1998
+ Rover withdrew from JV in 1996
# Also at Vrchlabí and Kvasiny
Source: Manufacturers' press releases, company reports and various articles.

**Figure 5.2   Location of foreign-invested car plants in CEE
(note that Elabuga and Togliatti are off the map)**

car components) in Hungary, 82% in Poland, and 67% in Czech Republic (Zemplinerová 1998 p.337). With the exception of Russia and Yugoslavia, car production came to be dominated by foreign capital. In Russia, continued domestic ownership was made possible by virtue of the large domestic market, the low purchasing power of the vast majority of car buyers and high import tariffs on car imports, whereas in Yugoslavia the situation resulted from the war and economic sanctions applied throughout the 1990s (see Pavlínek 2002a for details). Foreign investors have been attracted because of two major factors: first and foremost, the possibilities for low-cost production that could increase their competitiveness in lucrative Western European markets, and second, by the market potential of the region. Investment incentives and the avoidance of trade barriers (in the cases of

Figure 5.3   FDI inflows for car assembly in CEE, 1990–2003

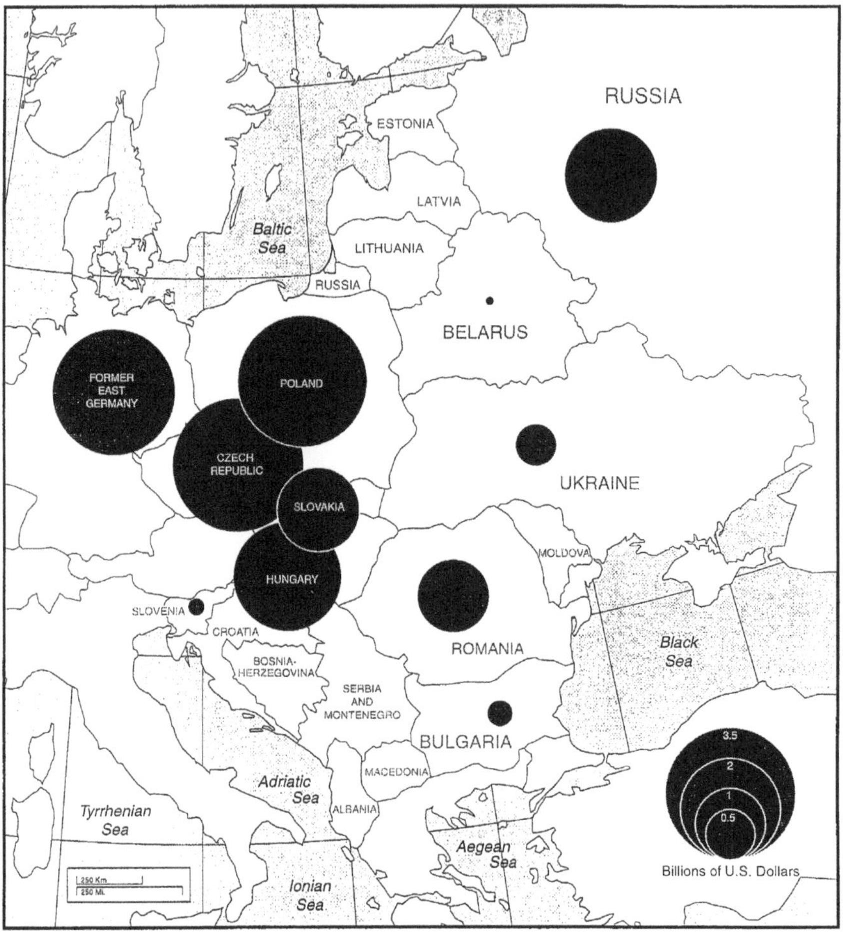

Czech Republc and Poland in particular) by producing cars locally were two additional factors (Havas 1997 pp.220–1). The principal reasons for setting up car industry subsidiaries in CEE may differ from country to country: see Havas (2000b p.124) for Hungary; Poznańska and Poznański (1996 p.78) for Poland, and Pavlínek and Smith (1998) for Czech Republic and Slovakia. In reality, there is always a combination of factors in which the cost-cutting strategies and/or market potential of the host country must be complemented by other necessary considerations such as good transportation and communication infrastructures, geographical proximity to major markets, incentives for foreign investors, and, overall, a positive business environment – including political stability. VW's expansion in CEE has been the most successful. In 1999, VW's 605,378 cars accounted for 40.3% of foreign-made cars produced in the region, followed by Fiat with 343.063 (22.8%), Daewoo with 221,239 (14.7%), Renault with 187,656 (12.5%) and four relatively small producers: Suzuki with 64,015 (4.3%), GM with 51,308 (3.4%), Ford with 25,551 (1.7%) and Subaru with 4,090 (0.3%) (MOTORSAT 2001).

One of the most important immediate changes in the foreign-owned firms and joint ventures was the introduction of new management practices, work organisation, and quality control. As argued in 1996 by Jaroslav Povšík, a trade union leader at Škoda, VW did not bring new skills but 'new work organisation and order'. Because the work discipline was poor in many companies – since the 'labour force was spoiled by state socialism', (according to an interview in 2000 with Pavel Pravec, Executive Board Chairman and Plant Manager of Barum Continental Otrokovice), the introduction of new work organisation and order often involved drastic changes in privatised companies. These included basic cost-cutting measures and increasing labour productivity brought about by minimising waste and improving the utilisation of the working day by employees. Using this simple strategy, which was also extended to component suppliers, VW substantially increased the quality of Škoda's Favorit model, produced in the state socialist era, without any investment during the two years after takeover (according to an interview in 1996 with Jaromír Šebesta of the Czech Republic's Industry and Trade Ministry). Similar changes were experienced in other foreign-owned companies and joint ventures elsewhere across the region (Swain 1998; Sadler and Swain 1994; Pavlínek and Smith 1998) and also in successful domestically-owned companies in the Czech automobile component industry (based on 2000 and 2001 interviews which refer to those conducted with 20 different component suppliers in Czech Republic). What this 'new organisation and order' means in practical terms is the replacement of the state socialist work organisation and shop-floor practices with capitalist hegemonic factory regimes. The anarchy of production associated with state socialism and typified by continual reorganisation of the labour process (Burawoy 1985 p.163) has been replaced with the planned organisation of production.[2] The considerable autonomy workers enjoyed on the shop floor and the control of production they exercised under state socialism (Clarke et al. 1994 p.181; Burawoy and Krotov 1993 p.53) was removed and replaced by the increased authority of the foreman and effective managerial control, as more efficient horizontal and vertical organisation of the work-place was introduced (Smith and Pavlínek 2000; see also Swain 1998 on greenfield investments in Hungary).

The effects of FDI on car output have been dramatic, but very uneven geographically, resulting in new geographies of production across the region in the 1990s (Figure 5.4). Although overall CEE car production declined by 400,000 units between 1990 and 1995 (2.1–1.7mln) it grew by 767,000 vehicles between 1995 and 2000 (excluding East Germany).[3] At the national scale, three basic production trends can be recognised. First, Czech Republic, Poland and Slovenia joined by Hungary and Slovakia – two newcomers to the business (Figure 5.5) – experienced the fastest growth during the 1990s (though Poland was falling out of this group

Figure 5.4   Car production trends in CEE, 1989–2002 (1989=100)

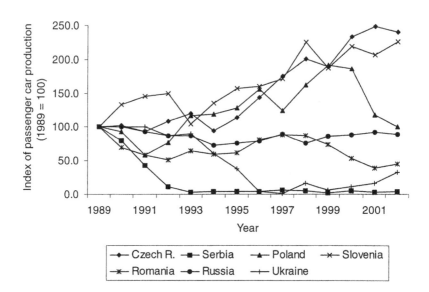

Source: National statistical yearbooks; national automobile associations; OICA 2001, 2002, 2003. Czech and Polish data for 1989–1996 include light commercial vehicles.

after 1999 as Figure 5.4 indicates). The dramatic growth in total output among these five ECECs (by 890,000 units between 1990 and 2000 and over one million if production in East Germany is included) resulted from FDI linked with radical restructuring and integration into European car production networks. Second, Russia and Romania experienced gradual production declines: 13% in Russia and 23% in Romania between 1990 and 2000 (indeed Romanian car production declined by 46% during 1989–2000). In Russia, the reasons for the decline included a continuing dominance of domestic producers focused on defensive restructuring strategies and the inability of foreign companies to set-up successful production operations and to win a significant share of the Russian market (see Meyer – 2000 p.137 – on the distinction between defensive and strategic restructuring). In the case of Romania, the inability of its largest passenger car maker Dacia to find a Western investor until 1999 (Renault) and the problems of a Daewoo-owned factory in Craiova led to a production decline in the 1990s. Third, car production collapsed in Ukraine and Serbia in the 1990s. In Serbia the collapse was caused by war and economic sanctions; in Ukraine it resulted from the failure of Daewoo to revive the collapsed domestic production successfully (see Pavlínek 2002a for details).

Figure 5.5 Car assembly in Hungary (below) and Slovakia (above), 1992–2002

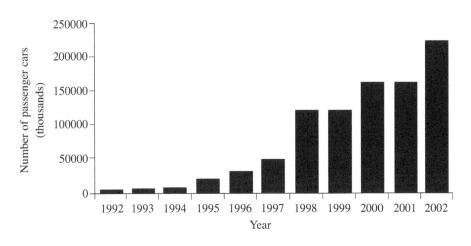

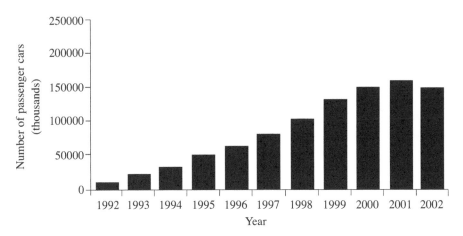

Source: AIA SR 2001; EIU 2001; OICA 2003.

Strategies of FDI-driven Car Industry Transformations

I have selected two criteria to classify these strategies in CEE in the 1990s: the mode of entry and local economy linkages. First, foreign investments could be divided based upon whether they went into the existing companies (brownfield investments) or into the formation of new sites (greenfield investments) (see also Havas 1997 p.236). In the case of brownfield investments, the bulk of investment went into the existing state socialist companies which had been purchased, restructured, and integrated into the existing production and distribution networks of multinational companies. In this sense, brownfield investments are path dependent. The primary motive for entry was typically market capture, combined with export potential. Examples of such investment include Fiat's purchase of Poland's FSM, VW's purchase of Škoda (then Czechoslovakia), Daewoo's purchase of Automobile Craiova – formerly Oltcit (Romania), Avtozaz (Ukraine) and FSO (Poland) and most recently Renault's purchase of Dacia (Romania). Greenfield investments are new production operations set up by foreign companies in host countries. In terms of CEE car manufacturing, this type of investment has been less common in car assembly but quite widespread in components production. In terms of assembly, East Germany, Hungary and Poland were major recipients of greenfield car assembly plants: Suzuki, Audi, and GM Opel set up assembly operations in Hungary in the 1990s (Havas 1997); in East Germany GM Opel and VW built new assembly plants in Eisenach, Mosel, and Dresden, and GM Opel built a new plant in Gliwice, Poland (whereupon car assembly ceased in Hungary in 1999). Additionally, BMW and Porsche also decided to build their new assembly facilities in East Germany, both in the city of Leipzig. Porsche launched its new facility for the annual production of 20,000 sport utility vehicles (SUVs) in 2002. BMW started the production of its new small model in 2003 with a capacity for about 200 000 cars annually. In December 2001, Peugeot and Toyota announced their decision to jointly build a new factory at Kolín (Czech Republic), to produce 300,000 small cars annually, starting in 2005. In January 2003, PSA Peugeot Citroën selected the Slovak city of Trnava for its new plant to assemble 300,000 small cars annually, commencing in 2006 (Table 5.1).

Embeddedness of foreign-owned car producers in local and regional economies is the second classification criterion. Linkages of foreign-owned plants with domestic firms are considered the most important mechanism through which technology transfer takes place, additional jobs are generated and new local enterprises are formed (Dicken 1998 p.251). However, the degree of embeddedness of foreign-owned companies in the local and regional economies of the host countries varies considerably. Following Dicken et al. (1994 p.38) who consider supplier relationships to be 'probably the most important single indicator of local embeddedness' of foreign investors, I focus only on supply linkages of foreign-owned car manufacturers. I am aware, that this focus on spillover effects and supplier linkages may not measure embeddedness of foreign companies in local economies very accurately because their participation in local social and service networks and other non-material linkages are ignored (Domański 1999 p.84).

Based on these two criteria we can recognise four types of FDI-related transformations in the CEE car industry in the 1990s: embedded path-dependent transformations; embedded greenfield operations; disembedded path-dependent transformations; and disembedded greenfield operations (Table 5.2). It is important to note that, although these categories were identified in the 1990s, they will not necessarily continue to be valid in the early 21st century as they are by no means

Table 5.2   Classification of FDI-driven transformations in the CEE car industry, 1990s

Embeddedness	Path dependency	
	Yes	No
High	Embedded path dependent transformations (Škoda, FSO, FSM, Dacia)	Embedded greenfield operations (Magyar Suzuki, Opel Polska)
Low	Disembedded path dependent transformations (VW Slovakia, Renault Revoz Slovenia)	Disembedded greenfield operations (Audi Hungária, Opel Hungary)

fixed, and the above strategies are only pursued as long as it is profitable for the parent companies. The local content of car manufacturing may substantially increase or decrease when the production strategy of the parent corporation, the scale of car assembly at a particular site, or the role of a particular plant in the corporate hierarchy changes. On the one hand, the local content of VW Slovakia may substantially increase because its rapid production growth may necessitate manufacturing of an increasing number of components locally for logistical and cost-cutting reasons. On the other hand, local content, and thus the degree of embeddedness, may decrease over time in other cases. This has been happening in the case of Škoda since the 'common platform' strategy was introduced by VW; this has led to an increased use of VW's concern-wide suppliers – at the expense of Czech suppliers.

Embedded Path-Dependent Transformations

Typically, brownfield investors also strive to rebuild the existing networks of component suppliers to exploit further low production costs, and existing industrial traditions, and skills in the region by encouraging their Western component suppliers either to form joint ventures with the existing domestic suppliers or to build new facilities. Such production-network transformations are successful as long as sufficient time is allowed by foreign investors for gradual change. These types of investments and associated 'offensive flexibility' strategies (Leborgne and Lipietz 1991 pp.41–6) tend to benefit local economies by integrating local component suppliers into a web of localised network suppliers, leading to relatively high levels of embeddedness. However, the nature of this embeddedness is often contradictory and uneven (Pavlínek 1998; Pavlínek and Smith 1998). As such, the brownfield investments by existing car manufacturers and their networks of component suppliers suggest elements of path-dependent transformations of the pre-existing CEE car industry, which result in forms of 'path-dependent embeddedness': meaning the situation in which former networks of an assembler's component suppliers that existed under state socialism survived and were activated and transformed to function under new conditions by learning new actions (Stark 1995 p.69; Smith and Swain 1998 pp.39–43). As noted above, these forms of path-dependent embeddedness are typified by the formation of joint ventures between the domestic part suppliers which existed under state socialism, and foreign component suppliers. Analysis of FDI in Czech Republic and Slovakia, and the case of VW-Škoda in particular, has shown that in this type of investment former industrial strengths and traditions are exploited and built upon by foreign investors because it is economically advantageous for them to do so (Pavlínek and Smith 1998; Meyer 2000).

A similar situation also existed in Poland and Romania. In Poland, Fiat was buying 76% of its component supplies from locally-based parts producers and suppliers in 1996 – twice as much as in 1992. Daewoo had to quickly increase the local EU content to at least 60% in its CEE plants to avoid tariff barriers on their EU exports. This goal was achieved in Poland in July 1999 after Daewoo set up 15 Polish-South Korean joint ventures making car components (Havas 2000b p.252; Koza 2000). In Romania, Daewoo also provided technical and financial support to South Korean component suppliers to locate in the country to bring the local content up to a planned 80% by 2000. Renault launched a similar process of supplier-network transformation after buying Dacia in 1999 to 'attain maximum local integration in the future' (Renault 2001 p.1). Romanian suppliers were pressed to improve quality quickly to achieve Western standards through quality assurance processes. In situ foreign suppliers have been subject to a new selection process, and major first tier multinational firms were encouraged to produce components locally. Companies such as Cabléa, Continental, Johnson Controls, Magnetto and Superplast have already established local production operations in Piteşti (Ibid pp.1–3).

However, it is important to realise that path dependency is not a constraint on investors' profit-seeking behaviour, and that path-dependent transformation is only pursued when it benefits foreign investors. In the case of the car industry, 'lean' production strategies implemented in the mass production of cars strongly encourage some suppliers to locate close to their customers to implement a JIT system efficiently. This is one of the reasons why car assemblers encourage the redevelopment of existing supplier networks. They also attempt to keep down costs by developing subcontracting links with companies located in CEE because they can also produce and supply less sophisticated components less expensively by using the same comparative advantage of low cost, but skilled labour. Any links between privatised car producers and the local economy were cut in cases where reliance on previous industrial strengths, traditions and producer networks was not considered important and were not believed to be beneficial to the new activities. In such cases, foreign-owned companies switched from domestic to foreign and from CEE to Western suppliers (Ellingstad 1997 p.13; Grabher 1994 p.182; Havas 1995 p.38). In other cases, car companies such as VW-Škoda and Suzuki replaced the existing Czech or Slovak (VW- Škoda) and Hungarian (Suzuki) component suppliers if they could not deliver the expected quality and timing of deliveries within a particular transition period determined by Western investors (1999, 2000, 2001 interviews).

Path-dependent strategies are therefore always combined with active path-shaping strategies (Nielsen et al. 1995 pp.5–8) by Western investors, which result in a number of restructuring strategies being pursued – even by the same investors – in different settings. These allow them to exploit particular local advantages which typically include combinations of production costs, governmental incentives, infrastructure, relative location, and other local strengths such as past industrial tradition and specific labour skills. For example, Table 5.2 shows that VW has successfully pursued very different investment and development strategies in the Czech, Slovak, and Hungarian auto industries leading to relatively deep forms of embeddedness in its Škoda plants in the Czech Republic, relatively low forms of embeddedness in its VW-Slovakia plant in Bratislava, and even lower in its Audi Hungária engine and car assembly plant in Győr (see Pavlínek and Smith 1998 pp.626–9 and Meyer 2000 pp.139–42 for details on Škoda Auto; VW-Slovakia and Audi Hungária are analysed below).[4]

Embedded Greenfield Operations

Greenfield investments are often set up with the specific aim of avoiding the potential problems associated with previous conditions, such as pre-existing traditions, labor conflicts, networks, and locations, that could constrain new operations. Magyar Suzuki is a typical example of such an approach. According to Péter Bod (1998 p.23) the former Hungarian Minister of Industry and Trade, who negotiated Suzuki's investment, Mr. Suzuki insisted on locating the factory in a place with no industrial tradition and no history of state socialist industry because, in his own words, he needed 'farmers and housewives, rather than spoilt workers' to assemble Suzuki cars in Hungary. The long and careful recruitment process by Magyar Suzuki to construct a docile and flexible workforce has been well documented (Swain 1998 pp.660–4; Sadler and Swain 1994 pp.399–400). This approach mirrors the strategies of Japanese transplants in Western Europe in the 1980s and 1990s (Hudson 1994 pp.338–9, 1995 pp. 81–3, 1997a pp.471–3) and it differs profoundly from the experience of brownfield investments such as VW-Škoda.

Suzuki's greenfield assembly plant at Esztergom is, however, a rare example of a greenfield investment that became quickly embedded in local economy. The investment achieved 68% of local content in 1998 (30% of Hungarian suppliers' content, up from 6% in 1992 when the plant was set up). As in the case of VW-Škoda, Suzuki played a very active role in 'sorting out' its suppliers and arranging joint ventures between Hungarian and Western component suppliers. Unlike VW, however, Suzuki was primarily driven by the EU requirement for non-EU companies to attain at least 60% of local content to avoid the EU's tariff barriers on their imports. About 55 of Suzuki's Hungarian-owned component suppliers focus on supplies of low-tech and lower-value-added components, while high-tech and high-value-added components continue to be imported from Japan. Magyar Suzuki also buys supplies from 35–40 EU-based component suppliers and three fully or partly foreign-owned ECE suppliers. The company has encouraged its suppliers to move their production facilities to an industrial park in Esztergom (Ellingstad 1997 pp.12–3; Havas 1997 p.227, 2000a pp.100–1, 2000b p.256; Hungarian Trade Commission 2001; Kapoor and Eddy 1998 p.44; Sadler and Swain 1994 p.398). However, the situation in which domestic component suppliers concentrate on the production of more simple and low-value-added components, while high-tech and high-value-added components are supplied either from Western Europe or from foreign-owned component suppliers in CEE, is not unique to Magyar Suzuki. In fact, this is typical of the situation that developed in the region's car manufacturing in the 1990s (2000 and 2001 interviews, see also Chomicka 1999 p.4 on Poland; Krecháč 2001 on Slovakia).

Opel Polska represents the second example of greenfield investment that has become quickly embedded in the local and regional economy. GM Opel's 1996 decision to build its greenfield assembly plant in Gliwice led to the rapid formation of joint ventures and construction of new plants in the Polish car component sector. By 1998, when the Opel Polska plant was launched, more than 70 joint ventures, each with investment exceeding $1mln, were formed among local component suppliers and foreign companies, and 27 greenfield factories were set up in Poland as a whole to supply foreign-owned car manufacturers (Domański 1999 pp.82–3). The local content of Opel Polska exceeds 50% (Opel Polska 2001). As in the case of Škoda and Magyar Suzuki, GM Opel also trained selected Polish component suppliers so that they could meet its quality standards and delivery schedules, and pressured some of its established Western suppliers to set-up local facilities in Poland (Chomicka 1999 pp.4–5). As in the case of Magyar Suzuki, Opel Polska implemented a careful selection of employees for its new plant and all new employees underwent intensive training in Opel and GM plants in Europe and the USA (General Motors 1998 p.2).

Disembedded Path-Dependent Transformations: The Case of VW-Slovakia

The VW plant in Bratislava (Slovakia) is an example of disembedded path-dependent transformation (Pavlínek and Smith 1998; Smith and Ferenčíková 1998). VW-Slovakia was originally established as a joint venture between VW and BAZ in 1991, but VW has fully owned the plant since 1994 (for fuller discussion see Ferenčíková 1995, 1996; Smith and Ferenčíková 1998 and Krecháč 2001). Prior to establishing the joint venture, BAZ existed as a classic branch plant of the Škoda network with little autonomous production and technological capability, and functioned as a component supplier to Škoda. As opposed to VW-Škoda, there has been no effort by VW to establish local development and the design of automobile products after their takeover of BAZ. Instead, the Bratislava plant has been used to cut production costs by transferring the assembly of gearboxes and specific niche-market cars (special editions of Golfs, Passats, Boras, Polos, and, most recently, the luxury SUV Colorado) from Germany in order to increase their competitiveness in Western markets. Therefore, the possibilities of low-cost production based on low labour costs (estimated to be one tenth of those in Germany) and the existence of a labour pool in the Bratislava area were major reasons for VW's investment. The previously limited role of BAZ in the Czechoslovak automobile industry has affected its position in the VW production hierarchy, as VW could not build up on any autonomous car production. Because BAZ was established only relatively recently under state socialism, the industrial strengths and traditions that existed in companies such as Škoda were largely absent in Bratislava. For example, the vast majority of current workers in VW-Slovakia have been recruited from a wide territory and had not worked in BAZ before the establishment of the joint venture, suggesting that pre-existing production skills were not very important to VW. They thus replaced the components role that BAZ had under state socialism with a globally integrated cost-cutting strategy to produce niche-market products and components for VW corporate networks. The path dependency of this experience revolves around building upon the marginal and latecomer role of BAZ to establish a strategy of replacement – a clear example of low-wage defensive restructuring (Pavlínek and Smith 1998 pp.628–9).

In contrast to the experiences of VW- Škoda, and reflecting the position of automobile production in Slovakia under state socialism, the local embeddedness of VW-Slovakia has been very limited. Approximately 85% of components used in the production are supplied directly from VW component producers in Germany. There are two reasons for the low local content of VW cars assembled in Slovakia. First, the company has focused on the low volume production of special models, which has not compelled Western suppliers of VW to establish production facilities close to the plant. Second, the existing domestic component-supplier network has been very limited. As a result, unlike the situation in Czech Republic, VW could not pressure Western component suppliers to form joint ventures with domestically located firms. This situation did not change even after VW decided to invest an additional DM600mln to expand production substantially at its Bratislava plant in order to accommodate the assembly of more than 100 cars per day. Despite the threefold growth in production, local content did not substantially increase (Figure 5.5). Instead, more than twenty fully-loaded freight trains with components arrive at VW-Slovakia every day, via a direct rail link established between VW-Slovakia and VW's consolidation center in Germany which receives shipments from VW suppliers and distributes them to various VW facilities throughout Europe. The same trains are used to export the finished products from the factory as almost 100% of assembled cars (98.6% in 1999 and 100.0% in 2000) are exported abroad (AIA SR 2001). Consequently, the regional economic effects of VW's investment in Bratislava have so far been very limited compared both with embedded path-dependent and with embedded greenfield investments.

However, recently this situation has started to change as VW-Slovakia entered a very different stage of development. In 1999, VW-Slovakia succeeded in a competition among 12 VW factories over the production of a new SUV – the VW Colorado. After this is launched, annual production is expected to exceed 200,000 vehicles at the Bratislava complex. The rapidly growing output, that is expected to reach 270,000 cars annually, compelled VW-Slovakia to rethink its strategy of supplying components almost exclusively from Germany. Consequently, in 2000 VW decided to build an industrial park that would accommodate 12–15 of VW's Western component suppliers. These suppliers will deliver component modules through a just-in-time regime for Colorado production (Bibza 2000). In May 2000, VW also opened a new greenfield factory in Martin, central Slovakia, to which it will transfer the production of gearboxes from its Bratislava complex to make space there for increased car assembly (Anon 2000). Nevertheless, to what extent these changes will actually enhance the links of VW-Slovakia with the local and regional economy remains to be seen, as VW's foreign suppliers are likely to be the suppliers of system components (or modules) assembled from imported components not manufactured by local firms. With this a likely scenario, the role of domestic suppliers will remain limited. According to Weidokal and Stagles (2001) VW-Slovakia had 1,200 suppliers in 2000, of which only two were Slovak firms supplying directly to the plant.

Disembedded Greenfield Investments

Disembedded greenfield investments represent the fourth type of investment strategy. These are typically export-oriented assembly branch plants, vertically integrated into their parent MNEs, which are set up to exploit cheap and often poorly organised labour in the region. Such assembly plants import components from a developed country, and the components are assembled into finished products which are then exported back to the core country(ies) or foreign-owned car factories in CEE. Typically, these branch plants are not integrated into local economies at all, or only very little, as they do not use any or only a few, local suppliers (Grabher 1994 p.191, 1997 p.126; Hardy 1998 p.650; Pavlínek 1998 pp.77–80; Pavlínek and Smith 1998 pp.628–9; Sadler and Swain 1994 p.398). Grabher (1994 p.191, 1997 p.126) has therefore labeled such plants as 'cathedrals in the desert' and the economic landscape they characterise as 'disembedded regional economies'. Export-oriented investments typify the new international division of labour in which simple and labour-intensive manufacturing activities are increasingly moved away from core to semi-peripheral and peripheral countries which offer abundant low cost labour. In this sense, CEE's cathedrals in the desert are very similar to the 'maquiladora' plants on the US-Mexican border, or foreign-owned factories in export-processing zones of less developed countries (Ellingstad 1997; Pavlínek 1998; Pavlínek and Smith 1998).

Audi Motor Hungária Kft, a greenfield engine and assembly plant built in the Hungarian city of Györ, close to the Austrian border, is a typical example of an export-oriented investment in the CEE automotive sector. The plant was located in Hungary to exploit its cheap and skilled labour, strong governmental incentives (including a ten-year profit tax holiday), relative location on the EU border, its communication infrastructure, and the possibility to employ lean production techniques that could not be introduced in Germany because of union resistance. The facility was opened in 1994, and by 2001 it was the only remaining Audi engine plant. In April 1998, the assembly of Audi's TT Coupé and Roadster began, with a capacity of 40,000 cars annually. Because of its success, the assembly of the two strongest and best equipped Audi A3 models, with an annual capacity of 15,000 cars, was launched in April 2001. Fully painted TT and A3 bodies, together with more than 90% of all components, are shipped by train from Ingolstadt to Györ where final assembly takes place. Finished cars are transported back to Germany to be sold mainly in the USA and Western Europe (Pál 2001).

The plant has been economically very successful and is expanding fast: in 2000 it imported 95% of the components for engine production and car assembly from Germany and, as a result, its linkages with local economy are virtually non-existent. Audi refuses to use local suppliers because they cannot meet its quality standards. In this sense, the role of Audi Hungária in VW's European and global production system mirrors that of nearby VW-Slovakia, as both are low cost production sites for components and niche market vehicles produced for export. Both of these investments were highly successful for VW, although their impact on local economies was limited. However, Audi Hungária will soon differ from VW Bratislava in one important respect. Audi built its first engine R&D centre outside Germany at Györ in 2001. As a result, some of the engine design and development work, which thus far has all been done in Germany, will be transferred to Hungary. The R&D center will employ eighty Hungarian engineers concentrating on the development of Audi, Seat, Škoda and VW engines (Havas 1997 p.235; Jancsó 2000; Kapoor and Eddy 1998 pp.44–45; Kiss 2001).

Other examples of similar automotive investments in Hungary include the Ford and GM plants that were also very slow to develop any local economy linkages (Sadler and Swain 1994 p.398). GM's Opel plant at Szentgotthárd (Opel is GM's European division) achieved only 5% of local content in the engine plant in 2000 (car assembly that had 8% of local content ended in 1999) and Ford's local content was also only five percent at its Székesfehérvár's plant in the mid-1990s (Ellingstad 1997 p.12; Havas 1997 pp.229–31; Jancsó 2000). Local content was gradually growing during the 1990s in attempts to cut costs, but the few existing local suppliers were typically second-tier or third-tier suppliers of low-tech and low-value components such as windshield wipers. Even the technologically most advanced local companies were unable to produce more complex and technologically advanced components of the desired quality. The car manufacturers therefore supported setting up of joint ventures between domestic and Western component suppliers or establishing Western subsidiaries in Hungary (Havas 1997 p.231). Overall, disembedded greenfield investments share many characteristics with disembedded path-dependent transformations. These are extremely efficient and profitable operations that capitalise on the comparative advantages of the cheap, skilled, flexible, and obedient labour force combined with governmental investment incentives and location close to the EU border. However, compared to embedded investments in the car industry, their contribution to local and regional economic development in the region has so far been very limited because of their very weak and only slowly developing supplier links with local companies.

The Role of the ECE Car Industry in the European and Global Division of Labour

Different strategies pursued by Western car makers in ECE have been closely related to this region's changing role in the division of labor in the industry across Europe. The FDI-driven integration of the ECE car industry into the periphery of the European production system concentrated on its two basic functions in the 1990s. First, ECE specialises on the mass production of small cars for domestic and export markets, to capitalise on its cheap and skilled labor and car manufacturing traditions. Second, the region has increasingly concentrated on the low-volume production of niche-market products, such as SUVs and high-priced sports cars, mainly for export to the EU (Havas 2000b pp.245–8; Humphrey et al. 2000 p.8; Humphrey and Oeter 2000 p.54; Lung 2000 p.24). Whereas the mass production of small cars is typically associated with high levels of embeddedness in local economies, the low volume production of special models is typified by low levels of embeddedness.

Mass Production of Small Cars

The mass production of small and cheap cars in ECE is, to a large extent, path dependent because the state socialist industry specialised in the production of precisely this type of car. However, path dependency was not the reason why this tradition continued in the 1990s. Instead, the combination of low labour costs, high quality labour, a manufacturing tradition, and the geographical proximity to Western Europe was ideally suited to the cost-cutting strategies of Western car producers. Because of the fierce competition in the market for small cars, and the small profit margins, car makers are forced to produce these cars as cheaply as possible. In addition, small cars are traditionally by far the best-selling passenger cars in ECE because of the lower purchasing power compared with Western Europe. Thus, a typical strategy of Western car makers in ECE has been to establish the best possible market position in a country where they produce and then expand their market share in the rest of the region and in Western Europe. VW in Czech Republic, Suzuki in Hungary, Daewoo, Fiat, GM Opel in Poland and Daewoo and Renault in Romania focus on the standardised mass production of small vehicles for domestic and export markets.[5] This regional specialisation will be further strengthened in the first decade of the 21st century as the Serbian Kragujevac Zastava car producer will most likely undergo a similar peripheral integration into the European system as other ECE producers have done in the 1990s. Negotiations with potential foreign investors were expected to start in mid-2002 and a strategic partnership at the beginning of 2003 (BBC 2001).

Low Volume Production of Special Models

Such production has developed in ECE as one of the Western car makers' responses to the fragmentation of the car market in Western Europe. It would be less cost-effective to commit the same kind of capital investment for a smaller series of special models as it would be for the mass production of standard cars, because this type of production requires an increased labour input and greater production flexibility. The manufacture of small volumes of a particular vehicle thus requires cost-cutting strategies in which labour costs and flexibility play an important role. In this type of production, shop floor flexibility is not derived from technologically-intensive flexible production methods, but from labour flexibility. Therefore, the cost of labour must be low enough to achieve efficiency and the labour must be highly flexible in order to accomplish rapid changes to new products. Young workers with a limited work history, who are thus less influenced by old habits that may be difficult to break, are typically hired for this type of operation. They can be quickly trained and are less likely to develop strong labor unions: the average age of Audi Hungaria's workers in 2001 was only 28 years (Sabatini 2000) which can be compared with the situation at Magyar Suzuki referred to by Swain (1998 p.661). A number of ECE factories such as Audi Hungária, Opel Hungary and VW-Slovakia proved to be very successful in this type of car manufacturing mainly as a result of the flexible labor strategies and lean production methods that would often be impossible to adopt in Western Europe because of union resistance or existing labour legislation.

For example, Audi Hungária, which assembles the expensive sports cars in Győr, subcontracts the handling of all materials. A subcontracted material handling and logistics firm unloads the train cars that bring in 90% of components from Germany and then moves the car bodies and other components to the plant. Also, Audi personnel have nothing to do with moving anything around the plant as all this work is subcontracted. All the components are supplied to the assembly line in small volumes by the logistics company. Most of the assembly of sports cars is done manually, which allows production of the coupe and roadster versions (and their front-wheel and four-wheel drive variants – with different rear suspensions) on the same assembly line, as adapting to the differences between the two models is relatively easy compared with the situation with automated assembly lines. The Audi Hungária lean production system became the basis for a worldwide implementation of the Audi Production System in 1999 (Sabatini 2000). The production system of VW Slovakia is based upon similar production strategies (Krecháč 2001; Pavlínek and Smith 1998 p.629). Opel Hungary pursued a similar maximum outsourcing strategy in its engine production as that of Audi Hungária.

Additionally, the company introduced a continuous 24–hour, seven-day production based on four shifts. VW-Slovakia introduced a floating working time system in 1997 to increase production flexibility further and cut labor costs. In the new system, an annual limit of work hours is set but the working time is unevenly distributed throughout the year – based on the immediate production needs (Krecháč 2001 p.81). This allows VW-Slovakia to react quickly to changing demand for its vehicles while, at the same time, paying only those workers that are necessary for the given production volume. These examples highlight another important role of the ECE car industry in the European and global automotive industry division of labour. ECE has increasingly served as a testing ground for new production strategies, what Grabher (1997 pp.127–9) has called 'experimenting-with-the-future' approaches. If successful, these strategies are then subsequently implemented in modified forms in parent factories of the particular car maker. In this respect Škoda's experimentation with strategic resourcing been well documented (Pavlínek and Smith 1998 p.627; Havas 2000b pp.250–1).

These different production strategies followed by Western car manufacturers in ECE suggest the existence of relationships between work organisation, type of car production, its embeddedness in local economies, and the division of labour within the European car industry. In both cases discussed above, ECE's peripheral position within the European car production system explains why these particular product specialisations emerged in the region in the 1990s. The strategies of mass production of cheap cars in ECE are most concerned with the overall production costs that compel car manufacturers to seek local suppliers of simple components while pressing their Western first tier suppliers to establish their operations in ECE, thus embedding car production in local and regional economies. The overall production costs, including the cost of labor and its skills, play a decisive role in this type of production. Although labour flexibility is important, it is not as critical a component of mass production strategies as of the low volume production of special models: the low volume production of niche market vehicles rests upon high labour flexibility – as shown above. Although labour costs play an important role in the decisions to develop this type of production in ECE, it is the high level and a particular kind of labour flexibility described above that has led to its development and location in ECE at the periphery of the European car production system. Such labour flexibility could hardly be achieved in Western Europe given its existing labour laws and the strength of its labour unions. The emphasis on labour flexibility rather than overall production costs, combined with low production volumes, also explains why foreign car makers in this type of production have not been so eager to establish strong supplier links within ECE and also why embeddedness of their operations has been low compared with that of the mass production of small cars.

Conclusion

The role of FDI in the CEE car industry transformation has been, and is likely to continue to be, crucial. FDI not only usually saved the existing car manufacturers, but also led to the increased production of cars and car components in the region. The effects of foreign capital at the level of individual car companies have also been significant, as FDI led to their profound, although uneven, transformations. With the exception of Russia, a comparison of car producers utilising foreign capital with those that have remained in local hands reveals that without the infusion of foreign capital it is likely that a collapse of the industry would have taken place. The integration of car manufacturers and component suppliers into transnational networks in an increasingly globalised industry improved their chances of survival by establishing a foothold in the global economy. Although such inclusion in the transnational economic networks is important both for the companies, the regions and the countries involved (Amin and Thrift 1997 p.155), the long-term positive regional economic effects of FDI have, in some cases, been limited and in many respects questionable.

FDI-driven car restructuring in parts of CEE led to the (re)integration of the industry into the periphery of the European car production system. In this I have shown that this integration may follow very different pathways, depending on distinct strategies pursued by Western car makers in the region, based on a combination of selective uses or a rejection of the path dependency and local embeddedness of automobile production. The effect of foreign capital on restructuring of the car industry has been uneven both sectorally and geographically. Whereas foreign capital has been heavily invested in car production and related component manufacture, its effects on the production of trucks, buses and motorcycles have been much smaller. FDI in the car industry has also contributed to the growing economic inequalities within CEE since it has been concentrated in ECE while the former Soviet Union – and indeed the southern part of ECE – has received only limited amounts of FDI. There seems to be a trend for the selective process of an integration of specific parts of CEE into European and global automobile production networks.

Within the European car production hierarchy, CEE's role has been twofold: a largely path-dependent specialisation in the mass production of small cars; and a newly developed specialisation in the assembly of low volumes of special purpose vehicles for export to rich markets. Low labour costs, high labour flexibility based on liberal labour legislation, and a high intensity of work have been instrumental in the development of this kind of car assembly, and have also allowed car makers to experiment with new lean production strategies in CEE. Although the mass production of small cars tends to be locally embedded, the local links of plants that assemble low volumes of niche automobiles tend to be very small, and these thus represent typical disembedded investments. In this paper, I have demonstrated that both of these specialisations are clearly related to the peripheral position of the region within the European car production system, and that there are links between the position of CEE on the European car production periphery and the type of car production developed and the work organisation employed by foreign car makers.

I have argued that the regional economic effects of FDI have been most profound in cases where forms of path-dependent embeddedness occurred. These forms of path-dependent embeddedness are largely based on the recreation and transformation of supply networks inherited from the state socialist period. However, it is important to keep in mind that these forms of path-dependent embeddedness only developed in cases when it was profitable for foreign investors, such as VW-Škoda. Locally embedded rather than disembedded investments in car manufacturing are more likely to represent a long-term commitment by foreign investors and have important implications for the future economic development and labor markets of a region. At the same time, it is important to realize that even this type of investment is problematic. Foreign ownership can lead to technological dependence on multinational corporations, and may intensify the CEE car industry specialisation in low cost and labour intensive production, preventint it from developing a more high-tech and more value-added production mix. It may also create strong dependence on a single corporation, such as VW in the case of Czech Republic and Slovakia, which entails significant risks as regional fortunes are linked to decisions made in a corporation external to the region, which has commitments to multiple products and markets.[6] Such risks recently materialised after Daewoo Motor, the parent company of three car factories in CEE (Daewoo-FSO Motor in Poland, Daewoo Automobile Romania, and Avtozaz-Daewoo in Ukraine) went bankrupt. These assembly plants face liquidation if no third party seller is found because they have been excluded from GM's tentative takeover of Daewoo Motor (Anon 2001b). In this case, foreign ownership failed to solve he fundamental problem for which foreign investment was sought in the first place: namely the long term survival of the car producers. Thus, in CEE it still remains to be seen what the long-term effects of car production transformation through FDI will be.

Notes

[1] This is an updated version of the paper originally published under the same title by Pion, London in 2002 in Environment and Planning A, Vol. 34, No. 9, pp. 1685–1709. I want to thank Pion Limited for the permission to republish the article. Support for this research was provided by two grants from the University Committee on Research at the University of Nebraska at Omaha. The author also wishes to thank the Slavic Research Center of Hokkaido University for their support during the preparation of this paper.
[2] The planned organisation of the way of appropriation and distribution of goods and services (what Burawoy 1985 calls relations of production) typical of state socialism was replaced by the anarchy associated with capitalism.
[3] Including the former GDR, total CEE car production declined by 277,000 units during 1990–5 and increased by 886,000 units during 1995–2000. Thus output in 2000 was 609,000 units greater than in 1990 (see Pavlínek 2002a for data on individual CEECs).
[4] VW produces its Bora, Golf, Polo – and most recently Colorado – models in Bratislava. The

plant produced 180,706 cars, 364,000 gearboxes and 7.94mln other components in 2000. VW invested more than DM1.0bln in the plant by 2000 and in 1999 it planned to triple car production there in three years. In 2001, VW planned to invest a further DM450mln in Slovakia (Václavík 2001). Similarly VW plans to invest an additional $330mln in its Audi engine plant in Györ, Hungary during 2000–3 and Škoda is investing $562mln in a new engine plant at its Mladá Boleslav production complex in Czech Republic.

[5] For example in addition to the Seicento, Fiat is producing its small Palio model only in Poland for its global sale and distribution network. Daewoo is focusing on the assembly of its small models Matiz and Lanos in its Polish Daewoo-FSO factory in Warsaw. GM Opel's new factory at Gliwice products 70,000 small Opel Astra models. Suzuki is focusing on the production of its small Swift models in Hungary of which about 75% are exported, and 66% of Škoda producted in 2000 were small cars (Felicias and Fabias). Škoda is exporting 80% of its production and 52.6% of Škoda sales were realised in Western Europe (Škoda Auto 2000 p.26). Renault-Dacia is also specialising in the production of such cheap entry level cars by 2010 and plans to produce more than 500,000 Dacias annually (Renault 2001 p.3).

[6] The VW-Slovakia-dominated car industry of Slovakia accounted for 19% of Slovakia's GDP in 1999 and 16% of the country's exports. In the case of Czech Republic, 14% of exports are attributable to VW-Škoda and its suppliers and some four percent of the Czech workforce are employed directly by VW-Škoda (Anon 2001 p.60; Kimentová 2000).

References

AIA SR (2001), *Štatistika – Ročenka Automobilového priemyslu Solvenskej Republiky 2001*, Zdruzenie atomobilového priemyslu Slovenskej republiky and Slovenská obchodná a priemyslná komora, Bratislava.

Amin, A. and Thrift, N. (1997), 'Globalization Socio-Economics Territoriality', in R. Lee and J. Wills (eds), *Geographies of Economies*, Arnold, London, 147–57.

Anon (1991), 'BAZ: Najednou bez dluhů!', *Lidove noviny*, May 29, 2.

Anon (2000), 'Nádej pre Martin', *Magazín MOT*, 7/2000 <www.mot.sk>.

Anon (2001a), 'Škoda Auto: Slav Motown', *The Economist*, 358, 60–2.

Anon (20001b), 'Discarded Daewoo Plants face Liquidation', *Financial Times*, September 28.

BBC (2001), 'Servian ministers detail accord on Zastava car producer', *BBC Monitoring Service*, July 30.

Bibza, S. (2000), 'VW-Slovakia v roku 2000', *Magazín MOT*, 5/2000 <http://www.mot.sk>.

Bod, P.A. (1998), 'The Social and Economic Legacies of Direct Capital Inflows: The Case of Hungary', in J. Bastian (ed), *The Political Economy of Transition in CEE: The Light(s) at the End of the Tunnel*, Ashgate, Aldershot, 13–44.

Bordenave, G. and Lung, Y. (1996), 'New Spatial Configurations in the European Automobile Industry', *European Urban and Regional Studies*, 3, 305–21.

Burawoy, M. (1985), *The Politics of Production*, Verso, London.

Burawoy, M. and Krotov, P. (1992), 'The Soviet Transition from Socialism to Capitalism: Worker Control and Economic Bargaining in the Wood Industry', *American Sociological Review*, 57(1), 16–38.

Burawoy, M. and Krotov, P. (1993), 'The Economic Basis of Russia's Political Crisis', *New Left Review*, 198, 49–69.

Chomicka, J. (1999), *Poland: Automotive Parts Market*, U.S. Foreign Commercial Service and Department of State, Washington, D.C.

Clarke, S., Fairbrother, P., Borisov, V. and Bizyukov, P. (1994), 'The Privatization of Industrial Enterprises in Russia: Four Case-Studies', *Europe-Asia Studies*, 46, 179–214.

Dicken, P. (1998), Global Shift: Transforming the World Economy, Guilford Press, New York.

Dicken, P., Forsgren, M. and Malmberg, A. (1994), 'The Local Embeddedness of Transnational Corporations', in A. Amin and N. Thrift (eds), *Globalization Institutions and Regional Development in Europe*, Oxford University Press, Oxford, 23–45.

Domański, B. (1999), 'Structure Regional Distribution and Selected Effects of FDI in Polish manufacturing in the 1990s', *Wirschaftsgeographische Studien*, 24/25, 71–88.

EIU 2001, *Vehicle Production Since 1990*, EIU, London <www.autoindustry.co.uk>.

Ellingstad, M. (1997), 'The Maquiladora Syndrome: Central European Prospects', *Europe-Asia Studies*, 49, 7–21.

Ferenčiková, S. (1995), 'Vstup zahraničného kapitálu do slovenskej ekonomiky: na príklade vybraných joint ventures', *Ekonomický casopis*, 43, 140–53.

Ferenčiková, S. (1996), 'Priame zahraničné investície: teória a prax v Slvenskej republike', *Ekonomické rozhl'ady*, 25, 17–35.

General Motors (1998), 'New Opel Polska Plant in Gliwice Based on the Lean Manufacturing Principles of Eisenach', *General Motors Press Release*, October 29.

Grabher, G. (ed) (1993a), *The Embedded Firm: On the Socioeconomics of Industrial Networks*, Routledge, London.

Grabher, G. (1993b), 'Rediscovering the Social in the Economics of Interfirm Relations', in G. Grabher (ed), *The Embedded Firm: On the Socioeconomics of Industrial Networks*, Routledge, London, 1–31.

Grabher, G. (1994), 'The Disembedded Regional Economy: The Transformation of East German Industrial Complexes into Western Enclaves', in A. Amin and N. Thrift (eds), *Globalization Institutions and Regional Development in Europe*, Oxford University Press, Oxford, 177–95.

Grabher, G. (1997), 'Adaptation at the Cost of Adaptability? Restructuring the Eastern German Regional Economy', in G. Grabher and D. Stark (eds), *Restructuring Networks in Post-Socialism: Legacies Linkages and Localities*, Oxford University Press, Oxford.

Granovetter, M. (1985), 'Economic Action and Social Structure: The Problem of Embeddedness', *Amerian Journal of Sociology*, 91, 481–510.

Håkansson, H. and Johanson, J. (1993), 'The Networks as a Governance Structure: Interfirm Cooperation beyond Markets and Hierarchies', in G. Grabher (ed), *The Embedded Firm: On the Socioeconomics of Industrial Networks*, Routledge, London, 35–51.

Hardy, J. (1998), 'Cathedrals in the Desert? Transnationals Corporate Strategy and Locality in Wroclaw', *Regional Studies*, 32, 639–52.

Hausner, J., Jessop, B. and Nielsen, K. (eds) (1995), *Strategic Choice and Path-Dependency in Post-Socialism: Institutional Dynamics in the Transformation Process*, Edward Elgar, Cheltenham.

Havas, A. (1995), 'Hungarian Car Parts Industry at a Crossroads: Fordism versus Lean Production', *EMERGO Journal of Transforming Economies and Societies*, 2(3), 33–55.

Havas, A. (1997), 'FDI and Intra-Industry Trade: The Case of the Automotive Industry in Central Europe', in D.A. Dyker (ed), *The Technology of Transition: Science and Technology Policies for Transition Countries*, Central European University Press, Budapest, 211–40.

Havas, A. (2000a), 'Local Regional and Global Production Networks: Reintegration of the Hungarian Automotive Industry', in C. von Hirschhausen and J. Bitzer (eds), *The Globalization of Industry and Innovation in Eastern Europe: From Post-Socialist Restructuring to International Competitiveness*, Edward Elgar, Cheltenham, 95–127.

Havas, A. (2000b), 'Changing Patterns of Inter- and Intra-Regional Division of Labour: Central Europe's Long and Winding Road', in J. Humphrey, Y. Lecler and M.S. Salerno (eds), *Global Strategies and Local Realities: The Auto Industry in Emerging Markets*, Macmillan, Basingstoke, 234–62.

Hudson, R. (1995), 'The Japanese, the European Market and the Automobile Industry in the United Kingdom', in R. Hudson and E.W. Schamp (eds), *Towards a New Map of Automobile Manufacturing in Europe? New Production Concepts and Spatial Restructuring*, Springer, Berlin, 63–91.

Hudson, R. (1997a), 'Regional Futures: Industrial Restructuring New High Volume Production Concepts and Spatial Development Strategies in the New Europe', *Regional Studies*, 31, 467–78.

Hudson, R. (1997b), 'Changing Gear? The Automobile Industry in Europe in the 1990s', *Tijdschrift voor Economische en Sociale Geografie*, 88, 481–7.

Hudson, R. and Schamp, E.W. (eds) (1995), *Towards a New Map of Automobile Manufacturing in Europe? New Production Concepts and Spatial Restructuring*, Springer, Berlin.

Humphrey, J., Lecler, Y. and Salerno, M.S. (2000), 'Introduction', in J. Humphrey, Y. Lecler and M.S. Salerno (eds), *Global Strategies and Local Realities: The Auto Industry in Emerging Markets*, Macmillan, Basingstoke, 1–15.

Humphrey, J. and Oeter, A. (2000), 'Motor Industry Policies in Emerging Markets: Globalisation and the Promotion of Domestic Industry', in J. Humphrey, Y. Lecler and M.S. Salerno (eds), *Global Strategies and Local Realities: The Auto Industry in Emerging Markets*, Macmillan, Basingstoke, 42–71.

Hungarian Trade Commission (2001), *More Car Factories to follow Suzuki: Opel and Audi to Hungary*, HTC, London <http://freespace.virgin.net/huntrade.london/>.

Jancszó, A. (2000), 'Automotive Industry to Lead the way to Prosperity?', *Business Hungary*, 14 <http://www.amcham.hu>.

Jessop, B. (1997), 'The Governance of Complexity and the Complexity of Governance: Preliminary Remarks on some Problems and Limits of Economic Guidance', in A. Amin and J. Hausner (eds), *Beyond Market and Hierarchy: Interactive Governance And Social Complexity*, Edward Elgar, Cheltenham, 95–128.

Kapoor, M. and Eddy, K. (1998), 'Two-Track Hungary: A Survey of Hungarian Industry', *Business Central Europe*, 6(50), 41–9.

Kiss, T.S. (2001), 'Audi Speeds along with Expansion Plan', *Budapest Sun Online*, 9(37) <http://www.budapestsun.com>.

Klimentová, S. (2000), 'Automobilová v´yroba nosn´ym motorom strojárskeho priemyslu', *Národná obroda*, 11(218).

Koza, P. (2000), 'Waiting for a Rescuer', *Business Central Europe*, 7, December <http://www.bcemag.com>.

Krecháč, J. (2001), *Priestorové súvislosti atomobilového priemyslu na Slovensku*, Unpublished Master's Thesis Prírodovedecká fakula UK, Bratislava.

Lagendijk, A. (1997), 'Towards an Integrated Automotive Industry in Europe: A "Merging Filière" Perspective', *European Urban and Regional Studies*, 4, 5–18.

Law, C.M. (1991), *Restructuring the Global Automobile Industry: National and Regional Impacts*, Routledge, London.

Leborgne, D. and Lipietz, A. (1991), 'Two Social Strategies in the Production of New Industrial Spaces', in g. Benko and M. Dunford (eds), *Industrial Change and Regional Development: The Transformation of New Industrial Spaces*, Belhaven Press, London, 27–50.

Lung, Y. (1992), 'Global Competition and Transregional Strategy: Spatial Reorganization of the European Car Industry', in M. Dunford and G. Kafkalas (eds), *Cities and Regions in the New Europe: The Global-Local Interplay and Spatial Development Strategies*, Belhaven Press, London, 68–85.

Lung, Y. (2000), 'Is the Rise of Emerging Countries as Automobile Producers an Irreversible Phenomenon?', in J. Humphre, Y. Lecler and M.S. Salerno (eds), *Global Strategies and Local Realities: The Auto Industry in Emerging Markets*, Macmillan, Basingstoke, 16–41.

Meyer, K.E. (2000), 'International Production Networks and Enterprise Transformation in Central Europe', *Comparative Economic Studies*, 42(1), 135–50.

MOTORSAT (2001), *International Automobile Statistics* <http://perso.club_internet.fr/motorsat/>.

Nielsen, K., Jessop, B., and Hausner, J. (1995), 'Institutional Change in Post-Socialism', in J. Hausner, B. Jessop and K. Nielsen (eds), *Strategie Choice and Path-Dependency in Post-Socialism: Institutional Dynamics in the Transformation Process*, Edward Elgar, Cheltenham, 3–44.

OICA (2001), *World Motor Vehicle Production by Country and Type 1997–2000*, <http://www.oica.net>.

Opel Polska (2001), *Opel Polska plant in Gliwice: Facts and Figures*, document provided for the author by Jacek Zarnowiecki of Opel Polska.

Pál, B. (2001), 'A3 Made in Hungary', *Business Hungary*, 15(6) <http://www.amcham.hu>.

Pavlínek, P. (1998), 'FDI in the Czech Republic', *Professional Geographer*, 50, 371–85.

Pavlínek, P. (2000), 'Restructuring of the Commercial Vehicle Industry in the Czech Republic', *Post-Soviet Geography and Economics*, 41, 265–87.

Pavlínek, P. (2002a), 'Restructuring the CEE Automobile Industry: Legacies Trends and Effects of FDI', *Post-Soviet Geography and Economics*, 43, 41–77.

Pavlínek, P. (2000b), 'Domestic Privatization and its Effects on Industrial Enterprises in ECE: The Evidence from the Czech Motor Component Industry', *Europe-Asia Studies*, 54, 1127–50.

Pavlínek, P. (2000c), 'The Role of FDI in the Czech Automotive Industry Privatization and Restructuring', *Post-Communist Economies*, 14, 359–79.

Pavlínek, P. and Smith, A. (1998), 'Internationalization and Embeddedness in ECE Transition: The Contrasting Geographies of Inward Investment in the Czech and Slovak Republics', *Regional Studies*, 32, 619–38.

Pickles, J. and Smith, A. (1998), *Theorizing Transition: The Political Economy of Post-Communist Transformation*, Routledge, London.

Poznańska, J.K. and Poznański, K.Z. (1996), 'Foreign Investment in the East European Automotive Industry: Strategies and Performance', *EMERGO Journal of Transforming Economies and Societies*, 3(2), 70–82.

Poznański, K.Z. (1993), 'Restructuring of Property Rights in Poland: A Study in Evolutionary Economics', *East European Politics and Societies*, 7, 395–421.

Poznański, K.Z. (1996), *Poland's Protracted Transition: Institutional Change and Economic Growth 1970–1994*, Cambridge University Press, Cambridge.

Ratajczyk, A. (1996), 'No Crime No Scandal – But Lots of Questions', *Warsaw Voice*, 7(382), 9.

Renault (2001), 'Renault set to make Dacia a Global Entry-Level Brand', *Renault Press Kit*, April 2 <http://www.renault.com>.

Sabatini, J. (2000), 'Assembling the Awesome Audi', *Automotive Design and Production*, 4/2000 <http://www.autofieldguide.com>.

Sadler, D. (1998), 'Changing Inter-Firm Relations in the European Automotive Industry: Increased Dependence or Enhanced Autonomy for Components Producers?', *European Urban and Regional Studies*, 5, 317–28.

Sadler, D. (1999), 'Internationalization and Specialization in the European Automotive Components Sector: Implications for the Hollowing-out Thesis', *Regional Studies*, 33, 109–19.

Sadler, D. and Swain, A. (1994), 'State and Market in Eastern Europe: Regional Development and Workplace Implications of FDI in the Automobile Industry in Hungary', *Transactions of the Institute of British Geographers*, 19, 387–403.

Sadler, D., Swain, A. and Hudson, R. (1993), 'The Automobile Industry and Eastern Europe: New Production Strategies or Old Solutions?', *Area*, 25, 339–49.

Škoda Auto (2000), *Škoda Auto Výroční zpráva 2000*, Škoda Auto, Mladá Boleslav.

Smith, A. (1998), *Reconstructuring the Regional Economy: Industrial Transformation and Regional Development in Slovakia*, Edward Elgar, Cheltenham.

Smith, A. and Ferenčíková, S. (1998), 'Regulating and Institutionalising Capitalisms: The Micro-Foundations of Transformation in ECE', in J. Pickles and A. Smith (eds), *Theorising Transition: The Political Economy of Post-Communist Transformations*, Routledge, London, 25–53.

Stark, D. (1992), 'Path Dependence and Privatization Strategies in ECE', *East European Politics and Societies*, 6, 17–51.

Stark, D. (1995), 'Not by Design: The Myth of Designer Capitalism in Eastern Europe', in J. Hausner, B. Jessop and K. Nielsen (eds), *Strategic Choice and Path-Dependency in Post-Socialism: Institutional Dynamics in the Transformation Process*, Edward Elgar, Cheltenham, 67–82.

Stark, D. (1996), 'Recombinant Property in East European Capitalism', *American Journal of Sociology*, 101, 993–1027.

Stark, D. and Bruszt, L. (1998) , *Postsocialist Pathways: Transforming Politics and Property in ECE*, Cambridge University Press, Cambridge.

Swain, A. (1998), 'Governing the Workplace: The Workplace and Regional Development

Implications of Automotive Foreign Investment in Hungary', *Regional Studies*, 32, 653–71.

Václavík, I. (2001), 'Autá motorom ekonomiky: pre automobilový priemysel pracuje v SR 130 firiem', *Hospodárske noviny*, August 24.

Weidokal, M. and Stagles, J. (2001), *Central and East European Automotive Overview*, Pricewaterhouse Coopers LLP, Washington, D.C.

Wells, P. and Rawlinson, M. (1994), *The New European Automobile Industry*, St. Martin's Press, New York.

Zemplinerová, A. (1998), 'Impact of FDI on the Restructuring and Growth in Manufacturing', *Prague Economic Papers*, 7, 329–45.

Chapter 6

Place Marketing for Foreign Direct Investment in Central and Eastern Europe

Craig Young

Introduction

Post-socialist CEE has seen a rapid adoption of 'Western' place marketing strategies. This chapter therefore constitutes an initial survey of the role of place marketing in attracting FDI into the region. Given the lack of any consistent evidence with which to evaluate the effectiveness of such strategies, the aim of the chapter is to present an overview of place marketing practice. First, the chapter discusses what place marketing for FDI is and the determinants of attracting such investment. Next the chapter outlines the changing context of FDI locational decision making in CEE and the implications of this for place marketing. The chapter then provides an overview of place marketing practice in the region by analysing case studies reflecting the variety of experience across the region and the spatial scales over which place marketing for FDI is undertaken. This leads to a concluding discussion of the key aspects of place marketing for FDI in post-socialist CEE. There is an inevitable bias towards the experiences of the more advanced post-socialist economies where place marketing is best developed, but this also reflects the lack of case studies of place marketing practice in the region and the lack of data which enable evaluation of that practice. There is a complete lack of any systematic cross-national evaluation on which to base evaluation of practice. Hence the need for further research to support the evaluation of place marketing and identify good practice to assist policy making by ensuring sustainable and embedded flows of quality FDI as a part of the region's development strategies.

Place Marketing for FDI

Theories of FDI focus on three groups of factors. These are: ownership-specific advantages (e.g. ownership of intellectual property or managerial skills); location-specific advantages (why companies seek to exploit ownership advantages internationally e.g. the ability to access local markets or cheap labour); and internalisation incentive advantages (why production takes place within the MNE in a variety of countries rather than devolving production to foreign companies e.g. avoiding transaction costs) (Hine 1999, following Dunning 1981). Although patterns of FDI into ECE question the applicability of such theories in this context (Du Pont 2000), FDI theory is significant for place marketing as locational attractiveness is one part of FDI decision-making that governments can influence. FDI is 'at the crossroads of a demand for localisation and a supply of localisation', and MNE willingness to invest abroad thus intersects with the ability of the host country to manage its attractiveness (Fabry and Zeghin 2002 p.292).

To achieve this post-socialist ECE rapidly adopted 'Western' place marketing strategies to compete for FDI by improving the competitive advantage of localities. Place marketing is a mix of: changes in the form and function of localities (the 'place product'); the use of financial incentives; the promotion of a new image of place; and changes in the way that places are governed. Thus localities try to make themselves more attractive than others to key economic decision makers to stimulate local economic development, and in CEE a key target is to attract globally mobile FDI. Thus the ability of a host country to market competitive immobile assets, such as skills, infrastructure, services, institutions assisting FDI and supply networks to complement the mobile assets of MNEs is argued to be increasingly important (Lall 2002). The ability to do this is often related to the capacity to create high levels of 'institutional thickness' arising from institutions sharing a common industrial purpose which are developing economic strategy through collaborative interactions. Institutional thickness, it is claimed, can 'hold down the global' (Amin and Thrift 1994), enhancing localities' economic competitiveness by embedding international investment.

There is no consensus over the effectiveness of attempts to influence FDI flows (Phelps 1999). Smith and Ferenčíková (1998 p.161) suggest that the uneven distribution of FDI into the region is partly due to 'the wholesale inability of governments (both local and national) to influence the location of inward investment'. However, states can attempt to influence FDI. The transaction costs of investing and producing vary between countries affecting their competitive position, market failures in information are important as the decision-making process of MNEs is subjective, and there is a need for co-ordination of immobile assets to meet MNE requirements (Lall 2002). Governments also determine the inward investment climate, through policy on education, training, industrial and regional policy, support for SMEs, establishing a credible business environment and investment infrastructure, and privatisation (Beyer 2002; OECD 2003). This is also significant for improving the embeddedness of FDI, as these factors influence the linkage behaviour of foreign investors (Meyer and Qu 1995). The key is to identify MNE requirements, change the place product to fulfil them and supply information about the locality. Government cooperation is also important in the establishment of supra-national multilateral investment rules to regulate MNEs at their scale of operation and to standardise national investment regulations (Young and Brewer

1999). Governance infrastructure is thus important in defining the investment environment (Du Pont 2000; Globerman and Shapiro 2002; Smekal and Sausgruber 2000), to influence the manner in which 'the activities of global corporations may be tied to particular territories' (Alden and Phelps 1999 p.275).

A number of studies have attempted to identify the location-specific variables which are determinants of FDI location (Table 6.1). The CEE investment context is characterised by problems with several of these variables, including risk, uncertainty, a poor legal framework and weak infrastructure, lack of reliable information and the inheritance of factors from communism. However, transformation has opened up opportunities for FDI (Collis et al. 1999). Institutional changes, the modernisation of the economic environment (particularly privatisation), clarification of property rights, adoption of bankruptcy laws and FDI legislation, price liberalisation, market growth, trade policy and the relative cost of inputs have enhanced the investment environment. However, there is no consensus over the relative importance of these determinants. Some studies have concluded that the key determinant of FDI location in the region is the maintenance of an environment conducive to sustained economic growth, with progress in economic reforms, particularly liberalization, privatisation, stabilisation and institutional changes (Claessens et al. 1998; Hunya 2000). However, Garibaldi et al. (1999) concluded that the legal and political climate rather than macroeconomic fundamentals have shaped FDI to transitional economies, and that macroeconomic stability without a business friendly environment is not enough to attract FDI. Du Pont's (2000) study of foreign investors in Poland demonstrated that only 40% of those surveyed were concerned with investment-climate factors.

Country level surveys also reveal no pattern in the relative importance of determinants. In Poland, a ranking by foreign investors suggested that 64% saw cost efficiency factors as the key issue, 57% market factors, 49% labour factors and 40% investment-climate factors (Du Pont 2000). In Slovenia (OECD 2002) access to local and adjacent markets (42% and 36% of investors respectively) was adjudged most important, with technology and know-how seen as important by 30% of investors, quality of labour (27%), financial support of potential joint-venture partners (25%) and securing materials and parts (11%). Low cost labour was only a factor for just two percent. Just about the only point of agreement in the literature is that financial incentives – though the most widely used strategy to attract FDI to the region – are the least effective method (Beyer 2002; Estrin et al. 1997; Hunya 2000; OECD 2003), though they may be important to small firms (Meyer and Qu 1995). This means there is no simple strategy for attracting FDI. The different study methodologies, country circumstances and FDI requirements make it impossible to identify a hierarchy of investor needs. Any of these factors could be crucial dependent on the stage of decision-making (Collis et al. 1999) and the specific needs of investors. Du Pont (2000 p.243) concluded that 'patterns and determinants of FDI are affected by factors that vary considerably in different countries [and thus] it is a mix of requirements structured in a competitive package matched to the specific needs of an inward investor which is crucial' (Christodoulou 1996 p.21).

Table 6.1 Location-specific determinants of FDI significant for MNEs

Labour	Infrastructure	Political Economy	Quality of Life
Skilled and educated labour at relatively low cost	Good transport	Sound and stable legal framework of industrial property rights	Cultural affinity with host country
High productivity	Suitable sites and premises, with services, development consent and expansion possibilities	Suitable economic and fiscal policies which create a conducive business environment	Relatively easy language to acquire
Flexible employment legislation	Good I.T. and telecomms	Economic and political stability	Good environment
Good industrial relations	Suitable local suppliers who are flexible and relatively low cost	Good place image endorsed by existing inward investors	Good housing, leisure and education
Effective vocational training and language skills	Relative cheapness of all of the above	Professional approach to investors and welcoming attitude	Low crime
Managerial expertise	Existence of business services	Incentives for FDI	Relatively low housing costs
	Access to technology and plant	Location within tariff walls eg. EU, export incentives and import barriers	
	Quality place marketing institutions ('one-stop shops')	Size of market, prospects of market growth and/or potentially strong margins on products	
		Potential for vertical integration of production and/or 'springboard' effect into neighbouring markets	
		Presence of related and supporting industry, track record in sector	
		Degree of economic transformation, privatisation etc.	

Source: adapted from Collis et al. 1999; Koźmiński and Yip 2000, and Estrin et al. 2000.

Investors follow a multiple set of market seeking, factor-cost advantage seeking and strategic motives, or may be influenced by a good opportunity, such as buying a company offered for privatisation (OECD 2002). FDI is a dynamic process and place marketing should take into account its forms and changing environment.

The Changing Context of FDI Decision-Making and the Implications for Place Marketing

Since 1989 there have been significant shifts in MNE corporate competition and FDI decision-making. These vary by sector and may be more significant amongst the leading edge corporate strategies of larger MNEs (Phelps 1999). The picture is complicated by the fact that FDI takes one of two main forms (following OECD 2002). Horizontal FDI is motivated by market access and typified by an orientation to sales in local markets. Vertical FDI is factor cost advantage seeking (i.e. resource seeking), efficiency seeking or strategic asset seeking. In this case there is geographical separation of different stages of the value-added chain. Each of these types of FDI implies different (though inter-related) decision-making issues within the changing context of FDI which is outlined below. Corporate strategy has shifted from truly global strategies to trans-national strategies centred on major regional markets, such as the EU, intensifying competition between MNEs within those regions (Phelps 1999). As CEECs gain accession to the EU in 2004 and 2007 and perhaps beyond, they will be drawn into this increasingly competitive environment for FDI, while those countries remaining outside of the EU will have to compete with new member countries. This regional focus has significance for horizontal forms of FDI wishing to exploit local CEE markets, but also for those concerned with gaining access to EU markets. Furthermore, MNE choice of location is increasingly driven by corporate efficiency and competitiveness which is less cost-based and more technology and innovation based (Alden and Phelps 1999). The place of subsidiaries in international production networks and the need for them to move to higher value-added products within the MNE corporate network is becoming crucial. New technology facilitates the fragmentation of the production process i.e. the division of the industry's value chain into smaller functions located in different countries. MNEs are instrumental in setting up supply chains that cut across national borders and CEE producers are increasingly becoming part of this global division of labour based on production fragmentation. The complex specialisation implicit in intra-industry trade extends the division of labour to parts and components of products within transborder supply chains i.e. intra-product trade.

This has considerable implications for FDI into the region. Trade within MNE networks has been the most rapidly growing component of world trade between 1980–2000, and CEECs have experienced a faster growth in trade in parts and components than in manufactures. More than one-third of ECE-EU trade relates to intra-product trade generated by fragmentation of the global production process

(Kaminski 2000). The development of ECE-EU trade from 1989 indicates the region's changing position in this international division of labour, with the EU emerging as the dominant trading partner. Between 1989–93 unskilled labour intensive goods, agriculture based products and industrial raw materials drove this re-orientation, but there has since been a shift in the composition of CEE-EU trade towards technology, capital and skilled labour intensive products. This indicates the region's growing participation in more sophisticated and higher value-added production activity. Foreign firms have emerged as the largest exporters in Czech Republic, Estonia, Hungary and Poland, and FDI in the region has been mainly responsible for the shift to skilled labour and technology based products.

Progress in economic transformation influences the nature of FDI and the ability of ECE countries to attract it. In the CECs privatisation is still significant for FDI flows but there is increasing greenfield and merger and acquisitions forms of FDI. Hungary made an early decision to privatise rapidly through opening strategic sectors to external investors. In Czech Republic, however, the system of mass voucher privatisation initially formed a barrier to FDI (Kaminski 2000). By 1998 the relatively promising privatisations were largely complete in these countries, with remaining privatisation limited to the more complex or politically sensitive sectors such as utilities, which are less frequent but larger. In these countries the link between privatisation receipts and FDI flows has been broken, and the challenge is to attract non-privatisation FDI (EBRD 2003).

In SEECs by contrast, FDI has been largely privatisation driven, with greenfield investments in manufacturing relatively rare. In Croatia, Romania and Bulgaria a slow but straightforward privatisation policy should still attract FDI for a few years, though delayed privatisation in Bulgaria contributed to a drop in FDI in 2002 (EBRD 2003). While Romania and Bulgaria have mainly attracted low wage, export oriented greenfield investments their growing economies and markets may attract such investments in consumer goods sectors. In the western Balkans, such as Albania, Bosnia and Hercegovina, Macedonia and Serbia & Montenegro, the risk factor is too great for investors, with export oriented investors rarely interested. Investment has been mainly privatisation or greenfield to serve the local market (Hunya 2002). In CIS countries FDI has been concentrated in countries rich in natural resources especially Azerbaijan, Kazakhstan and Russia. Reasonably solid macroeconomic performance, growing domestic markets and political stability have enhanced their attractiveness. However, there is a need to attract non-resource focused FDI which requires greater investor confidence and domestic growth (EBRD 2003).

FDI mode is influenced by the investment environment of countries. If a country is characterised by uncertainty then joint ventures are often a less risky vehicle for entering local markets. Where there are clearly defined property rights, low risk and good privatisation then a mix of greenfield investment, acquisitions and joint ventures is more common. This may also change over time as investors become more confident, with joint ventures forming an initial stage leading to majority-ownership or greenfield investment. The first wave of CEE investment comprised a significant level of joint ventures and acquisitions. While countries such as Czech

Republic, Hungary and Poland had a roughly even distribution between greenfield investments, acquisitions and joint ventures in the early years of transformation, the Russian Federation was dominated by joint ventures (Collis et al. 1999). In the SEECs greenfield investments in manufacturing have been rare (Hunya 2002). Thus the type and volume of FDI attracted varies considerably over the region, influenced by the complex relationships between economic reform, privatisation and previous record in FDI.

These trends have implications for place marketing, though these vary by type or mode of FDI, sector, country development and existing FDI. The region will have to be increasingly competitive to attract FDI. Place marketing must engage potential investors as early as possible to establish their needs, as strategies to attract FDI, for example, will vary with horizontal and vertical forms. FDI is becoming skill- and efficiency-seeking (Globerman and Shapiro 1999) as globalisation moves competition from price to quality (Fabry and Zeghni 2002) and CEE is increasingly integrated into the fragmented production chains of MNEs in the higher value-added stages. The implication of this is that quality factors (e.g. quality of labour or local supplier networks) have often become more important in locating FDI than cost factors like the availability of cheap labour (Christodoula 1996). There is also a desire to avoid 'race to the bottom' strategies (e.g. 'invest with us, we have the cheapest labour/costs or the weakest environmental/labour regulations') which would relegate the region to a low-wage, low-skill, low-standard periphery. These are problematic anyway since wage levels in many countries are relatively high compared to competing locations, and EU accession and MNE demands are driving up social and environmental standards. Thus the ability of a country to provide competitive, quality immobile assets to complement the mobile assets of MNEs is increasingly important. Localities which can help MNEs to establish globally competitive operations have more chance of attracting FDI (Lall 2002).

Place marketing agencies must thus develop more sophisticated strategies to capture FDI in an increasingly competitive field, and focus more on attracting quality, hi-tech and high value-added FDI. They must also attract investment which will become more embedded into local economies. FDI can take the form of 'cathedrals in the desert' i.e. large-scale capital investments with little spill-over into local networks of producer-suppliers thus lacking embeddedness into local economies and limiting its benefits (e.g. Hardy 1998; Smith and Ferenčíková 1998). To reinforce the point about quality, higher value-added elements of MNE production chains are potentially less mobile (Hunya 2000). Thus place marketing must attract higher quality, more embedded forms of FDI and ensure that such investments remain embedded. This is more likely to happen with investments which are tied into local supply networks and which benefit from aftercare programmes. This implies marketing strategies which are more focused on investors' needs and have a longer term commitment to investors. This can also be important to maximise ongoing FDI i.e. by keeping existing investors happy there is a greater likelihood of attracting more investment from the same parent company. Marketing using the testimony of existing investors and even word of mouth contact between existing investors and new investors are also reasons for maintaining longer term relations with inward investors. Overall this suggests that simple promotional campaigns to raise investor awareness of localities in the region are now quite limited in their potential impact and that place marketing must become a sophisticated and complex set of strategies.

Place Marketing Practice in CEE

This section presents an overview of place marketing practice in the region by presenting a set of case studies across a range of scales, reflecting the multi-scalar nature of governance throughout Europe. From this perspective (Brenner 1999) there are ongoing transformations of the territorial organisation of the state within which new institutions and regulatory forms are produced at supranational to subnational scales, while the role of national scale governance is being redefined in response to the current round of capitalist globalisation with which CEE is seeking to reintegrate itself. The end of state-socialism brought attempts to democratise former communist societies. This has led to increased supranational political cooperation, dramatic changes in national level governance, a re-regionalisation in many of these countries and the introduction of self-government with new powers at regional and local levels. This changing political context has important implications for place marketing as a form of multi-scalar governance in the region which is reflected in the case studies below.

Supranational Place Marketing: Euroregion Baltic

Place marketing has become supranational in scale as part of broader attempts to achieve cooperation in economic development within Europe, with one example being the activities of newly established Euroregions. Euroregion Baltic (ERB) was established in 1998, bringing together representatives of regional and local authorities from bordering areas of Denmark, Latvia, Lithuania, Poland, Russia and Sweden into a new trans-national body. The aims of ERB are to improve the quality of life of people in the area, promote mutual contacts, plan activities for sustainable development and promote actions aimed at cooperation between regional and local authorities. Several programmes have been started to try and further integrate the region. Using EU Interreg IIIb funding the SEAGULL-DevERB project aims to deepen and solidify regional cooperation. The Joint Transnational Development Programme has similar aims in line with EU objectives of social and economic cohesion, while the Swedish-funded Good Governance Programme established in 2002 seeks to strengthen the role of cooperation between municipalities and regions to enhance the attraction of large infrastructure projects and the position of ERB for EU regional funding <http://www.eurobalt.org; http://www.eurobalt.org/english/ pdf/ annualrep2002eng. pdf>. Although the ERB is not a place marketing organisation, establishing these stronger supranational links, albeit in a completely new regional alliance, provides the post-socialist localities involved with a medium for marketing themselves as part of a transnational entity, and specifically one with a European identity oriented to EU aims and goals.

The Lithuanian member, Klaipeda County, thus seeks to re-image itself as being well integrated into European transit corridors, having a young, educated and well qualified population, demonstrating economic growth, having already attracted FDI which distinguishes it from other counties, and offering quality standards and stability regulated by processes of accession to the EU and NATO. This image clearly attempts to incorporate notions of Klaipeda as a 'quality' location for investment by focusing on images of a quality workforce and well advanced economic restructuring. Included in its regional development plan are the aims of economic development including 'a favourable atmosphere for investors, to attract as much FDI as possible' <http://www.klaipeda.aps.lt>. The Kurzeme planning region in Latvia is represented by the Kurzeme Regional Development Agency, established in 1999 to facilitate socioeconomic development with one aim being 'to coordinate the attraction of financial resources from international and national financing institutions, EU programmes and funds' <http://www.kurzeme.lv>. Thus the ERB represents a new transnational entity which offers a vehicle for post-socialist locations to market themselves as part of an international body with a specifically European identity, offering wider regional cooperation but also to help distinguish themselves from other competing post-socialist localities.

Sophisticated Place Marketing at a National Scale: Czech Republic

At a national level some ECE countries have developed sophisticated place marketing strategies. Czech Republic has become a leader in attracting FDI. The 1998 National Investment Incentives Scheme offers tax incentives, job creation grants, retraining grants and site development support. A significant aspect has been the development of the Czech Agency for Foreign Investment, or CzechInvest. This is an important example of an promotional organisation which has developed a sophisticated approach to place marketing in order to promote the country internationally as 'one of the most advantageous areas in Europe in which to locate mobile FDI into the production and strategic services sectors' <http://www.czechinvest.cz>. Established in 1992, it is evolving into a national development agency, with a marketing role, that seeks to attract FDI by improving the investment environment and the comparative advantage of Czech industry in the global economy. 2003 interview material indicates that CzechInvest promotes FDI opportunities using media presence, targeted seminars, workshops and conferences, and company visits. Press releases on key developments in the country are targeted on carefully selected media which may have more impact on perceptions. Journalists and trade missions are introduced to key investors. Marketing events abroad, such as trade fairs, and specialist seminars for journalists and investors also help promote the country's image. The place image projected is of the Czech Republic as a quality location for FDI, reflected in hitech target sectors: strategic services; R&D and design centres; biotech; hitech chemicals; and (though less important) the automotive sector. A new target sector is electronics and semi-conductors. The marketing department develops investor leads through direct marketing activities especially overseas. CzechInvest manages the investment incentives programme, offering tax relief on corporate tax, job creation grants, retraining grants and low cost land. It also provides information on the business climate; brings together FDI with joint venture partners, suppliers, greenfield sites and production facilities; and helps investors with the regulatory environment.

Developing investor leads involves impressing potential investors with the quality of the Czech Republic as an investment location. Thus activities focus on improving the place product in four key areas: the local supply base; infrastructure; the labour force; and support services for investors. High-quality local supply networks and the availability of information about them can reduce barriers to inward investment (Estrin et al. 1997; Turnock 2001) and avoid 'cathedral in the desert' forms of FDI. CzechInvest's Supplier Development Programme upgrades the quality of local suppliers, improving the links between them and foreign-backed manufacturers. Local companies are provided with information and consultancy support following an initial evaluation. After one year they are assessed relative to an EFQM score (a not-for-profit organisation which provides models of 'excellence' against which to measure business management quality to help businesses make better products or deliver improved services) and then they upgrade to meet this standard. Thus by improving their technological capacities, management, quality control, and organisation they should better meet the needs of MNEs. Profiles of local suppliers are provided to foreign manufacturers. Upgrading local suppliers thus aims to improve the quality of the place product to attract FDI for embedding into local economies. New investors often require serviced industrial property. CzechInvest's Industrial Zone Support Programme helps municipalities and regional authorities in preparing sites, the construction of production halls, and accreditation of industrial zones. The latter point is a key quality indicator as it establishes qualifications and technical standards to ensure professional management.

Financial support is available for job creation and retraining which improves labour-force quality. Further efforts to improve the quality of the place product are represented by the 2002 framework programmes in support of strategic services projects and the establishment and expansion of technology centres which seek to attract high value-added FDI by offering subsidies for business activity and for training. These strategies combine attracting hi-tech, IT-based and high value-added FDI with improving the workforce to enhance the Czech Republic's image. The quality of investor support is also promoted. CzechInvest's ability to 'cut through red tape' is emphasised, further assisted by assigning project managers with sector-specific knowledge to each investor. From 2002 investors have benefited from an after-care team. On-site visits to investors gather feedback on the investment experience and improve information flows. These relationships with investors help attract further investment from their international parent companies by solving problems so that they expand investment. Testaments from satisfied investors are promotional tools and links to further investors. All these activities attempt to enhance the embeddedness of FDI and the quality of the investment environment. Enhancing the quality of the 'place product' is thus a significant strategy, and it is monitored by surveys of investors and potential FDI which 'tell us how the product is changing, to see if there are areas we need to change in the economy' (CzechInvest interview 2003). In this way policies enhance the Czech Republic as an inward investment location and help movement to more mature place marketing strategies concerned with improving the place product in response to the changing competitive environment for attracting FDI. While other CEECs have developed similar sophisticated approaches, many other states have still to address such issues in their place marketing activities.

City-Region Scale Place Marketing: Łódź, Poland

The regional scale is increasingly the favoured territorial scale among corporate investors. In many CEECs there has been a process of re-regionalisation, particularly to establish a new structure compatible with EU membership. In eastern Germany for example, Herrschel (2003) identifies the establishment of a completely new set of regions from 1990 which have been at least partly concerned with developing a new image which is externally projected to assist the region as a territorial destination of global economic interest. Though the relationship is complex there is often marketing of the city region, with different institutional arrangements being developed between the urban and regional scales. The Łódź city region in central Poland provides a good example. From the 1820s and then particularly under communism Łódź developed as a massive industrial region specialising in the production of cheap textiles, growing to become Poland's second largest city. With the collapse of its markets in 1989 the region entered a period of economic depression with high levels of unemployment as the SOEs shed labour during restructuring or bankruptcy. In addition, the city had since the 19th century suffered from a poor image as a grey, grim, polluted industrial city. From the early 1990s the city and other authorities became proactive in using place marketing as a means of attracting economic investment.

In the earlier years of transformation there was a strong emphasis on challenging the negative stereotype of the city. Campaigns in the 1990s continued to draw on Łódź's image as a site of industrial dynamism, using the exciting history of the city as the basis of a new image for the city as a place for brave, active and entrepreneurial people. Followed many other campaigns in ex-industrial European cities, stress was also laid on the centrality and connectedness of a large urban and regional centre, with its human resources, and economic dynamism (Young and Kaczmarek 1999). In the 2000s the emphasis has shifted, with less stress on industrial heritage and more on a rebranding effort for a 'City of Colour' to displace the grey, negative image <http://www.uml.Lodz.pl>. Though not entirely rejecting an industrial image, more emphasis is placed on the city's cultural role, its commercial and international role, and its attributes as an educational, research and medical centre. Thus in more recent years Łódź has also changed its marketing efforts to focus on improving the quality of the location for FDI. This includes changes in the organisational structure of local governance. In the 1990s marketing was the remit of the City Council's Department of City Strategy. In the 2000s further organisational development involved the creation of a dedicated Investor's Service Section (see website) offering detailed information about opportunities (including a list of available sites) and assistance over the stages of the investment process, including the preparation of investment applications to meet the approval of the city's mayor and the Minister of Internal Affairs & Administration. As is common with many other post-socialist locations, financial incentives are also included in the City Council's marketing package. These include exemption from real property tax where production is undertaken, customs duty allowances on imported machinery or components, and retraining grants and loans to assist the employment of Polish workers. Again there is more of an emphasis on upgrading the quality of the local workforce to meet the needs of FDI rather than stressing the cheapness of labour.

These activities also require organisational cooperation within the city council and with other external agencies to market the city region more widely. There are a range of organisations involved in promoting the Łódź region and there is to some extent a regional network of economic development organisations who also undertake place marketing. Two important organisations are the regional development agency and the regional (voivodship) self-government office (LVO). The agency was established in 1992 as a partnership between regional authorities, public institutions and entrepreneurs to cover a wide range of activities related to the economic development of the region. It is part of a wide network of national and international bodies and seeks to provide services for other regional institutions (assisting local self-government on development strategy) and Polish and foreign SMEs, including the economic promotion of the region in Poland and abroad <http://www.larr.lodz.pl>. Working with the city council and the LVO, it also promotes and implements regional development projects in line with EU standards, again suggesting an emphasis on upgrading the quality of the location for investors. The LVO also promotes the city region as a place for 'the dynamic development of entrepreneurship' <http://www. uw.lodz.pl>. Again there is an emphasis on imaging the region as one with good economic progress, a very good educational base producing a good local knowledge economy and with good cultural attractions (i.e. quality of life marketing). Being responsible for privatisation in the region, the LVO and its website is particularly focused on attracting investors for full or joint venture initiatives, stressing the region's connectedness, resources, infrastructure, the availability of special areas 'providing well equipped land ready for investment', a qualified labour force and investment partners. Thus again there is much emphasis on imaging the region for quality investment.

Place Marketing by Local Self-Government: Polish Communes

The establishment of local self-government at the lowest level was a key part of post-1989 attempts to introduce democracy. It also gave localities power in determining local economic development, with implications for place marketing at this level. In Poland, the 1990 Local Government Act gave communes ('gminas') the right to own property, collect certain taxes, manage their financial resources and act within a legal framework on all matters concerning the development and management of the commune. This gave them flexibility in developing strategies to attract investment. In a survey of Polish communes, 63% reported undertaking some form of promotional activity, while assisting inward investment – by developing the commune resources and improving the infrastructure – was even more significant. Just over half of commune governments had established dedicated promotion or marketing units, while 81% responded that attracting inward investment was a key LED strategy (Young and Kaczmarek 2000). Communes in the Łódź region have adopted a variety of means of attracting FDI (based on author interviews). Most communes reported problems with a lack of funds for marketing which inhibits advertising. But Rzgów commune, close to the city, has tried to identify alternatives to agriculture by stimulating economic growth and in the late 1990s a focus was developed on car sales when service centres were opened by Mitsubishi, Toyota and Scania. The key marketing strategy was advertising in Polish national newspapers offering lower taxes and commune-owned land for development.

Given the lack of resources an important strategy has been attempts at building institutional thickness through cooperation with other communes, both locally and internationally. The largely rural commune of Nowosolna formed a union with four other communes which has established a database of information from farmers about what space and equipment they have and how it might be useful for establishing small craft businesses. The union has an office in the city and was represented at a City Fair in Poznań, a promotional exhibition to attract investors. In Parzęczew commune attempts to develop marketing have involved twinning programmes with local governments in France and Germany, and also with the Polish community in Lithuania, on the grounds that it will offer direct contact with investors and know-how about promotion. Another strategy is to develop schoolchildrens' holiday exchanges to build these links. Communes can also reduce local taxes as a financial incentive. Several communes have areas in the Łódź Special Economic Zone, one of several demarcated by the Polish government to attract inward investment, particularly through financial incentives. Communes are limited in their ability to manipulate local taxes to setting the level of real estate tax: when Nowosolna commune limited taxes at below 30% of normal rates to attract investors, it reduced income to the commune budget. However, Finnish capital invested in textile production after seeking out a suitable investment site while a Swedish/Danish joint venture producing furniture for Ikea has a longer association with the area. Ozorków reduced taxes for investors to zero for two-three years. In Parzęczew commune 50% tax reductions attracted inward investment into textiles

(50 jobs) and Chinese investment into pre-prepared meals on a site sold to them by the commune created a further 20.

Communes have also been active in improving the quality of the local infrastructure. In many cases this was demanded of local government by their citizens in order to improve their quality of life, but this has also created resources for marketing. In Konstantynów improving the quality of the place product was achieved through a strategy of 'privatisation of the communal infrastructure' since the policy of the commune is to 'take care of the roads, the communal infrastructure, keep the taxes reasonable and allow the rest to take care of itself'. Parzęczew commune has also upgraded the quality of the infrastructure and used as marketing features the easy accessibility of phones and the fact that nearly 90% of houses have tap water (such basic improvements being very necessary in the early years of local government). Andrespol commune invested in water systems to attract investors on the basis of pure water for manufacturing with proper waste water treatment afterwards. Meanwhile, in Ozorków commune an important focus has been on upgrading the quality of the environment and in particular the purity of the soil so that state certification of soil quality has attracted investment into the processing of locally-grown vegetables. Further strategies seek to improve the quality of the labour force. Parzęczew commune introduced a target of producing well-qualified school graduates in the hope that they would return to assist development, and to this end foreign language teachers have been introduced in primary schools and languages made a compulsory element of the syllabus in order to upgrade the labour force. Such is the variety of experience with innovative strategies of place marketing, although many communes are hardly proactive. Indeed throughout CEE there is much geographical differentiation in local development pathways in general and place marketing in particular.

Circumventing Marketing Institutions: Proactive Local Leadership in Poland

Lack of sophisticated place marketing is a legacy of the communist era that has left local and regional administrative structures with a lack of institutional thickness. This enhances the role of key individuals, such as local mayors, in directly negotiating FDI; a point demonstrated by the industrial town of Płock, located northwest of Warsaw with a population of 40,000. Lacking experience in economic development in the early 1990s, the executive council appointed a mayor from a business management background, and generally left economic development strategy to his vision. Due to the structure of the executive authority, and the mayor's vaguely-defined authority, he was able to take over the vacant second vice-president's office (responsible for urban planning and zoning) and thereby controlled almost every city agency concerned with economic development. In 1991 the city authority was successful in attracting Levi Strauss to invest $20mln in a jeans factory, creating 1,200 jobs. The success of this relocation was due to the mayor's business experience which he used to create a positive investment climate. He personally negotiated the deal having persuaded the council to grant him extensive powers to cut through red tape. To accelerate finalisation of the deal, which usually takes more than six months, he brought together the six relevant government ministers in one room and reached an agreement on the spot. Finally, the mayor raised a $0.7mln no-interest loan through business contacts to buy the site, allowing a tax write-off of the same amount to the lender. What is significant here is the role played by one personality (the mayor) in locating the investment in Płock in the absence of other economic development intermediaries in the private sector at that time (Owen 1994). Similarly, proactive local mayors were significant in attracting investment into the Wrocław region of Poland by using their informal networks to circumvent official networks and thereby cut through red tape (Hardy 1998). Thus where there are competent officials, the absence of institutional thickness and a formal place marketing infrastructure may not hinder FDI, but this further adds to the differentiation in place marketing.

Weak Place Marketing: Ukraine and Zaporizhya

The above case studies of place marketing at a range of scales have focused on examples where there are proactive attempts to market localities. However, there are many localities which do not have a well developed place marketing infrastructure. Despite its size and potential, Ukraine is one country which has failed to successfully engage FDI. This poor record is blamed on lack of progress with privatisation and a poor investment climate. FDI legislation was adopted in 1992 but the lack of clear legislation and its erratic implementation has deterred investors. Lack of clarity over private land ownership and legal protection of business interests do not assist FDI, while there are problems with export barriers which restrict market opportunity, and privatisation has tended to exclude foreigners (Ishaq 1999). Problems have also been experienced at the local level. The city of Zaporizhya suffered intense deindustrialisation after 1991 but has generally failed to attract FDI to assist restructuring. 'Institutional thinness' has led to a lack of effort to make the city attractive to FDI by producing information, improving infrastructure or clarifying land purchasing procedures (Van Zon 1998). Even a proposed $1.3bln six-year joint venture investment by Daewoo into the local car producer Autozaz has proved highly problematic as the company has struggled to modernise production lines, and the Ukrainian government took time to introduce import levies on the second-hand car import market. The authorities had to make considerable concessions to gain the investment. These included scrapping import taxes on spare parts, a ten-year break in profit taxes, tax-free land status for future development, an $80mln debt write-off for Autozaz and the banning of the import of cars more than five years old. These changes were opposed by local car dealers and by the EU. So far the development plans have been restricted with more emphasis on the pre-assembly of US and Korean cars – at Illichivsk near Odessa – rather than those with a high local content which would enhance embeddedness. The outcome of the venture illustrates the difficulties that Ukraine faces in attracting FDI (EBRD 2003; Hawrylyshyn and Sheremeta 2000) and highlights its need to develop more sophisticated place marketing.

Conclusion

This chapter has presented an overview of the changing context for attracting FDI into CEE and the nature of place marketing practice as a response. As MNE corporate strategy becomes more concerned with competitive strategies based on factors other than just the availability of cheap production factors, and seeks to address more quality led issues to ensure competitiveness, then place marketing to attract such investment must respond to address this quality issue. In addition, place marketing also needs to attract FDI that is embedded in localities, which provides good linkages into local economies, and is sustainable in the long term. Another issue is to avoid being drawn into a fight for inward investment which simply leads to some localities being losers in development terms. These issues suggest that place marketing in the region must develop more sophisticated marketing strategies to attract, retain and further develop FDI. Place marketing has to proceed beyond simple re-imaging strategies to improve the quality of locations for investment and address investor-specific needs in the long-term, but as the case studies illustrate it is only certain localities which are developing this level of expertise.

A crucial factor in such efforts is to recognise the multi-scalar nature of FDI and place marketing as a form of governance. Such sophisticated place marketing strategies require coordination of marketing at a range of scales. As the examples above show supra-national forms of governance assist with raising investor awareness and the transfer of good practice. They are also significant in regulating the environment for investment at a transnational scale. Developing the sophisticated place marketing strategies requires coordination between scales. This involves international coordination of direct marketing activities, national-level regulation of the business and investment climates as well as national policies which affect investment directly (e.g. customs regulations, business and investment regimes) and which seek to upgrade the place product, such as investment in education, training and infrastructure. National scale coordination of regional development policy and of subnational governance is also important to implement such changes. Thus although there is an ongoing rescaling of governance and a rise in the significance of supranational and subnational governance structures, it remains the case in CEE for place marketing that the actions of the nation-state remain highly significant (Smith and Pickles 1998), though in a context in which the national scale of accumulation and state regulation have been decentred (Brenner 1999). Moreover, while place marketing in the region varies in its practice and effectiveness across geographical scale, there is also a considerably uneven geography of practice across the region. As the case studies above illustrate, while some locations (at a range of scales) have well developed, sophisticated place marketing practices which address the 'quality' issue and the needs of FDI, many have not yet been grasped the quality agenda, while others suffer from institutional thinness and a lack of resources with which to undertake any meaningful place marketing activities.

That place marketing practice varies considerably over space in CEE reflects its place in the complex transformation of the region. Smith and Pickles (1998) note that transition programmes failed to appreciate that capitalist economies are regulated through complex sets of institutional structures which are produced over long time periods; also that there has been considerable diversity in the ways in which new sites of accumulation are constituted in the transformation process. The adoption of place marketing in the region has largely been a straightforward and uncritical adoption of strategies from the experience of Western capitalism. However, the diversity of local responses in post-socialist transformation arises out of the interrelation of previous sets of socio-economic relations with new forms of regulation and accumulation rendering transformation a path-dependent process. Place marketing is an important process which selectively brings together existing socio-economic resources with changes to the place product (e.g. upgrading the quality of a locality) and which plays a role in combining old sets of productive relations with new ideas introduced through FDI. As Smith and Pickles (1998) also note, marked regional differentiation is emerging in CEE between sets of regional economies experiencing dramatic restructuring and globalised growth related to their ability to mobilise their global connectedness particularly through their ability to attract FDI, and sets of marginalised economies increasingly left behind in the process of capitalist restructuring. Though the relationships are by no means straightforward, the ability to develop institutional thickness in the multi-scalar governance of place marketing and FDI are a further significant set of processes in this production of uneven development in the region.

References

Alden, J. and Phelps, N.A. (1999), 'Conclusions: Multinationals and Global-localisation – Some Themes', in N.A. Phelps and J. Alden (eds), *FDI and the Global Economy*, The Stationery Office/Regional Studies Association, London, 269–80.

Amin, A. and Thrift, N. (1994), 'Holding down the Global', in A. Amin and N. Thrift (eds), *Globalisation Institutions and Regional Development in Europe*, Oxford University Press, Oxford, 257–60.

Beyer, J. (2002), '"Please Invest in our Country": How Successful were the Tax Incentives for Foreign Investment in Transition Countries?', *Communist and Post-Communist Studies*, 35, 191–211.

Brenner, N. (1999), 'Globalisation as Reterritorialisation: The Rescaling of Urban Governance in the EU', *Urban Studies* 36, 431–51.

Christodoulou, P. (1996), *Inward Investment: An Overview and Guide to the Literature*, British Library, London.

Claessens, S., Oks, D. and Polastri, R. (1998), *Capital Flows to CEE and the Former Soviet Union*, World Bank, Washington, D.C.

Collis, C., Berekeley, N. and Noon, D. (1999), 'Attracting FDI to East European Economies: Can Lessons be Learned from the UK Experience?', in N.A. Phelps and J. Alden (eds), *FDI and the Global Economy*, Regional Studies Association/Stationery Office, London, 139–56.

Du Pont, M. (2002), *FDI in Transitional Economies. A case study of China and Poland*, Macmillan, Basingstoke.

Dunning, J.H. (1981), *International Production and the Multinational Enterprise*, Allen and Unwin, London.

EBRD (2003), *Transition Report Update*, EBRD, London.

Estrin, S., Hughes, K. and Todd, S. (1997), *FDI in CEE: Multinationals in Transition*, Pinter, London.

Estrin, S., Richet, X. and Brada, J.C. (2000), 'A Comparison of FDI in Bulgaria, the Czech Republic and Slovenia', in S. Estrin, X. Richet and J.C. Brada (eds), *FDI in Central Eastern Europe: Case Studies of Firms in Transition*, Sharpe, New York and London, vii-xxxi.

Fabry, N. and Zeghni, S. (2002), 'FDI in Russia: How the Investment Climate Matters', *Communist and Post-Communist Studies*, 35, 289–303.

Garibaldi, P., Mora, N., Sahay, R. and Zettelmeyer, J. (1999), 'What Moves Capital to Transition Economies?', *IMF Conference 'A Decade of Transition'*, Washington, D.C.

Globerman, S. and Shapiro, D.M. (1999), 'The Impact of Government Policies on FDI: The Canadian Experience', *Journal of International Business Studies*, 30, 513–32.

Globerman, S. and Shapiro, D. (2002), 'Global FDI Flows: The Role of Governance Infrastructure', *World Development*, 30, 1899–1919.

Hardy, J. (1998), 'Cathedrals in the Desert? Transnationals Corporate Strategy and Locality in Wrocław', *Regional Studies*, 32, 639–52.

Hawrylyshyn, B. and Sheremeta, P. (2000), 'Ukraine: Europe's frontier', in A.K. Koźmiński and G.S. Yip (eds), *Strategies for CEE*, MacMillan, London, 264–83.

Herrschel, T. (2003), 'Regulating Post-Socialist Space: Territory Representation and the "New Regionalism" in eastern Germany', *Association of American Geographers' Conference*, New Orleans.

Hine, R. (1999), 'Globalisation versus Regionalism? Trade Investment and Multinationals', in N.A. Phelps and J. Alden (eds), *FDI and the Global Economy*, Regional Studies Association/Stationery Office, London, 31–50.

Hunya, G. (2000), *Integration through FDI*, Edward Elgar, Cheltenham.

Hunya, G. (2002), *FDI in SEE in the early 2000s*, WIIW, Vienna.

Ishaq, M. (1999), 'FDI in Ukraine since Transition', *Communist and Post-Communist Studies*, 32, 91–109.

Kaminski, B. (2000), *How Accession to the EU has Affected External Trade and FDI in Central European Economies*, World Bank, Washington, DC.

Lall, S. (2002), 'FDI and Development: Research Issues in the Emerging Context', in B. Bora (ed), *Foreign Direct Investment*, Routledge, London, 325–45.

Meyer, S. and Qu, T. (1995), 'Place-Specific Determinants of FDI: The Geographical Perspective', in M.B. Green and R.B. McNaughton (eds), *The location of FDI*, Ashgate, Aldershot, 1–13.

OECD (2002), *FDI in Slovenia: Trends and Prospects* <http://www.oecd.org>.

OECD (2003), *Policies Towards Attracting FDI* <http://www.oecd.org>.

Owen, C.J. (1994), 'City Government in Płock: An Emerging Urban Regime in Poland', *Journal of Urban Affairs*, 16, 67–80.

Phelps, N. (1999), 'Introduction: The Corporate and Institutional Dynamics of FDI in the Global Economy', in N.A. Phelps and J. Alden (eds), *FDI and the Global Economy*, Regional Studies Association/Stationery Office, London, 1–12.

Smekal, C. and Sausgruber, R. (2000), 'Determinants of FDI in Europe', in J.R. Chen (ed), *Foreign Direct Investment*, Macmillan, London, 33–42.

Smith, A. and Ferenčíková, S. (1998), 'Inward Investment Regional Transformations and Uneven Development in Eastern and Central Europe', *European Urban and Regional Studies*, 5, 155–73.

Smith, A. and Pickles, J. (1998), 'Introduction: Theorising Transition and the Political Economy of Transformation', in J. Pickles and A. Smith (eds), *Theorising Transition*, Routledge, London, 1–24.

Turnock, D. (2001), 'Location Trends for FDI in ECE', *Environment and Planning C*, 19, 849–80.

Van Zon, H. (1998), 'The Mismanaged Integration of Zaporizhzhya with the World Economy: Implications for Regional Development in Peripheral Regions', *Regional Studies*, 32, 607–18.

Young, C. and Kaczmarek, S. (1999), 'Changing the Perception of the Post-Socialist City: Place Promotion and Imagery in Łódź, Poland', *Geographical Journal*, 165, 183–91.

Young, C. and Kaczmarek, S. (2000), 'Local Government, Local Economic Development and Quality of Life in Poland', *GeoJournal*, 50, 225–34.

Young, S. and Brewer, T. (1999), 'Multilateral Investment Rules Multinationals and the Global Economy', in N.A. Phelps and J. Alden (eds), *FDI and the Global Economy*, Regional Studies Association/Stationery Office, London, 13–30.

Chapter 7

Foreign Economic Relations and the Environment: A Historical Approach

Jonathan Oldfield and Andrew Tickle

Introduction

The purpose of this chapter is to explore the environmental implications of foreign economic relations in CEE. As such, it takes a broad historical approach in order to highlight the changing nature of this relationship during both the socialist and post-socialist periods. Consonant with the overall theme of this book, our focus is on both the former Soviet Union (FSU) and its socialist satellite states in ECE (together, the Soviet bloc) and their joint and varied relationships with each other and the so-called 'West'. We include FSU not only because the book includes some examination in this domain but also because of its hegemonic political and economic role in the region after the Second World War. This underscores our belief – shared in no short measure by many other geographers (e.g. Carter 1994) – in the importance of historical influences in the region, which can be equated partially with the contemporary concept of path-dependency. Using similar historical antecedents, we also hold firm to the view that foreign economic relations have long been implicated in the region's environmental situation and cannot be solely associated with late 20th century processes currently described as globalisation (see Fagin and Tickle 2002 p.41).

Despite the isolationist tendencies of the FSU and – to a lesser extent – the other countries within the bloc, foreign economic links were maintained throughout the socialist period, although they waxed and waned in tandem with the prevailing political climate. Political and economic crises, from changes in the Soviet leadership to the later exhaustion of the socialist economic model, were often significant markers heralding or indeed forcing altered attitudes to economic engagement with the West and elsewhere. Nevertheless, in all periods, the prevailing set of foreign economic relations, broadly characterised by the FSU's dependence on the export of natural resources and the import of industrial technology, had notable environmental ramifications. The final economic and political crisis, which precipitated the post-1989 fall of socialist regimes throughout the region, necessitated rapid and fundamental changes to economic relations with the West. Western firms saw the opportunity to gain from low production costs and a vastly under-provisioned consumer market whilst the countries in the region hoped to gain from an influx of investment to aid the restructuring process. Burgeoning East-West economic linkages have a range of environmental implications for the region and these can be conflicting in nature. For example, FDI flows are typically credited with having the potential to improve the region's environmental situation through the 'greening' of indigenous manufacturing production. At the same time, the footloose characteristics of FDI may also result in the establishment of environmentally damaging activities which undermine national regulatory frameworks.

In trying to explore some of the issues surrounding the environment and foreign economic relations (which we use here to connote a range of East-West economic linkages including FDI, trade flows, multilateral development bank loans to bilateral and other aid packages), we look first at the early years of the FSU and the associated patterns of resource utilisation that can still be seen today. We then examine how these patterns were subsequently imposed and replicated (not always successfully) in central and eastern Europe from 1945. This foregrounds a more detailed analysis of foreign economic relations (now dominated by FDI plus loan and aid deals) and related environmental issues since 1990. We conclude by identifying a number of issues of concern, both in relation to the potential for ongoing environmental impacts and opportunities for further academic enquiry.

East-West Economic Relations during the Socialist Period

According to Kornai (1992 p.339), the classical socialist economy was to be characterised by self-sufficiency and a generally closed relationship with the non-socialist world. Nevertheless, as intimated above, while the relationship between the two ideologically-opposed regions was certainly limited this should not disguise the instances of interaction and contact. In particular, Western technology and know-how was able to penetrate via state-mediated trade links and associated special exchange programmes. This relationship was underscored by a noted 'import hunger' for high quality technical goods within the Soviet Union and the socialist bloc more generally (Ibid p.346). Furthermore, it can be argued that an appreciation of the nature of the relationship between the bloc and the West is important for a full understanding of its resulting environmental situation. While careful not to become apologists for the policies of Soviet and allied regimes or trivialise the complexity of their environmental problems, it should be remembered that the FSU's industrialisation drive was predicated on a desire to 'catch up and overtake' the West in the 'race' for high modernity. Furthermore, this ideologically-based aspiration was a factor in maintaining the gap between socialist (Marxist-Leninist) environmental policy and practice. A range of nature protection and environmental laws were introduced by the Soviet authorities with some issued during the early 1920s (e.g. Lisitzin 1987; Pryde 1991). Legislation covering nature protection and pollution control were introduced in other socialist countries from the 1950s (e.g.

Cole 1998; Moldan 1990). Nevertheless, these laws were unable to prevent the emergence of a strained and often worsening environmental situation within the region by the end of the 1970s: for a detailed critique of the environmental limitations of Marxism-Leninism see DeBardeleben (1985).

The Cold War stand-off was also an important factor since it encouraged the development of an extensive defence sector which was responsible for some of the FSU's worst environmental issues. The deleterious environmental effects were shrouded in the veil of secrecy surrounding all Soviet and Warsaw Pact military initiatives. The environmental legacy of defence sector activities included the disruption of large areas of land and extensive petrochemical and nuclear pollution (e.g. Bøhmer et al. 2001). In the West, similar environmental legacies also exist in the USA, Australia and the former French Polynesian islands where, for example, nuclear-related activity had, and continues to have, significant environmental and health implications for local communities (e.g. Dalton et al. 1999). Unintentionally, the Cold War period also promoted some environmental benefits, with large areas of land throughout the region (particularly in the 'iron curtain' border areas with the West) either remaining undeveloped or reverting to near wilderness – usually with positive implications for the conservation of biodiversity (Tickle 2000).

Early Russian/Soviet Economic Models

While an appreciation of the ideological confrontation between East and West provides a general background on which to begin to understand environmental conditions within the Soviet bloc, more specific historico-economic relationships are also worthy of further exploration. As mentioned earlier, the nature of East-West economic relations ebbed and flowed during the course of the socialist period, largely in response to the prevailing political environment within the FSU. Even before the revolution, Russia was 'half empire and half colony' – being strongly dependent on foreign capital, with westerners owning 90% of its mines, 40–50% of chemical and engineering plants and over 40% of banking stock (Deutscher 1967 p.12). But dividends were not re-invested in the country, thus fomenting the Bolsheviks' desire for revolution and to nationalise and socialise the means of production. Nevertheless, by the early to mid-1920s the Soviet economy was in a chaotic state. The combination of civil war and societal revolution severely undermined the productive capacity of the domestic economy. Furthermore, the newly-forming administrative and regulatory networks were too weak and disorganised to provide the necessary guidance or control. As a consequence, capitalistic activities were actively supported during this period within the framework of the New Economic Policy in order to provide some respite from the prevailing difficulties. In particular, Lenin believed that there were important economic benefits to be gained from allowing foreign capital to invest in the Soviet Union's natural resources (Deutscher 1967 p.66; Nove 1992 p.84). But the number of concessions established with foreign investors was generally limited and together accounted for under 1.0% of industrial output by the end of the 1920s (Nove 1992 p.84). Nove goes on to suggest that the wariness of foreign investors was the main reason behind the small number of concessions and – as such – mirrors strongly the contemporary situation in the Russian Federation and other FSU states.

Trade relations developed with capitalist countries during this period based largely on the export of natural resources. For example, Soviet timber procurement plans were partly determined by the need for foreign currency to purchase Western machinery and equipment in order to drive the urgent restructuring and industrialisation process (Nove 1992; Weiner 2000 p.149). Animal products such as furs also formed a source of hard currency during the early Soviet period (Weiner 2000 pp.152–4). The perceived utility of such natural resources was also a driving force behind the Bolshevik's support for conserving vast areas of the FSU in a system of nature reserves ('zapovedniki') which were to be strictly protected from inappropriate development (Weiner 1999; 2000). But with Stalin's rise to power towards the end of the 1920s, capitalistic relations were increasingly frowned upon and foreign concessions suffered as attention turned towards rapid industrialisation and collectivisation in order to begin in earnest the task of constructing 'socialism in one country'. The 1930s were characterised by a significant decline in the extent of economic relations with the West as domestic production increased markedly (Nove 1992 p.231). The FSU had now implemented a de facto 'war economy', in which vast resources (natural, human and political) were mobilised on a seemingly permanent basis (Kaldor 1990). Nevertheless, even during this period of ideological fervour, foreign expertise was still sought in order to assist the industrial drive. For example, Kotkin (1995 pp.37–71) recounts the involvement of US and German firms in the design and construction of the giant Magnitogorsk metallurgical 'combinat' during the first FYP. Magnitogorsk (located in the southern Urals) was in many ways meant to symbolise the FSU's decisive leap towards its socialist future, and yet it was founded in no small part on imported capitalist technology and know-how.

It was not until after the Second World War that the volume of Soviet foreign trade began to increase steadily once again (TsSU SSSR 1972 p.491). Unsurprisingly, a significant percentage of this foreign trade activity was with the new socialist states of ECE. At the same time, Comecon was formed to facilitate and control trade within the new 'internal market' and this became increasingly important following the death of Stalin in 1953. Soviet trade with capitalist countries also grew significantly during the Brezhnev period due to a combination of more constructive political relations between East and West (détente) and the rising price of Middle East oil (Nove 1992 p.391). While the extent of trade with capitalist countries remained below that of other socialist countries, this relationship was nevertheless important for the Soviet Union.

Developments in ECE and Comecon

While much of the analysis so far has focused on the FSU, it is important to recognise the diversity of economic relations with the West that existed amongst the socialist countries of ECE. Part of this is rooted in the peripheral 19th century relationship of these 'marchlands' to the West European core, although some areas like the Czech Lands are seen as semi-peripheral (Dingsdale 2002 p.83). By the early 20th century, key industrial areas had been built up, particularly within southern Germany and northern Austria-Hungary with the first instances of environmental degradation being noted in the 1920s (Stoklasa 1923). By the 1930s, the external trade of the new states of Austria, Hungary and Czechoslovakia was dominated by Germany who they supplied with raw materials in return for high value (often consumer) goods (Dingsdale 2002 p.90). By the 1950s this resource base (in respect of both natural resources and capital and labour) had been forcibly colonised by the Soviet Union with the inception of Comecon. This brought a new form of relative autarchy to the region and the imposition of a largely extensive model of socialist industrialisation on the satellite states, some of which were largely undeveloped (Smith 1994).

At this point the Western and Eastern models of industrial development began to diverge strongly with the Eastern model characterised by an undue bias towards outdated heavy industry and ideologically-driven decisions in respect of location and markets. The seeds of environmental degradation, which became evident by the 1970s and 1980s (e.g. Carter and Turnock 1993), were now sown in earnest in ECE by the imposition of a new set of 'foreign' (FSU-satellite state) economic relations. At the root of this relationship was the Soviet provision of fuel supplies (Ziegler 1991). This seemingly plentiful supply of cheap energy also encouraged and perpetuated an inefficient production process. At the same time, the newly-imposed ruling ideology of Marxism-Leninism sought to claim that whilst 'improper natural resource exploitation is a result of the capitalistic mode of production … a socialist economy necessarily pursues the wisest possible use of natural resources' (Pryde 1972 p.136). Nevertheless, this did not mean that academic and political elites in both the FSU and ECE were not aware of, or indeed concerned about, the socialist system's evident environmental deficiencies (DeBardeleben 1985; Tickle and Vavroušek 1998).

At the same time, varying forms of socialist economy evolved across ECE, leading to significant differences in both relative degrees of economic isolation and environmental degradation (Carter and Turnock 1993). The likes of Hungary, Romania and Yugoslavia were characterised by legislation permitting foreign investment in national production units (Lavigne 1974, p.345) long before the FSU's 1987 law on joint ventures (IMF et al. 1991 p.76). Furthermore, a significant number of bilateral economic cooperation agreements (varying markedly in both nature and extent) existed between socialist and capitalist industrial enterprises and these facilitated the transfer of technology and know-how from West to East (Lavigne 1974 p.344). More specifically, Yugoslavia's alternative model of socialism, based on decentralisation and autonomous 'units of production', also retained much private land as well as earlier Austro-Hungarian laws on nature and forest protection. This led to a far less degraded environment than, for example, in former Czechoslovakia where highly unsustainable patterns of development were underlain by almost total nationalisation of production units and the collectivisation of agriculture (Tickle and Clarke 2000).

By the 1970s both ideology and the economies of the ECE socialist states had become moribund and entered a phase of decomposition. Workers saw living standards rising in the West and their own standing still. In addition, the economic shocks associated with the fuel crises from 1970 onwards worsened the situation. The result was an economic opening to the West, seeking large-scale investment as a means of modernising both management and technology and increasing welfarist measures to combat popular disquiet (Schopflin 1993 pp.165–6). Unfortunately, the political dominance of the heavy industry lobby meant that most of the newly available funds went into the chemicals, metallurgy, construction and energy production sectors. Thus a second period of 'heavy industrialization' (Ibid) occurred with a concomitant increase in resource depletion and pollution impacts over the region (e.g. brown coal extraction in the 'black triangle' of North Bohemia, Saxony and Silesia).

At the same time, the FSU and its satellite states were opening politically to the West. Aware that their economies (especially in the Soviet case) were massively and negatively skewed by the weapons race, they sought détente as a way of addressing their failing economies. In addition the Soviet bloc wanted to catch up in the new field of computer technology. Negotiations towards both these ends began under the auspices of the OSCE and became known as the 'Helsinki process'. Although stuttering progress was made on arms reduction, the West had little to offer on hi-tech transfer. Fearing stalemate, a new and less contentious area of negotiation was opened on the environment. By 1979, these discussions had progressed sufficiently for a full international convention to be signed at the UN ECE in Geneva, setting out a framework agreement to curb transboundary air pollution. As a result of such developments, in addition to unintentioned initiatives in international diplomacy, environmental issues and policy were now opening to the West. During the 1980s, economic relations with the West European states had burgeoned to such a degree that the European Community invited states such as Czechoslovakia and Hungary to sign 'association agreements', offering and regulating increasingly open trade relationships. In contrast, the Soviet Union shied away from greater economic engagement with the West and maintained its traditional export-import relationship.

By the early 1980s, exports of fuel and electricity comprised approximately 70% of the FSU's exports to the capitalist West (Bradshaw 1995 p.134). In contrast, 'machinery and equipment' and grain dominated imports to the socialist bloc from the West. In the Soviet case, Western imports of machinery and equipment were typically used to facilitate the further development of industrial branches such as chemicals, hydrocarbons and forestry. Importantly, these branches had a distinctive geography, tending to be located in the Volga, Urals and West Siberian regions. It is therefore of little surprise that these parts of the FSU were subsequently characterised by acute environmental problems (Peterson 1993; Pryde 1991; Saiko 2001). Following on from such spatial patterns of investment, resource utilisation and consequent environmental impacts, the relationship between the FSU, its satellite states and border countries also required a physical infrastructure for both internal (Comecon) and external trade. A typical example with environmental consequences is energy supply where distribution systems for both electricity and fuel (oil and gas) were patterned politically rather than in relation to resource geographies or market factors. Although this could mean that border zones with the West were sometimes spared environmentally-damaging infrastructure developments, a highly inconsistent geography of trade networks was revealed when state socialism suddenly collapsed in 1989–1990. The spatial mismatch between the two systems obviously spelt profound difficulties for future integration of economic and environmental infrastructure.

East-West Economic Relations in the Post-Socialist Period

The Importance of FDI

The fall of socialist regimes heralded a fundamental change in both the nature and extent of economic relations between East and West. Much of the subsequent restructuring can be understood within the framework of neoliberal economic transformation with an emphasis on creating freely operating markets open to the world economy. Indeed, the importance of global economic connections is encapsulated in the economic dogma of those early years which stressed, amongst other things, the need to internationalise economic activities and relations. Furthermore, the structural development plans sponsored by the World Bank and other international organisations (for a critical assessment see Smith 2002) placed considerable emphasis on the importance of free trade and the attraction of FDI (EBRD 2001 pp.53–54; World Bank 2002). These Western-sponsored structural development programmes have some commonality with the rhetoric and positioning of earlier programmes for Africa and Latin America and ensure that the ECE/FSU region is progressively embedded within the uneven 'power relations' of the global economy. It should also be noted that the influence of these structural development programmes is reinforced by the apparent acceptance of Western-inspired concepts such as ecological modernisation and sustainable development (e.g. Jehlička and Tickle 2004; Oldfield 2001; Oldfield et al. 2003). These concepts are premised on the asserted synergy between economic growth and environmental improvement and are therefore concerned with notions of 'greening' and improved efficiency while in pursuit of material growth and profit. As such, they provide limited means to challenge the underlying nature of the global economic system and, more specifically, they help to justify an apathetic approach towards environmental issues in contrast to the urgency of economic concerns.

In some instances, old economic associations and ties were maintained during the early post-socialist period in order to facilitate the successful negotiation of an economically uncertain period. At the same time, entirely new geographies of economic association and dependency were allowed to emerge as former barriers disappeared or else were more easily circumnavigated. The reorientation of ECE trade to the EU is indicative of this trend (Gibb 1998 p.61; Gower 1999 pp.3–7). Furthermore, countries such as Hungary and Poland – characterised by nascent capitalistic relations during the socialist period – were better able to take advantage of the new economic opportunities presented by the expanding market. In contrast, Russia tended to fall back on familiar trade activities and exported natural resource exports to the West and elsewhere in order to guarantee short-term economic stability. This type of trading pattern mirrors that of many developing world countries and can lead to structural economic problems in the medium to long term and the over-exploitation of the natural resource base (Bradshaw and Lynn 1998; Markandya and Averchenkova 2000). Additional conduits of economic assistance have developed on a bilateral basis with economically-advanced countries due to the reinvigoration of historical ties and allegiances and these relationships can result in preferential economic relations and grant aid packages. Examples include the links between the Baltic States and Scandinavia (in particular Estonia and Finland); between Romania and both Canada and France; and to a lesser extent between the Czech Republic and Germany. In addition some of the weaker former members of Comecon (marginal ECE countries and small republics of the CIS) have often retained reciprocal trade ties.

Significant volumes of investment have also been channelled into the region through the intermediary of international organisations and lending agencies such as the World Bank Organisation, EBRD and GEF. For example, the World Bank committed approximately $40bln to the region during 1991–2002 (see http://www.worldbank.org/ accessed in April 2003). Russia has been the main recipient of loans accounting for approximately 30% of total flows during this period. By the end of the 1990s, the EBRD had invested billions of Euros in the region while at the same time stimulating additional investment from third parties (see http://www.ebrd.org/ accessed in April 2003). The EBRD was established at the beginning of the 1990s with the intention of facilitating movement towards market economies in the region through the encouragement of private enterprise. Its activities are stated to be based firmly on the concept of sustainable development and thus market-based economic restructuring is considered compatible with environmental improvement. Investment in municipal and environmental infrastructure in addition to power and transport networks is a key priority of the EBRD's activities (EBRD 2000 p.7). Indeed, the investment portfolios of both the EBRD and the World Bank are important in preparing the ground for future rounds of investment with their focus on large-scale infrastructure projects such as roads and energy provision, a trend also highlighted by Escobar (1995 p.165) with regard to the activities of the World Bank in developing world countries.

Another main source of investment is the aid provided by the EU. Two key regional programmes are PHARE and TACIS. PHARE has emerged as a main tool supporting the EU accession countries and accounted for approximately €13bln of technical assistance during 1990–2001. Since 1997, the programme has been wholly focused on the pre-accession process and supplemented by newly formed financial programmes such as SAPARD (with an agricultural and rural development focus) and ISPA (dealing with transport and environment). In contrast, TACIS is focused predominantly on the FSU states and has facilitated the transfer of approximately €4.0bln of technical assistance to the region during the 1990s (see http://europa.eu.int/comm/external_relations/ceeca/index.htm accessed in April 2003). It aims to support the establishment of competent market and democratic infrastructure (though a sectoral breakdown of funds indicates a focus on specific areas such as energy, environment, telecommunications and transport). Some concern has been raised over the effectiveness of these aid programmes in respect of the environment with suggestions of ambiguous outcomes and operational difficulties (e.g. Baker & Jehlička 1998; Fagin 2001) and more critical work is required here.

Allied to these important economic linkages, private FDI flows are proving to be an influential, albeit variable, factor in the ongoing transformation. Indeed, the attraction of FDI is a main policy aim of former socialist countries and is intimately bound up with the neoliberal project manifest within the region and a main policy aim of former socialist countries (Swain and Hardy 1998). The importance of FDI is further advanced by the evident relationship between the region's relative economic 'winners' and the attraction of this type of investment (World Bank 2002 pp.6–7). Per capita figures (based on net inflows recorded in the country's balance of payments) for the period 1989–2000 provide an indication of the extent to which per capita FDI flows vary within the region (EBRD 2001 pp.68–9). For example, Czech Republic ($2,102) and Hungary ($1,964) are clear leaders, although the Baltic countries of Estonia ($1,400) and Latvia ($1,056) are also prominent. Elsewhere within the FSU, Kazakhstan and Azerbaijan score $577 and $464 respectively compared with just $69 for Russia. While FDI is not always embedded firmly within the local and regional economy (Smith and Swain 1998 pp.43–5; Swain and Hardy 1998 p.589), it nevertheless has the potential to influence production systems and associated networks of production as well as consumption patterns.

The Environmental Implications of FDI

The opening up of former socialist economies to global economic flows has a range of implications for the region's environment. During the late 1980s and early 1990s, there was a strong conviction, especially amongst international organisations such as the EBRD and World Bank, that the adoption of a liberal democratic development model would ensure a significant improvement in the region's environmental situation (Fagin and Tickle 2002 pp.42–5; Herrschel and Forsyth 2001 pp.574–6; Pavlínek 1997 p.103). It should come as little surprise that those institutions engaged in the construction of new economies within the region tend to underline the beneficial environmental consequences of neoliberal reform allied to the conceptual framework of ecological modernisation. In particular, FDI is often seen as a key means for both facilitating and implementing the move away from the extensive, highly polluting Soviet development model towards a more intensive and rationalised production system through the introduction of environmentally efficient technologies and relevant technical know-how (Fagin and Tickle 2002 p.44). In relation to this, considerable attention has been devoted subsequently to an assessment of the extent to which production and pollution indicators have been decoupled since 1989/1991 in the light of economic restructuring. It would seem that the ECE countries have been far more successful in this respect than those of the FSU (OECD 1999 p.17; Zamparutti and Gillespie 2000 p.333). At the same time, organisations such as the OECD recognise the ambiguous nature of FDI and its potential to engender negative environmental trends within the region (OECD 1999 p.161).

The FDI flow has varied not only with respect to its extent but also with regard to its function and purpose. For example, while some FDI flows have focussed on existing brownfield sites, others have sought out virgin greenfield sites (Pavlínek 2002). New greenfield investments can bring manufacturing activity to previously non-industrial regions and may result in the loss of environmentally valuable or sensitive land. In contrast, brownfield investments must contend with past pollution issues and the question of environmental liability (Kolaja 1995; OECD 1999 p.162). Nevertheless, where new production units seek out local suppliers, typically building on pre-existing supply networks, there is potential for the dissemination of environmental 'good practice' down the supply chain. In addition, the quality control of inputs can put pressure on local suppliers to improve the efficiency of their own operations associated with the introduction of new technology and management practices. Pavlínek and Smith (1998 pp.626–7) draw attention to Volkswagen's investment in Czech Republic. In this case, the relatively poor quality of local components resulted in encouragement for indigenous suppliers to seek out foreign partners (Pavlínek 2002 pp.1697–8). Similarly, investment in the building sector Czech Republic and Slovakia by the Swedish company Skanska led to the early introduction of environmental management systems (EMS) before they were required by European law (Alizadeh pers.comm. 2003).

Nevertheless, while foreign investment has the potential to improve the environmental effectiveness of indigenous production systems, it is also important to acknowledge that such activity may result in environmentally damaging consequences. For example, the seemingly straightforward relationship between FDI and technical improvement needs questioning. In some instances the low cost of labour within the region is a prime pull factor for manufacturing FDI and this can encourage reduced investment in expensive new production machinery (Pavlínek and Smith 1998 pp.628, 631; Pavlínek 2002 p.1705). As far as pollution control is concerned, both investors and governments have often been more concerned with 'end of pipe' solutions than radical overhauls of the production system itself (Pavlínek and Pickles 2000 p.242; Turnock and Carter 2002 p.425). More fundamentally, the reworking of foreign economic relations in the context of neoliberal restructuring can raise a range of environmental concerns (e.g. Pavlínek

1997 pp.104–10; Pavlínek and Pickles 2000 pp.242–85). Indeed, many of the former socialist countries hold a relatively weak position within the context of the global economy and this has numerous actual and potential consequences for the state of the region's environment. Those countries not involved in the EU accession process and dependent on natural resource exports are particularly vulnerable to the vagaries of the global economic system. There is also a risk that the region as a whole may find itself with a comparative advantage in highly polluting production activities, a situation exacerbated by relatively weak environmental regulation and economic necessity (e.g. Pavlínek 1997 p.108; Pavlínek and Pickles 2000 p.284). The Russian Federation's apparent willingness to import spent nuclear fuel for processing would appear to support this premise (Oldfield 2002). This trend closely parallels the observations of Shiva (1999 pp.53–9), noting India's relative weakness in the face of attempts by industrialised countries to export their waste. The recent exponential increase in crushed rock export from Czech Republic is seemingly indicative of the reduced environmental constraints offered by ECE more generally, where protected areas have been exploited in a comparatively aggressive manner (Jehlička and Cowell 2003).

While a significant percentage of FDI is aimed at export-led production, some investment is focussed on capturing part of the region's domestic market. The truncated development of consumer production under socialism left a legacy of pent-up consumer demand within the region which foreign multinationals are eager to exploit. The developing service sector, while helpful in shifting the regional economies away from their dependence on heavy industry, is also giving rise to new environmental pressures such as increased motor vehicle emissions and domestic waste production (Oldfield 1999; Pavlínek and Pickles 2000). It is also necessary to recognise the influence of foreign investment activity in other sectors of the economy such as agriculture and natural resource extraction. For example, agricultural regions in countries such as Hungary have attracted Western investors due to the cheap cost of land and the perceived lack of environmental constraints. The restructuring associated with Western investment (usually through farm mergers and intensification of agricultural practices) has the clear potential to worsen rural environments with output gains achieved at the expense of lost or fragmented habitats and the overall socioeconomic fabric. The recent extension of a limited set of EU agri-environmental packages to the accession countries has the potential to offset such impacts.

In an allied manner Russia provides an ideal focus, at least in theory, for those multinationals interested in natural resource extraction. The recent flurry of FDI activity associated with the Caspian Sea oil and gas developments, of which Russia is a major player, is illustrative of this point (EBRD 2001 pp.75–90). A significant percentage of Russia's natural resources is located in the country's Far East region and includes forest and mineral (energy and precious metal) resources. Furthermore, the hydrocarbon reserves offshore of Sakhalin have attracted the attention of multinationals such as BP, Exxon-Mobil and Shell and major investment is planned (Bradshaw 1999 p.29). It should be noted, however, that the general difficulties and uncertainties associated with investing in the region continue to depress foreign investment flows (Ibid). At the same time, there is also increasing evidence to suggest that some foreign firms operating within the Russian Far East have been able to take advantage of the confused regulatory situation in order to further their own interests while simultaneously undermining the integrity of the region's natural resource base. This trend is particularly evident in the forestry sector which is characterised by the involvement of companies based in China, South Korea and other countries of South East Asia (BROC et al. 2000 pp.15–6; Newell 2004). The Russian Far East's fishing resources are also being undermined by unregulated and illegal fishing activities. While much of this activity is being carried out by national operators, the situation has been aggravated by the influx of large, modern, Western-built trawlers – typically underwritten by foreign companies (Newell 2004). The involvement of foreign multinationals in Russia's economy has also raised issues concerning the social and environmental consequences of natural resource development for local communities (Wilson 2000, 2002).

The relative weakness of Russia's indigenous environmental regulatory regime draws attention to some of the difficulties the countries of the region can face in trying to monitor the activities of foreign companies. As intimated above, the profit motive ensures that foreign investors are not necessarily concerned with lowering levels of pollution emissions and can actively seek to reduce their environmental commitments in order to cut costs. Furthermore, the fluid and footloose nature of foreign investment can undermine the effectiveness of national and regional environmental regulation. This is most apparent when the investment is export-oriented with limited connections to the local economy and regulation thus tends to take place at the supranational scale (Bridge 1998 pp.224–7). Importantly, the emergence of a sustainability discourse within the arena of transnational economic activity is proving influential with respect to the regulation of footloose, globalised capital (Ibid). Furthermore, this discourse has permeated the work practices of both corporate and state bodies and helped to shape the lending practices of international organisations such as the World Bank and EBRD as well as the technical assistance programmes of the EU (Macnaghten and Urry 1998 pp.212–9). It is open to debate how effective such discursive developments are in addressing the environmental situation within the region. Even if they do provide a necessary regulatory role, indigenous environmental regulatory frameworks are nevertheless denied the opportunity to develop meaningful relationships with local firms and businesses, which, it can be argued, are an integral aspect of any maturing regulatory system. At the same time, there is a marked variability in the effectiveness of national regulatory frameworks across the post-socialist region. This unevenness is further aggravated by the EU accession process which demands the extensive reworking of existing legislative and policy frameworks amongst the accession countries to bring them into line with the EU norms (Oldfield and Tickle 2002). Evidence from the Czech Republic also suggests that EU-led ecologically modern discourses are often taken up uncritically by regulators (Jehlička and Tickle forthcoming) though other studies indicate some pockets of stronger indigenous policy communities surviving from the socialist period (Jehlička and Cowell 2003; Tickle 2000).

Conclusion

This chapter has focussed on the enduring economic relationship that have characterised the interaction between East and West during the course of the last 80 years. It has argued that these relationships have been influential in determining a distinctive set of environmental geographies that emerged within the region during the second half of the 20th century. Moreover, these relationships continue to play an integral role in shaping environmental futures throughout the region. Since the fall of the socialist regimes, the region has been progressively exposed to a more complex range of foreign economic pressures, although vestiges of pre-1989 East-West relationships remain formative factors in certain instances. These new influences range from the structural aid and loan packages of the EU and multilateral development banks such as the World Bank to private flows of FDI.

Cumulatively, these different influences amount to a significant influx of investment aimed at advancing and implanting capitalistic relations within the region. At the same time, the nature and extent of investment flows has varied considerably across the region. In general, the ECE countries have received a significant proportion of the available investment with those in line for EU accession being particularly prominent. In contrast, the FSU (excluding the Baltic States) has tended to lag behind in the investment stakes due to the reluctance of potential investors. These varied performances have implications for the environmental consequences of foreign economic linkages. At a general level, the influx of FDI, loans, and other payments is certainly encouraging a concomitant 'greening' of the production process within the region, although this process can throw up ambiguous results and more critical analysis is required here. This is particularly relevant to those countries not partaking in the EU accession process. More fundamentally, these flows of capital are abetting the progressive embedding of the region in the 'asymmetries' of the global economy, a process with the potential to undermine the environmental integrity of countries within the region to a greater or lesser extent. Those countries with a relatively high dependence on natural resource exploitation such as Russia are particularly vulnerable to the vagaries of global capital flows.

The relationship between foreign capital flows and the environment in the region is clearly complex as well as being spatially diverse – both between and within countries. However, this relationship is in need of further exploration in order to move beyond the generalised and often anecdotal nature of the current literature and instead provide the foundations for a more sophisticated conceptual framework sensitive to the regional particularities of the post-socialist landscape. In order to abet the development of such a framework, it would seem expedient to engage more freely with the experiences of other world regions. For example, there would seem much to gain from comparing the experience of resource-rich developing countries with the contemporary experience of the Russian Far East region. As it is, the current lack of empirical, case-led evidence and concomitant understanding is certain to undermine effective responses within the region to the challenges posed by foreign capital flows over the short to medium term.

References

Alizadeh, I. (2003), *personal communication*, Section Manager, Environmental Management System, IPS Skanska a.s., Prague.

Baker, S. and Jehlička, P. (1998), 'Dilemmas of Transition: The Environment, Democracy and Economic Reform in ECE – An Introduction', *Environmental Politics*, 7(1), 17–23.

Bøhmer, N., Nikitin, A., Kudrik, I. et al. (2001), *The Arctic Nuclear Challenge*, The Bellona Foundation, Oslo.

Bradshaw, M.J.B. (1999), *The Russian Far East: Prospects for the New Millennium*, Royal Institute of International Affairs Discussion Paper 80, London.

Bradshaw, M.J.B. and Lynn, N.J. (1998), 'Resource-Based Development in the Russian Far East: Problems and Prospects', *Geoforum*, 29, 375–92.

Bridge, G. (1998), 'Excavating Nature: Environmental Narratives and Discursive Regulation in the Mining Industry', in A. Herod, G.O' Tuathail and S.M. Roberts (eds), *An Unruly World: Globalization, Governance and Geography*, Routledge, London, 219–43.

BROC (Bureau for Regional Oriental Campaigns), Friends of the Earth-Japan, and Pacific Environment and Resource Center (2000), *Plundering Russia's Far Eastern Taiga*, Friends of the Earth-Japan, Tokyo.

Carter, F.W. (1994), *Trade and Urban Development in Poland: An Economic Geography of Cracow from its Origins to 1795*, Cambridge University Press, Cambridge.

Carter, F.W. and Turnock, D. (eds) (1993), *Environmental Problems in Eastern Europe*, Routledge, London.

Cole, D.H. (1998), *Instituting Environmental Protection: From Red to Green in Poland*, Macmillan, London.

Dalton, R.J., Garb, P., Lovrich, N.P. (1999), *Critical Masses: Citizens, Nuclear Weapons Production, and Environmental Destruction in the United States and Russia*, MIT Press, Cambridge, Mass.

DeBardeleben, J. (1985), *The Environment and Marxism-Leninism: The Soviet and East German Experience*, Westview, Boulder, Col.

Deutscher, I. (1967), *The Unfinished Revolution: Russia 1917–1967*, Oxford University Press, London.

Dingsdale, A. (2002), *Mapping Modernities: Goegraphies of Central and Eastern Europe, 1920–2000*, Routledge, London.

EBRD (2000), *The EBRD: Its Role and Activities*, EBRD, London.

EBRD (2001), *Transition Report 2001: Energy in Transition*, EBRD, London.

Escobar, A. (1995), *Encountering Development: The Making and Unmaking of the Third World*, Princeton University Press, Princeton, N.J.

Fagin, A. (2001), 'Environmental Capacity Building in the Czech Republic', *Environment and Planning A*, 33, 589–606.

Fagin, A. and Tickle, A. (2002), 'Environmental Movements Nation States and Globalisation', in F.W. Carter and D. Turnock (eds), *Environmental Problems of East Central Europe*, Routledge, London, 40–55.

Gibb, R. (1998), 'Europe in the World Economy', in D. Pinder (ed), *The New Europe: Economy Society and Environment*, Wiley, Chichester, 45–65.

Gower, J. (1999), 'EU Policy to CEE', in K. Henderson (ed), CEE and the EU, UCL Press, London, 3–19.

Herrschel, T. and Forsythe, T. (2001), 'Constructing a New Understanding of the Environment under Postsocialism', *Environment and Planning A*, 33, 573–87.

IMF, The World Bank, OECD and EBRD (1991), *A Study of the Soviet Economy Volume 2*, OECD, Paris.

Jehlička, P. and Cowell, R. (2003), 'Czech minerals policy in transformation: the search for legitimate policy approaches', *Environmental Sciences: Journal of Integrative Environmental Research*, 11(1), 79–109.

Jehlička, P. and Tickle, A. (2004), 'Environmental Implications of Eastern Enlargement: The End of EU Progressive Environmental Policy?', *Environmental Politics*, 13(1), in press.

Kaldor, M. (1990), *The Imaginary War: Understanding the East-West Conflict*, Blackwell, Oxford.

Kolaja, T. (1995), 'Environmental Liabilities and Privatisation', *Environments in Transition: The Environmental Bulletin of the EBRD*, Spring, 1–6.

Kornai, J. (1992), *The Socialist System: The Political Economy of Communism*, Clarendon Press, Oxford.

Kotkin, S. (1995), *Magnetic Mountain: Stalinism as a Civilization*, University of Los Angeles Press, Berkeley and Los Angeles, Calif.

Lavigne, M. (1974), *The Socialist Economies of the Soviet Union and Europe*, Martin Robertson, London.

Lisitzin, E.N. (1987), 'The Union of Soviet Socialist Republics', in G. Enyedi, A.J. Gijswijt and B. Rhode (eds), *Environmental Policies in East and West*, Taylor Graham, London, pp. 311–333.

Macnaghten, P. and Urry, L.J. (1998), *Contested Natures*, Sage Publications, London.

Markandya, A. and Averchenkova, A. (2000), 'Transition and Reform: What Effect does Resource Abundance Have?', *Environmental Planning D: Planning and Design*, 27(3), 349–63.

Moldan, B. (ed.) (1990), *Životné prostredie České republiky: vyvoj a stav do konce*, Academia, Prague.

Newell, J. (2004), *The Russian Far East: A Reference Guide for Conservation and Development*, Daniel and Daniel, McKinleyville, Calif.

Nove, A. (1992), *An Economic History of the USSR 1917–1991*, Penguin Books, London.

OECD (1999), *Environment in the Transition to a Market: Progress in Central and Eastern Europe and the New Independent States*, OECD, Paris.

Oldfield, J. (1999), 'Socio-Economic Change and the Environment: Moscow City Case Study', *Geographical Journal*, 165, 222–31.

Oldfield, J.D. (2001), 'Russia, Systemic Transformation and the Concept of Sustainable Development', *Environmental Politics*, 10(3), 94–110.

Oldfield, J.D. (2002), 'Russian Environmentalism', *European Environment*, 12(2), 117–29.

Oldfield, J.D. and Tickle, A. (2002), Environmental Policy in a Wider Europe', *European Environment*, 12(2), 61–3.

Oldfield, J.D., Kouzmina, A. and Shaw, D.J.B. (2003), 'Russia's Involvement in the International Environmental Process: A Research Report', *Eurasian Geography and Economics*, 44(2), 157–68.

Pavlínek, P. (1997), *Economic Structuring and Local Environmental Management in the Czech Republic*, Edwin Mellen Press, Lewiston, N.Y.

Pavlínek, P. (2002), 'Transformation of the CEE Passenger Car Industry: Selective Peripheral Integration through FDI', *Environment and Planning A*, 34, 1685–1709.

Pavlínek, P. and Pickles, J. (2000), *Environmental Transitions: Transformation and Ecological Defence in Central and Eastern Europe*, Routledge, London.

Pavlínek, P. and Smith, A. (1998), 'Internationalization and Embeddedness in ECE Transition: The Contrasting Geographies of Inward Investment in the Czech and Slovak Republics', *Regional Studies*, 32(7), 619–38.

Peterson, D.J. (1993), *Troubled Lands: The Legacy of Soviet Environmental Destruction*, Westview, Boulder, Col.

Pryde, P. (1972), *Conservation in the Soviet Union*, Cambridge University Press, London.

Pryde, P. (1991), *Environmental Management in the Soviet Union*, Cambridge University Press, Cambridge.

Saiko, T. (2001), *Environmental Crises: Geographical Case Studies in Post-Socialist Eurasia*, Pearson Education, Harlow.

Schopflin, G. (1993), *Politics in Eastern Europe 1945–1992*, Blackwell, Oxford.

Shiva, V. (1999), 'Ecological Balance in an Era of Globalization', in N. Low (ed), *Global Ethics & Environment*, Routledge, London, 47–69.

Smith, A. (1994), 'Uneven Development and the Restructuring of the Armaments Industry in Slovakia', *Transactions of the Institute of British Geographers*, 19, 404–24.

Smith, A. (2002), 'Imagining Geographies of the "New Europe": Geo-Economic Power and the New European Architecture of Integration', *Political Geography*, 21, 647–70.

Smith, A. and Swain, A. (1998), 'Regulating and Institutionalising Capitalism: The Micro-Foundations of Transformation in Eastern and Central Europe', in J. Pickles and A. Smith (eds), *Theorising Transition: The Political Economy of Post-Communist Transformations*, Routledge, London, 25–53.

Stoklasa, J. (1923), *Die Beschadigung der Vegetationen durch Rauchgase und Fabriksexhalationen*, Urban and Schwarzenberg, Munich.

Swain, A. and Hardy, J. (1998), 'Globalization Institutions Foreign Investment and Reintegration of East and Central Europe and the FSU with the World Economy', *Regional Studies*, 32, 587–90.

Tickle, A. (2000), 'Regulating Environmental Space in Socialist and Post-Socialist Systems: Nature and Landscape Conservation in the Czech Republic', *Journal of European Area Studies*, 8(1), 57–78.

Tickle, A. and Clarke, R. (2000), 'Nature and Landscape Conservation in Transition in Central and South-Eastern Europe', *European Environment*, 10(5), 211–9.

Tickle, A. and Vavroušek, J. (1998), 'Environmental Politics in the Former Czechoslovakia', in A.Tickle and I.Welsh (eds), *Environment and Society in Eastern Europe*, Longman, Harlow, 114–45.

TsSU SSSR (1972), *Narodnoe khozyaistvo SSSR 1922–1972: Yubileinyi statisticheskii ezhegodnik*, TsSU SSSR, Moscow.

Turnock, D. and Carter, F.W. (2002), 'Conclusion', in F.W. Carter and D. Turnock (eds), *Environmental Problems of East Central Europe*, Routledge, London, 419–32.

Weiner, D.R. (1999), *A Little Corner of Freedom: Russian Nature Protection from Stalin to Gorbachev*, University of California Press, Berkeley, Calif.

Weiner, D.R. (2000), *Models of Nature: Ecology, Conservation and Cultural Revolution in Soviet Russia*, University of Pittsburgh Press, Pittsburgh, Pa.

Wilson, E. (2000), *North-Eastern Sakhalin: Local Communities and the Oil Industry*, University of Birmingham Russian Regional Research Group Working Paper 21, Birmingham.

Wilson, E. (2002), 'Est' zakon, est' i svoi zakony: Legal and Moral Entitlements to the Fish Resources of Nyski Bay, North-Eastern Sakhalin', in E. Karsten (ed.), *People and the Land: Pathways to Reform in Post-Soviet Siberia*, Dietrich Reimer Verlag, Berlin, 149–68.

World Bank (2002), *Transition: The First Ten Years. Analysis and Lessons for Eastern Europe and the Former Soviet Union*, World Bank, Washington, D.C.

Zamparutti, T. and Gillespie, B. (2000), 'Environment in the Transition towards Market Economies', *Environment and Planning B: Planning and Design*, 27, 331–47.

Ziegler, C.E. (1991), 'Environmental Protection in Soviet-East European Relations', in J. DeBardeleben (ed), *To Breathe Free: Eastern Europe's Environmental Crisis*, Woodrow Wilson Center Press, Washington D.C. and Johns Hopkins University Press, Baltimore Md., 83–100.

Chapter 8

Regional Development with Particular Reference to Cohesion in Cross-Border Regions

David Turnock

Introduction

There is nothing original about cross-border studies in ECE given the dramatic territorial changes experienced during the last century, but in the past interest has usually arisen out of the process of demarcation in what have invariably have been highly contested situations and the problems arising from boundaries acting as substantial barriers to movement. There is no distinguished history of cooperation in frontier regions even where a shared infrastructure made this almost essential, as in inter-war Upper Silesia (Hartshorne 1933). With frontiers as seemingly impenetrable obstacles leading to major dislocations, ethnic anomalies were conventionally approached through population exchanges while typical border regions would highlight marginality, as in the case of the Upper Soča valley of Slovenia (Moodie et al. 1955). However in sharp contrast to the conflicts and disputes of the past, much border study today is inspired by the European model of cross-border cohesion to the point where frontiers pose as a mere administrative formality rather than an inevitable barrier to intercourse. This developing field of study has now generated several collections of papers dealing with ECE (Baran 1995a; Deica et al. 1998; Dubcova and Kramerekova 2002) as well a range of global studies. The theme of transformation 'from conflict to harmony' has been discussed very thoroughly with respect to the upper Adriatic borderland (Bufon and Minghi 2001) and has produced numerous other studies profiling the Italian-Slovenian border with particular emphasis on social integration, which also includes the Croatian-Slovenian frontier in the area south of Trieste where Slovenia seeks control of the maritime approaches to Koper (Klemenčić and Gosar 2000). Indeed, cross-border cooperation (CBC) has made significant progress to the point where frontier regions are often seen as acceptable locations for FDI. It is therefore worth examining the potential from the viewpoint of a wider dispersal of investment which is widely seen as desirable in a situation where certain countries and locations attract a disproportionate share of the region's FDI. Frontier regions are chosen for study here because they have attracted considerable attention at a time when regional policy in the region has been generally weak (Bachtler et al. 2000). Hence the experience may provide pointers as to the directions in which regional policy should move in the future.

The Communist Legacy

Under communism there was no formal regional planning process in which local communities could participate. Administrative regions were essentially passive in the face of central government development programmes that were the haphazard outcome of sectoral decisions (the clout of some regional leaders notwithstanding) (Mihailovič 1972). Border regions were usually marginalised in the process, but the situation in ECE under communism surely provides an extreme case of minimal international contact in the context of a generally well-developed infrastructure and a high density of population (Dawson 1997). While the very nature of an autarkhic central planning process would have limited commercial and cultural contact, the region was also obliged to shoulder highly restrictive Soviet norms for cross-border relations, especially during the early part of the communist era. Boundaries between communist states were subject to complex bureaucratic regulation with relatively few local arrangements to facilitate family contact or daily commuting. Many cross-border routes endowed with road and/or rail links were closed; a situation which affected not only the heavily fortified 'Iron Curtain' but also the boundaries between communist states, including those shared with the USSR. Of course some post-1945 boundaries were newly demarcated – like the inter-zonal boundary in Germany, the Oder-Neisse frontier between the GDR and Poland and Poland's frontier with the USSR – and in these cases there would inevitably have been many closures of local lines of communication (as was the case after 1918 when the boundaries of Trianon Hungary were imposed and Banat was divided into Hungarian, Romanian and Yugoslav portions). But discussing the border between Hungary and the then Czechoslovakia, Podhorsky (1995 p.56) remarked that the USSR 'was remarkable for distrust even towards its own allies'. Where local facilities remained, use might be restricted to certain days of the week and additionally subject to quotas; thereby encouraging clandestine activity (Boar 1999). The result was isolation and the reduction of cross-border contact to a minimum. Szekely (1995) used the concept of circular cumulative causation to emphasise the negative effects of a closed frontier on the economic development of border regions, with limited employment prospects – beyond a relatively extensive agriculture characteristic of fringe areas – prompting selective out-migration and population decline; noted in Slovenia where Horvat (1992) identified 'demographically-endangered settlements'.

The situation was made worse by bureaucratic actions to depopulate border regions. Expulsion of the German population had a particularly depressing effect on the northwestern borders of Czechoslovakia where consolidation of agriculture was placed in the hands of large state farming enterprises (Bičik and Stepanek 1994).

Population decline in turn discouraged investment in infrastructure, while jobs in manufacturing (usually handicrafts, food processing and extractive industries) provided few opportunities for the well-qualified (Mikus 1986). In a bid to reduce illegal emigration ('Republikflucht'), the former GDR established a security zone on its western border zone – closed to non-residents – while the strained relations that followed Tito's expulsion from the Cominform had dramatic consequences in Romania where all unreliable elements were cleared from the border strip and transported to the Bărăgan steppe until 'normality' was restored in the mid-1950s. The 'Wisla' operation in Poland dispersed the Ukrainian population and the bid to homogenise the ethnically diverse borderlands of eastern Poland succeeded in emptying significant areas of the Bieszczady frontier region where a closed military zone existed during 1948–56 (Kisielowska-Lipman 2002). Until the 1975 Helsinki process imposed obligations on the communist states to observe basic human rights it was often extremely difficult for families divided by the frontiers to exchange visits. This affected East Germans wishing to travel to West Germany (and East Berliners going to West Berlin), Hungarians wishing to visit Romania, Slovakia and the former Yugoslavia – complicated in the former case by the Ceauşescu regime's controls on accommodation in the private houses of relatives – and also smaller groups like Romanians in Maramureş and northern Moldavia with relatives in Ukraine. The isolation of Poland during the Solidarity era gave rise to some highly insensitive planning by the country's neighbours in view of the pollution hazards associated with the GDR projects for nuclear power at Lubmin, oil refining at Schwedt and metallurgy at Eisenhüttenstadt; also the Czechoslovak plans for the Stonava coking plant and a hotel on Snieska mountain (Ciechocinska 1992).

There were however several measures to moderate isolation and neglect. Some domestic train services were allowed to transit short sections of foreign territory e.g. GDR trains passed through the Varnsdorf area of Bohemia, while Polish trains from Krosno to Przemyśl passed through Soviet territory (albeit with the carriage windows painted over in the latter case!) while Soviet freights entered northern Romania between Campulung pe Tisa and Valea Vişeului. The policy of full employment meant that work was provided not only in agriculture but also in light industry specially organised in border regions to make use of the labour available e.g. small textile enterprises in the villages of the Harz Mountains. But the shortage of labour in the GDR led to significant daily commuting from Poland and (to a lesser extent) Czechoslovakia: the Neisse frontier was opened for Polish workers who crossed daily from 1972 until 1980 when the Solidarity 'threat' closed the border again until 1989 (Buusink 2001). There was some cooperation over joint economic projects: notably hydropower projects on the Drava (Hungary-Yugoslavia); Danube (Czechoslovakia-Hungary, Romania-Yugoslavia and – as a planning exercise – Bulgaria-Romania); and Prut (Romania-USSR). Meanwhile environmental interests brought Czechoslovakia and Poland together over the Tatra National Park – an area where the frontier was particularly restrictive for Polish tourists – and the Baltic Sea Environmental Protection Committee was set up in 1960 with the aim of reducing pollution and creating a taskforce for programme implementation. Another special institution was the Alps-Adria Working Community founded in 1978 which included parts of former Yugoslavia from the start and was extended into Hungary during 1986–9 (Horvath 1993).

Some border 'loosening' occurred in Hungary in the 1970s through cooperation between Austria's Burgenland province and the Hungarian counties of Györ-Sopron and Vas in Hungary over a joint planning committee. It was in 1967 that Hungarian officials started contacts with Austrian Area Research and Area Planning Society in Vienna for joint frontier planning. Joint inspections of frontier areas in 1969–70 led on to the 1971 announcement initiating joint planning for transport, industry and the regulation of Lake Fertö/Neusiedlersee. Travellers between the Burgenland villages of Siegendorf and Morbisch were allowed to use the direct route through Hungary, while greatly increased travel by Austrians into Hungary increased supermarket turnover in Sopron and recreation use of facilities of Lake Fertö. Hungary also sought cooperation in border areas with Czechoslovakia and Yugoslavia. In 1970 a permanent committee on urban and regional planning was established in Belgrade with the aim of joint economic/regional development. Inspection tours were made in that year and in 1972 plans were drawn up for industry and water projects with new opportunities envisaged when Adria Pipeline crossed the border. In 1971 a working committee on regional development was set up in Bratislava to deal with the Czechoslovak-Hungarian frontier with coordinated development with regard to water supply; building (use of Hungarian components in housing projects in the Czechoslovak side); industry (links between the Košice and 'Lenin' metallurgical industries); and agriculture (through joint use of machine stocks, given that the main harvesting campaign began later in Czechoslovakia).

Czechoslovakia followed its expulsion of the German population from 'Sudetenland' with belated attempts to regenerate the borderlands. The 1966–70 FYP provided regional subsidies for agricultural cooperatives and industries, leading to the 1969 plan recognising three categories of living conditions (extremely-unfavourable, difficult and average) over an area of 17,000 sq.kms with a population of 0.7mln. Triggered by the need for more agriculture (in the interest of greater national self-sufficiency in cereals) and the consequent revival of village life to avoid uninhabited wilderness relieved only by afforestation, measures taken in 1970 to ensure 'regional proportionality' secured higher wage levels and subsidies for up to 70% of estimated construction costs for projects in agriculture, industry and the tertiary sector. But despite a substantial budget to finance production and services during the 1971–5 FYP – to improve economic and social life and discourage further out-migration – there was no great commitment by the industry, trade and tourism ministries despite the expectation that a labour shortage in the interior would force development. Meanwhile, Bulgaria supported border tourism in 1976 (with promotion of Bozhentsi as a fashionable resort) while a decree in 1982 aimed at bringing the two lowest order settlement categories up the national average in order to curtail out-migration. Steps were taken to develop small industrial enterprises and strengthen sheep rearing and forestry enterprises to ensure year-round employment. Private and semi-private enterprises (plus further electrification) were to improve services in the villages and in mountain regions the conventional brigade unit were combined with small/family groups and individual piecework systems. Settlement grants were offered to people who moved in for at least ten years and wage bonuses were paid from 1983. There was also a special role for the 'Komsomol' (youth organisation) to recruit 4,500 young workers/specialists – 2,500 for the northeastern Strandzha-Sakar region (where a 'Republic of Youth' was proclaimed) and 2,000 for other border areas, with the additional incentive of free travel to their place of birth twice a year. There were few reported successes but these examples indicate an awareness of the weakness of regional policy in some countries.

The Post-Communist Model of European Integration

Bureaucratic restrictions have been removed and the opening of the formerly-closed communist economies to global business has greatly increased cross-border freight and passenger flows. Border crossing points were increased in number, with more through roads and railways now available, while travel formalities were simplified by the abolition of visas (in many cases), although international travel continued to be restricted by the high cost of travel and accommodation. A new 19km railway connecting the Hungarian and Slovenian systems without the need to use a short border section in Croatia helped to provide 'missing links' in the international transport system (Maggi and Nijkamp 1992). Major changes also occurred in the Carpathians where few crossing points were previously available: for example, the rebuilding of the road bridge over the Tisza at Záhony in 1996 helped to improve the road link between Budapest and Lviv, Rovno, Zitomir and Kiev. However, while capacity at frontier crossing points has increased and there is some polarisation of development along transit routes, there has not necessarily been a boost to activity in border regions in general: rather, the impact of transformed international relations may be felt most strongly in capital and provincial cities located far from the frontiers. For foreign investors (particularly those on the western frontier of the region in Austria, Germany, Greece and Italy) business opportunities are now available throughout the region with respect to home market penetration and export. In the case of Greece, Petrakos (1996 p.18) sees a regional market of up to 80 million people in the Balkans since Albania's agricultural bias – with a population dynamism not found elsewhere in Europe – complements Bulgaria's manufacturing and Greece's strong tertiary sector; so the impact of Greek investment is felt throughout the region. And with regard to Italy's relations with Slovenia, Gosar (1996) notes that Italian investors do not generally choose Slovenia's border regions and the few enterprises that have been located in these areas are not usually owned by investors from the other side.

Laying the Foundations for CBC

Border regions have benefited considerably through the reconstruction of civil society which enables individuals and organisations (both governmental and non-governmental) to take initiatives. It is also significant that the EU and CoE have long championed CBC which means that candidate countries are throwing their weight behind the concept and using some of the blueprints already available. Signing the CoE convention on transfrontier cooperation (Hungary, Poland and Ukraine were among the first) signals recognition that border regions may experience varying degrees of demographic crisis and require support as fringe areas. Thus Slovakia approved CBC in 1998 by signing the 'Memorandum on Understanding and Establishing Organisational Structures for Bilateral Programmes of CBC' and went to approve Euroregion (ER) activities in the country the following year. Finance is provided through the EU Interreg Community Initiative of 1990, extended to ECE in 1994 through PHARE CBC with the aim of improving living standards in border areas: a sub-programme of a scheme that was first made available in 1989 for Poland and Hungary to enhance cooperation between local government, business and the community as well as regional development in the context of both economic development (agriculture, human resources and transport) and environmental protection. CBC projects – which may involve the various levels of government and regional development agencies – are allowed to take a tenth of the PHARE budget compared with just two to three percent of structural funding allocated to Interreg inside the Union. And on most borders up to a tenth of the money may go on small projects (under €50,000 – excluded under normal PHARE economic and social cohesion schemes) allowing regional and local actors to become increasingly involved. However, borders not on the EU limits were eligible for funding only in December 1998 when PHARE was extended to all border regions in ECE and CREDO funds became available. Moreover TACIS (Technical Assistance for the Commonwealth of Independent States) started in 1996 – with Finland's membership of the EU a key factor in the decision – and the programme now extends to the land borders of Belarus, Moldova, Russia and Ukraine. Management of CBC funds is undertaken by national governments and by EU but some use is made of institutes in the region e.g. Siauliai Business Advisory Centre (Lithuania) was contracted to manage the Lithuania-Poland CREDO programme in 1997, working with border committees. There are other donors such as the World Bank (whose help was crucial for the Tisza bridge at Záhony already mentioned) and some charitable organisations in the USA are also involved.

There is no question of the need for these programmes to achieve greater cohesion and maximise economic potential after communist neglect; not to mention the humanitarian issues arising from the fact that many of the region's frontiers derive from the partition of the Habsburg Empire in 1918 and the break-up of functionally-coherent areas: many Hungarians found themselves isolated from market centres like Arad, Oradea, Satu Mare and Timişoara which had passed under Romanian control (Berenyi 1992; Kovacs 1990), while the new border with

Czechoslovakia cut through well-developed functional zones, including the Rimamurany-Satgótarán heavy industrial system. And although alternative employment was eventually provided in new factories built in the towns of the border regions (Košice, Lučenec and Rimavská Sobota) the 'exiled' Hungarian minority regarded the border as an unacceptable constraint on movement. So it is entirely appropriate that a wide range of exercises have been undertaken to reduce the detrimental effects of state boundaries through reinforcing regional identity, promoting neighbourly links and developing economic potentials in order to reduce unemployment and raise/equalise living standards on both sides. The private sector has been stimulated through the growth of SMEs taking advantage of cheap labour, with the logic deriving from the complementary roles of each national territory (rather than structural homogeneity).

Reference should also be made to cross-border spatial planning and regional policy as well as common transport and telecommunication infrastructures – e.g. in the Bratislava-Vienna area (Korec 1997) – and similar arrangements for water, sewage and waste management. There is also activity through school partnerships and enterprises in the field of occupational training; and cooperation through the mass media, fairs-exhibitions and cultural, sporting and tourist initiatives. Finally, the need for environmental actions has given rise to quality integrated management of national parks and protected landscape areas as well as joint efforts to reduce industry-related environmental damage. The potential for interaction is variable as indicated by a substantial amount of research. These studies have profiled border regions involving pairs of countries (LACE/AEBR 1997) and the varying potentials were demonstrated, as in other subsequent works, promoting individual ERs (referred to below) such as those concerned with the Carpathians (CER) (Helinski 1998) and the Neisse (NER) (Gałęski et al. n.d.), and the group of ERs in Poland (Bojar 1996). Dokoupil (2002) classifies border regions according to development opportunity (positive, neutral or negative) and the state of cross-border relations (conducive to integration, cooperation or simply coexistence). The potential will clearly be higher where the population is substantial, communications are good and there is a degree of complementarity underpinned by good inter-ethnic relations (Dobraca and Ianoş 1998). A good example is Hodonin on the Czech-Slovak frontier where there is cross-border movement to work, shopping and school (Vaclav et al. 2002) – favoured by the special regime for Czechs and Slovaks who can cross the frontier freely wherever they wish.

Developing Organisations

Many organisations have appeared and they are in themselves a significant achievement. They include groups of municipalities such as the Union of Municipalities of Upper Silesia and Northern Moravia which brings together Jeseník with Nysa, Opava with Racibórz and Katowice with Ostrava and supports both new north-south highways and enhanced cultural relations for the ethnic minorities like the Poles of Těšín/Cieszyn (Kubik 1994). The partnership between

Frankfurt a.d. Oder and Słubice is significant through Viadrina University which has establishments in both places (see below); while Košice and Miskolc have grounded their collaboration in a shared interest in metallurgy. An Association of Danube River Municipalities presides over intensive links between Giurgiu and Ruse and Calafat-Vidin, with more modest associations elsewhere. Rural areas are involved in the Assembly of Border Localities established between Bihor (Romania) and Hajdú-Bihar (Hungary) clustered around the border crossing point on the edge of Oradea; also on the Polish-Slovakian frontier through a Slovak microregion for Biela Orava – formed in 2000 with interests in water/sewage systems and tourism promotion – complemented in Poland by the Beskid Intercommunal Association and the Union of Resorts of Sądecki Beskid. There is cross-border activity on either side of Babia Góra in respect of tourism (assisted by cycle routes) as well as culture, sport and SMEs – supported by PHARE and the Ekopolis (Jantárová čestá) foundation. Cultural groups are also being formed e.g. the Cultural Association of Ethnic Hungarians in Ruthenia, some of whom are now settling over the border (in Hungary) due to economic hardships at home.

On the economic side, the Carpathian Border Region Economic Development Association (EDA) was set up in 1994 though links between Hungary's Zemplen Local Enterprise Development Foundation in Sátoraljaújhely (where the main office is situated), and both the Carpathian Society of Hungarian Intellectuals at Užhorod (Ukraine) and the Nagykapos Regional Enterprise Development Centre (Slovakia) where small offices are maintained (Danko 1996). There are plans to extend the organisation to Poland and Romania. The council of 12 brings together mayors, professional businessmen and regional planners who collect and disseminate business information, while looking at the obstacles to CBC and the setting up of possible special economic zones. There has been help from the US Peace Corps (delivering economic expertise in such fields as privatisation), CESO (Canada), the Soros Foundation's Open Society Institute and the American 'Centre for International Private Enterprise'. The EDA publishes a 'Carpathian Business Review' which emphasises the critical need to stimulate SMEs through a network of offices where entrepreneurs can get business information, counselling, education and training; and financial support: note the three regional information offices in Sátoraljaújhely, Velké Kapušany and Užhorod. Enterprises need help in establishing contacts; securing transport services (including warehousing and multimode activity); and in matters concerned with technology and innovation generally. Partnerships with other countries have been encouraged through exhibitions and regional fairs. Other groups with economic interests include the Gorzów-based Polish-German Society for Economic Support, a joint stock company started in 1993 combining Berlin, Brandenburg, Mecklenburg-Vorpommern and Saxony with Szczecin, Gorzów, Zielona Góra and Jelenia Góra. The aims are to support investment plans (with contributions to joint ventures); provide advice and information (including an investors' database in liaison with Viadrina University) and organise exhibitions and fairs. In the same border region there is progress in the textile industry through collaboration involving the Centre for Innovation and

Technology (in Guben) and German/Polish textile producers; with the aim of specialising in quality ladies clothing, including model collections produced. Some 50 joint ventures have been set up with finance available through a Burgschaftsbank. Finally the Chamber of Commerce and Industry (CCI) in Borsod-Abaúj-Zemplén has established an Association of Carpathian Chambers; while cross-border agreements have been entered into by the CCIs of Bratislava and Győr; Košice and Miskolc; and Lučenec and Salgótarján. Incubators (or 'entrepreneur houses') operate at Salgótarján (Hungary) and Velké Kapušany (Slovakia), the latter using old buildings previously belonging to 'Jednota' consumer cooperative.

Euroregions (ERs)

The most conventional vehicle for CBC in specific areas is the ER which has been extended across ECE during the past decade. In 1992 the Polish Foreign Minister (K. Skubiszewski) called for such arrangements to extend cooperation without eroding the scope for individual cultural identification, after ERs were first recognised in Western Europe in the 1960s with the support of the EU Regional Development Fund and the CoE. Experiences were shared through 'Linkage Assistance and Cooperation for the European Border Regions' (LACE) an EU project administered by the Association of European Border Regions (AEBR) created in 1971 in the Rhineland. And further endorsement came from the EC's recognition in 1991 that border regions were disadvantaged because they lay at the extremities of transport systems planned on a national basis. In such areas trade was often distorted, while services might be wastefully duplicated and mobility hampered by differences in language, taxation, employment practices and welfare systems. Despite the legacy of the closed borders and a pronounced lack of cohesion, the ER model was successfully applied in ECE through progress along the eastern frontiers of Germany in 1991–3 following treaties with Poland in 1990–1 (Freiherr von Malchus 1997), while the CER emerged in 1993 as the first such region in ECE that did not include an EU member state (Corrigan et al. 1997). Since then ERs have continued to proliferate especially in the northern half of the region (Figure 8.1). They are part of the political culture underpinning the process of EU enlargement: highlighting the settlement of disputes between neighbouring states and enlightened treatment of ethnic minorities some of which are dispersed among several countries like the Lemko/Rusyn elements in Carpathian countries which can now work for a more coherent identity. Creation of permeable frontiers and the removal of bureaucratic blockages against socioeconomic progress (Suli-Zakar 1992) may now be seen as a significant part of a global world with enhanced mobility and integration. ERs are becoming formalised within the EU system of National Units for Territorial Statistics (NUTS) through the combining of basic units (NUTS III) into transfrontier regions which may then be recognised at the NUTSI or NUTSII levels (equivalent to the German Länder and Italian regions respectively).

ERs may develop from community groupings, for the Association of Rhodope

Figure 8.1   Euroregions in ECE

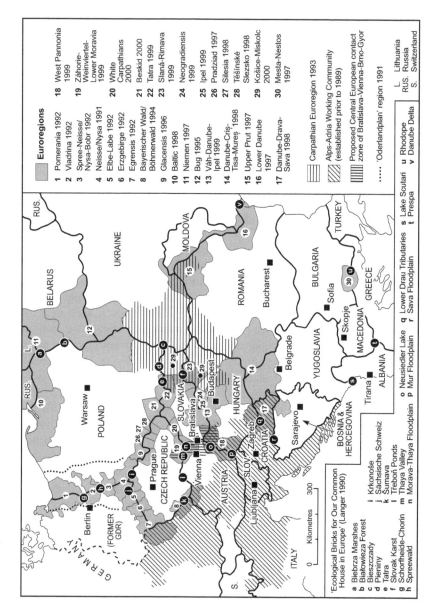

Municipalities spawned the Mesta-Nestos ER in 1997; while the Union of Municipalities of Upper Silesia and Northern Moravia (referred to above) led on the Silesian ER in 1998 and the Košice-Miskolc Association generated its ER in 2000. There may sometimes be coherence through an international transport axis: the Danube-Drava-Sava ER covering Baranya, Osijek-Baranya and Tuzla is significant in terms of the Budapest-Pécs-Osijek-Sarajevo axis. Most ERs combine territory from just two countries but the maritime approach of the Baltic ER brings together Denmark, Latvia, Lithuania, Poland, Russia (Kaliningrad) and Sweden. ERs usually have a significant establishment: in the case of Euroregion Baltic (ERB) a council, praesidium, executive board, secretariats and three working groups (regional development and spatial planning – with spatial planning centres in each of the countries; environment protection; and social problems/cultural exchange). Funding is again available through Interreg and there is a growing consensus that ERs are a force for stability and proof of good neighbourly relations which is helpful for EU candidate countries. Moreover, with the prospect of tighter administration on the borders of the EU, ERs may offer concessions with regard to visa formalities which may help to maintain intimate links with former Soviet states, especially where there strong cultural reasons for doing so, as between Moldova and Romania.

Integrating CBC into the Planning Process

CBC does not operate in isolation but is a significant element in the planning process. In 1989 there was no Western-style regional planning: expertise was lacking and given the desperate shortage of funds government was slow to get involved beyond ad hoc allocations of funds to maintain local government. In turn, local government had no experience of planning on the basis of growth potential through SWOT-type methodologies involving strengths, weaknesses, opportunities and threats. Of course, local authorities were free to build identity and promote their perceived attractions to boost investment, create jobs and thereby improve living standards and there were some notable successes for place marketing especially in the northern countries. But the basic municipalities are small, with limited resources and competence, while larger regions – the boundaries of which have often been contested in recent years – may not be fully accountable. Slovakia's regions were without elected councils until 2001 while the Hungarian county level 'is still fragmented and state-dominated and cannot fulfil its role as an intermediary between the central government and the lowest tier administrative units' (Szigetvari 2001 p.299). So typically, there was decentralisation, yet also fragmentation and weak institutional capacity with local budgets dependent on central decisions. There has also been limited ability to absorb funds and PHARE projects have sometimes overloaded administrative capacities. Moreover, despite the efforts of some inter-ministerial councils, central governments have been slow to develop West European-style strategic regional planning although control procedures are gradually being introduced to process planning applications – notably those concerned with FDI – and major projects are now subject to environmental and social impact assessment. In Poland, 'regional policies have been extremely weak and subordinated to sectoral policies of the national governments' (Gorzelak 2001 p.323). In the Czech Republic, as elsewhere, there were ad hoc interventions by government and international bodies but these were fragmented and uncoordinated. However, the Czech Republic and Hungary provided some help to areas experiencing the most difficult problems of economic and social cohesion – while Poland and Romania identified areas with major environmental hazards. The Czech Republic also introduced a regional policy that offered assistance to depressed border regions (Horvath 1996).

During the 1990s CBC programmes provided an element of regional policy and some included regional planning as a specific objective. The need was demonstrated by the former 'inner German' border (Wild and Jones 1993) and the dramatic changes in the transport network (Scharman and Tietze 1990). Thus Petrakos (1996 p.128) referred to 'a wide strip of land along the Greek-Albanian and the Greek-Bulgarian borders with lower levels of economic activity and development' and isolation from development axes to call for new planning systems in marginal areas that will overcome national frontiers (e.g. a unified approach to the Drina valley on the Bosnian-Serbian border). There are implications for the development of local democracy and decentralisation in favour of local administration. Momentum has come largely from Europe with a vision of cooperation on spatial policy. CoE has also become involved through a 1997 'European Conference of Ministers Responsible for Spatial Policy' (CEMAT) to establish guidelines for future spatial development (Kennard 2000a) linked with the Interreg concept of seven large overlapping regions of which two are relevant to ECE: Baltic Sea Region (BSR) and Central Adriatic Danubian and South-Eastern European Space (CADSES). Both have funds to allocate and these have gone to both EU member states and others e.g. Baltics Projects Facility (1996) to encourage cooperation networks across the BSR with provision for common actions with the TACIS programme.

In the case of the Baltic, the Baltic Sea Environmental Protection Committee dates back to 1960 and aims at reducing pollution and creating a taskforce for programme implementation. A new convention agreed in 1994 provided for a wider programme to run for 20 years. A Spatial Planning Commission was established on the German-Polish border in 1992 involving three Länder, four voivodships, the Bundesministerium für Raumordnung Bauwesen und Städtebau and Instytut Gospodarki Przestrzennej i Kommunalej. 'Spatial Policy Perspectives' were produced and adopted in 1995: seeking improved living conditions by minimising separation through CBC and environmental rehabilitation of damaged areas; also decentralised urbanisation through improved infrastructure (Storm-Pedersen 1993). Moreover, further planning in 1994 among the Baltic countries produced a concept of spatial development for an 'Oder' region rather more extensive than the ERs already referred to: the Berlin and Dresden areas of Germany and with Szczecin and Zielona Góra in Poland (Van de Boel 1994). The Pomeranian coastlands are attractive to Scandinavian investors, while German investment in Szczecin/Świnoujście reflects the import of building materials, chemicals and timber through ports that are closer to Berlin than Rostock. A multipurpose terminal (supplementing the coal handling facilities with container and Ro-Ro facilities) could also attract Scandinavian help, bearing in mind the potential of the 'Baltic Hanse'. As the Baltic's largest port complex (handling 25mln.t/yr), Szczecin provides a major centre with improved road access projected through a coastal motorway. Unfortunately, the city is poor for higher education, science and technology and is one of the less-competitive urban agglomerations (Markowski and Stawasz 2003). Moreover, local road improvements are needed and while are some new border crossing points – and the Altwarp-Nowe Warpno ferry – a link between Usedom and Wolin islands is needed through Garz-Świnoujście, crossing the Regalica River. It is also necessary to acquire 400ha to extend the port, while work is needed to modernise the quays with development of silos and improvement of the duty-free zone along with a computer-controlled shipping system and a port rail/road link through bridges over the Parnica as a port link; while a major international sailing and leisure centre is projected for Lake Dąbie. However, there are good rail links with Berlin, Prague and Vienna; also a link with West European waterways via the Oder-Havel Canal. In 1996, Poland launched its 'Programme to Develop the Oder Waterway by 2005' with links to the European waterway system with 20mln.t.yr carried on the waterway alone. The rehabilitation of potential waterway rings in the Baltic States, Kaliningrad and Poland involving the Augustów and Masurian canals and the coastal system from Gdańsk to Klaipeda has not been significantly discussed but there is a related Elbe project which involves German-Czech cooperation (Jansky 1997). Finally, through the Baltic arena Poland is also involved in cooperation with neighbouring countries (Belarus, Baltics and Russia i.e. Kaliningrad) in the 'Green Lungs of Europe' project – growing out of an initial 'Green Lungs of Poland' initiative – to coordinate activities in networks of national parks and other protected areas.

In the case of CADSES, a 'Vision Planet' is generating planning strategies through 'commonalities' and the identification of transnational action areas. The first results were provided in 1998 and all projects are to include both EU and non-EU members (even though Interreg and PHARE/TACIS application conditions and procedures are completely separate). There is a clear focus on improved north-south routes of the kind contemplated under communism but never achieved on a comprehensive basis. Now the Eurocorridors are gaining some funding priority and they provide a basis for cooperation on the borders of Germany and Poland with Czech Republic and Slovakia and in turn with Hungary and the SEECs. Several border routes have been improved including the Tisza crossing at Chop (Hungary-Ukraine). In the case of Poland-Slovakia, a project for a 'Cross Border Communication System' involves a comprehensive information flow to stimulate enterprise and intensify cooperation in respect of administration, scientific research, culture and training. PHARE has supported restoration of the Esztergom-Štúrovo bridge between Hungary and Slovakia, first opened 1895 but destroyed 1944 and reopened only in 2000; as well as the new Koroszczyn border facility for the Kukuryki crossing between Poland and Belarus (under construction during 1996–9). As regards the waterways the Danube is a Eurocorridor in its own right and connections still mooted with the Elbe and Oder, not to mention more local links e.g. with Timişoara through a possible revamping of the Bega Canal. The region also provides a context for improvement of east-west links across the Balkans e.g. from Athens and Thessaloniki to both Sofia-Bucharest and Skopje-Belgrade; also Patras-Ioannina-Tirana; Alexandropolis-Burgas-Varna-Constanţa; and Istanbul to Igoumenitsa and/or Durrës (via Sofia and Tirana) 'to improve the coherence of Balkan economic space' (Petrakos 1996 p.265). Some new railways are desirable e.g. between Bulgaria and Macedonia and former links might well be reopened such as the direct connection between Szeged and Timişoara that is part of the trilateral programme for DCMTER during 2004–6.

These projects will transform the potentials for CBC which are currently very variable. Tvrdon (1996) mentions nine areas of CBC potential for Slovakia and her neighbours with primacy falling to Bratislava-Vienna as the core of a Work Community of Danubian Countries ('Arbeitsgemeinschaft Donauländer'). Rechnitzer (1998 p.89) perceives a wider 'Wirkungsbereich' for Vienna, St.Polten and Wiener Neustadt (Austria); Brno (Czech Republic); Bratislava and Nitra (Slovakia); and Budapest, Györ, Székesfehérvár and Nagykanizsa (Hungary). The logic of an integrated Bratislava-Vienna city region is underpinned by the Danube (and its possible canal links with the Elbe, Oder and Vistula) as well as by cooperation over airports and the restoration of railway links now that the city region is no longer suppressed by closed frontiers (Altzinger 1997). A special economic zone is emerging in Bratislava's Petrzalka suburb to attract more Austrian investment; while Brno and Györ are also attracting Austrian investment and have plans for motorway links with Vienna. Several organisations have emerged, beginning with the 1991 agreement leading to the Coordination Study of the Development of the Slovak and Polish border region. Reference should also be made to an Austro-Hungarian Cross-Border Regional Council founded in 1992 (comprising representatives of Györ-Moson-Sopron, Vas and Burgenland) which has addressed Austria's concerns over prostitution, crime and drugs arising from more open borders (Rechnitzer 1997). Also public utility maps were harmonised to facilitate joint planning of infrastructure including sites for solid waste disposal. This provided the springboard for the ER founded in 1999 and also a 'West Gateway' Area Development Association for the Austria-Hungary-Slovakia tri-frontier area (Szorenyine-Kukorelli et al. 2000 p.227). In the run-up to EU enlargement CBC issues are now appearing in all levels of strategic planning and there is a good basis here for a joint vision of an integrated regional economy, social cohesion and good neighbourly relations. As a result of high cooperation between Austria and Hungary, CBC issues are appearing at regional and national levels of strategic planning so as to maximise fundraising potential and harmonise development priorities. In the run-up to EU membership there should be a joint vision of an integrated regional economy, social cohesion and good neighbourly relations. However effective regional planning for cross-border territories will depend on binding systems to ensure the necessary priority for implementation.

National Planning

Progress has been driven by applications for EU membership and consequent exposure to EU regional policy which seeks to redress the main regional imbalances within the EU – promoting economic and social cohesion and creating an area without internal borders; with reinforcement in the Maastricht Treaty of 1992 providing for a Cohesion Fund for environmental projects and Eurocorridors – with democratic control through the European Parliament. The ECEs are now being drawn into a uniform system with a vision of sustainable development: 'an environmentally respectful economic development which maintains today's resources and their functional capacity for future generations'. It is evident that in a wider Europe – without internal boundaries – traffic will increase and exert more pressure on some natural areas – hence the need for more planning with the European Spatial Development Perspective adopted 1997 as a basis for further discussion. However, it is only in the present 2000–6 period that EU candidate countries are producing national and regional plans as a basis for cohesion funding. Starting from 2000 in order to access PHARE funds, candidate countries must have National Development Plans (NDPs) for 2000–2. These will need to highlight key sectoral and regional priorities highlighted and harmonise with Regional Development Plans prepared by counties and Regional Development Agencies. So in 1999–2000 each country prepared its NDPs for 2000–2 mapping out its national strategy for economic and social cohesion to which all government and PHARE money (including ISPA and SAPARD pre-accession funding) will be linked through Regional Development Funds. All countries are also formulating integrated regional programme – including special provision for specified 'less-favoured areas' (LFAs) – for each NUTS II unit which will either be a top-tier administrative region or a groups of such regions (in which case plans will have to be broken down so as to achieve consistency). NER leaders think it will be 'good to create mutual medium-term cross border regional plans which are directly binding for all public planning offices as elements of cross border planning into which all local plans would gradually be introduced' (Gałęski et al. n.d. p.12).

Much has therefore been accomplished in recent years. Poland created a new Ministry for Regional Development in 2000 – leading to procedures for supporting regional development with state and EU funds: a national strategy for 2001–6 with a government support programme, to which the regions have responded with their own strategies in order to produce a final 'contract' (though other countries may place less emphasis on regional responses). Enhancement of absorption ability is a priority at all levels of government in order to secure the maximum benefits from PHARE funds (Bachtler et al. 2000 p.372). It will clearly take time to build up the necessary expertise and establish procedures to blend national, regional and local aspirations in line with EU requirements. The PHARE CBC Regulation of 1998 laid the basis for a better integration of such programmes into the candidate countries' overall regional development policy. Support provided by PHARE CBC should be consistent with support for investment in economic and social cohesion under the national PHARE programmes. Thus PHARE CBC programmes will also have to harmonise with the relevant national and regional programmes and their specific provisions for the coordinated development of border regions. Hungary's regional plan now urges the coordinated development of border regions especially since most border regions are underdeveloped and are 'exposed to the highest social stress and strain' (GHR 1997 p.130). It will be necessary to open new border crossings and improve existing facilities through better infrastructure, including information and Internet access. Also border areas need development through joint ventures, industrial parks and entrepreneurial zones, with attention to environment protection and control of illegal activities. Romania also sees CBC an element in regional development policy and encourages local actors to get involved in projects to meet local community needs and national development plan priorities. Thus Romania's South Region sees logic in CBC with Bulgaria launched from Călăraşi and Giurgiu (both centres of LFAs) while the North East region highlights CBC with Moldova and Ukraine; noting the importance of a new road from Dorohoi aiming at the Herta area of Ukraine where there is a Romanian minority (see below).

CBC Studies: Frontiers on the Edge of EU-15 with Significant Industrial Development

There has been much activity on the EU borders through FDI to exploit the 'wages precipice' through labour-intensive industry (Domański 2000). Writing on the Vimperk area of Czech Republic with local industry restricted largely to wood processing, Novotna (2002 p.51) is quite unequivocal is referring to the growth of branch plants whereby 'Bavarian businessmen use our cheap labour force'. German investors have not diverted capital from southern Europe to ECE where their interest lies in market penetration, but they are using border locations to reduce costs while retaining close organisational control through cross-border joint ventures. Here they also have informational advantages, cultural affinities and the support of both government and banking the systems: expansion into ECE 'is a natural move for German firms' compared with Greece and Portugal where investment has been sparse (Flockton 2000 p.124). Much the same applies in Austria and Italy: hence the 400ha cross-border industrial park opened at Szentgotthárd-Heiligenkreuz (1996) with a working commission for unified management including single energy, waste and emergency arrangements (e.g. for coordinated flood protection). Up to 80 firms are expected to provide 2,300 skilled jobs. In the case of Slovakia there is an industrial park at Jarovce-Kitsee and a Science and Technology Park in Bratislava. And for Slovenia there is collaboration along the Graz-Maribor axis impacting on the town of Maribor – location of the

Styrian Technological Park – where restructuring of the engineering and textile industries has generated the highest unemployment in the country.

Austrian assistance has also financed an Information and Business Support Centre in Kranj and regeneration at Jesenice. Reference should also be made to a tourist border zone project – which includes interest in special production activities like fruits, rare vegetables and cheese/sheep products – while the agricultural sector enters the picture through cooperation over viticulture. Environmental actions focus on the Lower Morava as a model area for CBC between Slovakia, Austria and the Czech Republic, with potential for tourism (especially agrotourism), agriculture and small industries (Seffer et al. 1999). Special attention is being given to the Morava meadows and NGOs from the three countries – including Austria's Distelverein (Union for the Morava River), Veronica (Czech Republic) and Daphne (Slovakia) – along with the government of Lower Austria and the environment ministries of the three countries, have been encouraged by WWF to conserve some 44sq.kms of Morava floodplain meadow where the rich flora provides a habitat for many rare and endangered bird species. Meanwhile. Italian involvement in Slovenian agriculture started with a pilot project (1997) by the Friuli-Venezia Giulia Agricultural Development Agency in respect of a 'Centro Pilota per la Vit i Vinicoltura' at Gorizia, aiming at enhanced economic integration through CBC in the context of Italian-Slovenian Interreg. The centre for experimentation is linked with a state research institute at Nova Gorica in Slovenia ('Knetijsko Venerinarsky Zavod' – receiving PHARE CBC funding) and Ljubljana University, with the aim of improving wine quality and developing aromatic characteristics. Environmental action can be seen within the Alps-Adria organisation with respect to the Timavo river which carries polluted water from Italy to Slovenia.

Greek investment across the border has the additional stimulus of a Greek population in the Albanian border region: thus ethnic minorities may be seen as 'assets that increase cultural diversity and bring nations closer to a truly international society' (Petrakos 1996 p.261). There is an emphasis on business networks which may emerge through the encouragement of SMEs by local CCIs, while Bulgaria and Greece have researched selected industries and Greece and Macedonia have organised trade fairs to attract Greek investors and proposed free zones in Bitola and Gevgelija (also Skopje). However political stability is a precondition for closer economic and business ties. In the Bulgaria-Greece frontier area a translation centre was established to facilitate contacts and a bilingual guide published. EU support for projects in both the Albanian and Bulgarian border regions is focused on agriculture, handicrafts, renewable energy and tourism (note 'Egnatia Tourism' for the Albanian-Greek border with attractive scenery and cultural heritage – originating in the work of the Ioannina-based Egnatia-Epirus Foundation during 1991–4). There is also an emphasis on business networks which may emerge through the encouragement of SMEs by local CCIs.

Germany's Border with Czech Republic and Poland

The most detailed study has been made of Germany's borders with ECE: Czech Republic and Poland – especially the latter which could become a central hub for much of the future EU (Barjak and Heimpold 2000; Kratke 1999). For Germany, Czech Republic and Poland are tied up with the 'East-Central European boomerang' – an axis of intensive transformation processes – with implications for the peripheralisation of remoter border regions. FDI in Czech Republic has been encouraged by free trade zones in Cheb and Ostrava, while centres of innovation (Economic and Technology Parks) have been established in České Valenice and Cheb and university faculties have been set up in some border towns (České Budějovice, Cheb, Frýdek Místek, Jindřichův Hradec and Ústí nad Labem). Tax holidays and other financial incentives are available for selected private firms and there are also grants to improve technical infrastructure. Meanwhile, locations in Poland have been earmarked where there is state-owned land and an adequate infrastructure e.g. close to Zielona Góra airport as well as railway junctions (Zbąszynek) and former Soviet airfields in the area where hangars, repair shops and garages are available (Szprotawa and Żagań). The Poles have the opportunity of developing land belonging to the Agricultural Property Agency at Porajow where a few hundred meters of Polish territory intervenes between the borders of Germany and the Czech Republic on the edge of Zittau.

'Cheap manufacturing in the border area can only be an initial phase on the way of Polish enterprises towards long-term growth of their technological competence and an improvement in the quality of cooperation' (Stryjakiewicz and Kaczmarek 2000 p.51). Crude exploitation of cheap labour is no help to Germany since it deepens the post-unification collapse of her eastern industry and creates further unemployment. Fortunately there are also increasing numbers of joint ventures and successful collaboration over textiles after the German industry was much reduced in Guben after the war despite a new chemical fibre plant built in the GDR era (subsequently modernised by West German investors with the workforce cut 9,000 to 800!). CCIs in Cottbus, Teplice and Zielona Góra are improving the base for trade, with most interest in metals, electrotechnics and foodstuffs. Meanwhile the ERs have played their part; being in the best position to create communication/cooperation structures between the two sides. Pomerania is encouraging the transfer of technical expertise and supporting SMEs – with a focus on electronics and engineering – through innovation centres; while Elbe-Labe (ELER) offers a bilingual information service for SMEs; and Spree-Neisse-Bóbr (SNBER) provides training facilities in Germany for Polish apprentices. Kratke (1996, 1998) reports genuine cooperation among firms in the region – e.g. the production of plastic windows through joint ventures between Stahlleichtbau of Frankfurt/Oder and Stilon of Gorzów – in contrast to outside firms that cross the wages precipice to offer low wage work on goods for export. This case demonstrates the need for regional development based on quality production and networking extending to both sides – not the cheap labour syndrome. However a new stage of transborder relationships involving free flows of capital, technology, information and people – forming a more integrated border region – is still largely a matter for the future. Areas like Liberec remain a testing ground for CBC. After isolation through closed frontiers with modernisation of the industrial structure further constrained by the policy of expanding the manufacturing base in Slovakia, the reopening of the frontier with Saxony and access to the German market (especially in Berlin) offers the possibility of a better future for light industry (Barlow et al. 1994). And Liberec will have the advantage of adjusting to the West European market without the burden of the communist legacy of heavy industry dominance. If the area (including Jablonec) does gain momentum, attention can shift to the sub-regional scale and the correction of divergent tendencies in fringe areas: positive perhaps in Rumburk-Varnsdorf close to Germany but less favourable in Frydlant and Česká Lipa with the burden of former uranium mining.

There has been an impressive development of commerce – from 'rucksack trade' to foreign trade enterprises and international fairs – which can support half the population in some areas. In 1995–6 the 16 largest bazaars on the Polish border generated 16.4% of official Polish exports to Germany. The largest – Cedynia, Gubin, Kostrzyn, Łeknica, Słubice, Świnoujście and Zgorzelec – averaged almost PZL200mln/yr, thereby reducing local unemployment and stimulating German companies to built cheap hypermarkets in Poland to compensate for lost trade (Wallace 1999). German people shop in Poland since prices are relatively low and the Institute for Market Economy Research calculates that border area trade is worth $6.0bln annually (about a tenth of GDP) and provides thousands of jobs, plus income to the local authorities. At Kostrzyn 1,200 jobs were created in the town's municipal bazaar and the large German clientele (including many from Berlin) patronised the petrol stations and catering establishments as well. as the local market. A lively bazaar trade also developed on the Poland's eastern frontier with Białystok hosting ECE's largest outdoor market in the mid-1990s. 'In many towns and communes with especially advantageous locations, lobbies of small private commercial and service businesses have grown so powerful that they are able to block the plans of large-scale economic projects' aiming at industrial or R&D investments (Stryjakiewicz 1998 p.204). However crime, corruption and mafia influence are also present. Higher tariffs for imported cars as opposed to parts stimulated not only assembly plants in Poland but also small networks handling damaged cars that could be broken down, taken into the country in pieces (using different customs posts) and then reconstituted.

Some innovations may be noted regarding shared use of services. Viadrina European University includes a Polish College and five student hostels in Słubice to complement the main complex in Frankfurt and there are common academic programmes for German and Polish students. There is a common sewage plant on Usedom while Guben-Gubin have twinned and evolved a common concept of urban development with a common sewage plant in Gubin and school project in Guben; also common participation in the World Exhibition 'Expo 2000' and an improved image that produced the 'Euro-Model Town Guben-Gubin Award' at the end of the

1990s. In ELER a cross-border project for joint electricity supply (exchanging at 110Kv) started in 1993 as a result of long term cooperation between Energieversorgung Sachsen Ost AG and EV Děčín. Pomerania ER evolved the idea for cross-border vocational training colleges in respect of car electronics, environment and building to facilitate restructuring. On the Czech-German border a medical centre emerged at Dippoldiswalde (near the Zinnwald-Cinovec border crossing on the E55) in a tourist region offering winter sports. Meanwhile, agriculture features in both ELER and NER through a structural development plan covering Upper Mandau-Spreequell. including links between Děčín Agricultural Chamber and Sächsische Schweiz/Osterzgebirge e.V. regional association. The farming cooperative 'Agrargenossenschaft Sächsische Schwiez' wants to build up a herd of ewes in Saupsdorf and strengthen cooperation with the Czechs on the restructuring of farms and cooperatives and landscape conservation methods. Some 350 Merino sheep are now being grazed on 80ha in a remote part of the national park and 240ha of a surrounding conservation area: commercialisation of the lamb's meat by the cooperative will follow. Students from Děčín will have hands-on experience at Saupsdorf and herds of up to 50 sheep will be grazed on the Czech side of the border. In SNBER a feasibility study has revealed that a rape seed oil plant in Guben could draw part of its supply from Poland with the waste returning to Poland for pig fodder (especially if better road links could be provided).

Pressing environment issues arise in the Czech-German-Polish border area known as the 'Black Triangle' through pollution by heavy metals as well as advanced soil acidification, retarded nutrient cycling and a major decline in biodiversity linked with atmospheric pollution and spruce monocultures. There is a need to modify the local economy with clean technology, alternative sources of energy, tourism, agriculture and forestry to provide the base for sustainable development in protected areas like the Karkonosze (Raj and Petrikova 2001). Meanwhile ELER is active over pollution control, nature protection and tourism development with a planned environmental study of Osterzgebirge to evaluate and protect the border area. A cross-border agency is envisaged for nature protection envisaged with input from regional trusts while both Czech Republic and Germany have identified strong development opportunities in tourism and leisure facilities (also food processing) and related environmental protection. Tourism also has a cultural base evident in the link between the open air Lusatian Museum at Buszyny near Łeknica, with its programme of special events, and a large agrotouristical farm. Indeed after experiencing some difficulty over integration with communism's border of 'eternal friendship' in the 1980s, the Lusatian Sorbs are building something of a capital in Żary where the monthly 'Przeglad Wschodnio-Luzycki' (East Lusatian Review) is published. Lusatian consciousness is also fostered by the Polish Society for Touring and Country Studies (which in Zgorzelec has taken the name 'Lusatian Land') and the Catholic Deanery of Zgorzelec organises Easter or Corpus Christi celebrations together with Catholics throughout Upper Lusatia.

In some respects the German-Polish border and its ERs is a genuine showpiece for CBC after the ups and downs of the communist era. There is some highly effective coordination e.g. in the 'small triangle' Zittau-Bogatynia-Hrádek n.N. Reference should be made to the 1991 Görlitz-Zgorzelec Cooperation Contract with its task forces to deal with the economy, construction and town planning (covering shared Internet presentation and transborder passenger transport), public health, social affairs and environment. Innovations include a Neisse river cleanup; emergency medical cover and work on drug addiction; culture, tourism and sport; and security through border patrols and camera surveillance in public squares. Adoption of the name 'European City of Görlitz/Zgorzelec' in 1998 – with a total population of 100,000 (62,000 in Görlitz and 37,000 in Zgorzelec) – is an indication of good progress towards integration. The zone has also been used to settle poor Carpathian Roma living on social benefits, begging and petty crime; also a small Greek community: the remnants of some 12,000 Greeks and Macedonians who arrived during 1948–50 as refugee communist insurgents (including 3,000 orphans) and remained until the 1970s when the Greek dictatorship fell. Forming almost two-thirds of Zgorzelec's population in the 1950s, most worked on state farms in Bogatynia but could not be integrated into the villages because they were both irreligious and over-enthusiastic about communism! There are also small communities of Jews, Slovaks – 'Čadca Highlanders' who arrived in 1946–7 – and Germans who worked in the lignite mines and power stations but lacked their own associations until after 1990 when the magazine 'Sorauer Heimatblatt' appeared. However, wider potential networks like Bautzen-Görlitz-Zittau-Jelenia Góra-Liberec remain poorly developed.

CBC Studies: Other Frontiers, with Limited Industry but Environmental and Tourism Potential

Here the emphasis is rather different and the strength of environmental interests creates an unfortunate impression that industry is a problem when in fact it is desperately needed. Environmental concerns arise out of pollution (already referred to in the case of the Black Triangle) evident in the rivers entering Hungary from the north: the Bodrog, Hernád and Sajó (linked with industries in Slovakia) but above all the Tisza threatened by toxic waste (as noted below). In the past Hungary has been interested in joint hydropower schemes but such is the opposition now maintained by environmental groups that these have become bones of contention – especially the Gabčikovo-Nagymaros programme agreed with Czechoslovakia in 1977. When Hungary's Antall government tried to withdraw in 1989, work resumed unilaterally on the Slovak side at Gabčikovo since water could be diverted at Cunovo, above the point where the river became the boundary between the two states (Fitzmaurice 1995). Since Hungary's withdrawal was deemed illegal by the International Court the remainder of the project (Nagymaros in Hungary) is still technically 'on' although there is still no consensus in the country in favour of resumption. Meanwhile, Hungary is also upset by another project on its border with Croatia. And although opposition is not as strong as in Austria, there is concern over the continuation of the nuclear power problem in Slovakia which resulted in the opening of the first reactor at Mochovce near Levice in 1998 (Nagy 2000). However there is a more positive basis for cooperation over the management of protected areas: the Danube-Drava National Park (with Croatia) and the Aggtelek karst region with Slovakia. Hungary also collaborates with Slovakia over nature heritage and conservation in the Danube-Ipoly area (including Naszaly mountain near Vác with a wide range of habitats).

In 1990, A. Langer wrote on 'Ecological Bricks for our Common House in Europe' and produced a list of frontier areas where action was merited. No integrated programme has emerged from this initiative, but several of the areas are now being more adequately managed. In addition, the Stability Pact – bringing together the SEECs in the light of recent inter-ethnic tension – has given rise to a Regional Environmental Reconstruction Programme (REReP) which has been taking shape since the Kosovo War. It includes a Swiss-funded programme, for implementation by the Regional Environment Centre, the World Conservation Union (IUCN) and the World Wildlife Fund for Nature (WWF) and other organisations with respect to five biodiversity-rich transfrontier areas requiring cooperative management. The shortlisted areas are the Belsitza Mountains (Bulgaria, Greece and Macedonia), Lake Shkodra (Albania and Montenegro), the Neretva River (Bosnia and Hercegovina and Croatia), Šar Planina (Macedonia and Kosovo) and Stara Planina (Bulgaria and Serbia). Prespa Park is already an important international project involving Albania, Greece and Macedonia.

The White Carpathians

Major concerns arose in the Carpathians where there has been much NGO activity since the start of the transition in a region which embraces six countries and has been badly affected by the closed frontiers of communism: not because of excessive levels of development but rather the creation of artificial barriers preventing a unified view over the conservation of the ecosystem. This has its origins in the White Carpathians (Bilé Karpaty/Biele Karpaty) on the Czech-Slovak border, recognised as an area of ecological deterioration – with additional socioeconomic problems of ageing and out-migration causing a breakdown of the sustainable community-environment relationship based on careful agricultural landuse. With the demise of the Czechoslovak federation, the area – once centrally located and mutually developed – became a frontier region at the very moment of transfer to a competitive market economy with administrative decentralisation and an expansion of private ownership. Constraints arose through the Protected Landscape Area (PLA) designation and an inadequate rural infrastructure arising from the assumption that communities would die out in favour of urban life based on industries which developed first in the inter-war years through Bat'a in Zlín and were extended under communism through the spread of engineering and electronics including the 'Tesla' branch plant at Rožnov pod Radhoštem, the 'Igla' needle factory at Valašske Klobouky and 'Tryodyn' at Brumov-Bylnice (Figure 8.2).

Figure 8.2   The White Carpathian Euroregion

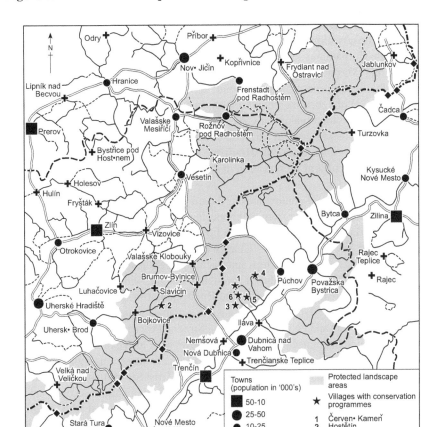

There are two interlocking strands to the programme. The first concerns the deterioration of the mountain meadows which require regular grazing in order to retain their distinctive flora and avoid a trend towards scrubland. The PLA has been working with farmers to retain an appropriate level of agricultural activity and the White Carpathians were one area examined by an EU-funded research programme on 'Central and Eastern Europe Sustainable Agriculture' to evaluate both the incentive scheme – which offers subsidies for agricultural practices that will avoid abandonment and thereby safeguard wildlife – and the work of the agriculture and environment ministries which is not always harmonious (Krumalova and Ratinger 2002). There has been input from the White Carpathian branch of Czech Union of Nature Conservationists based at Veselí n.Moravou – interested in meadow management, grassland restoration and fruit conservation – which has been instrumental in gaining the support of the PLA administration and the town council for an education and information centre. Košenka – a member of the Czech Union for Nature Conservation – has also emerged to care for rare orchids in the meadows of the White Carpathians around its base at Valašske Klobouky, while the regional organisation 'Veronica' in Brno has also contributed its own expertise on sustainable development. The second strand concerns the protection of communities. With encouragement of an inspired environment minister (Josef Vavroušek) before the breakup of Czechoslovakia a non-profit-making Society for Sustainable Living was created and it now has branches in the region at Trenčín on the Slovak side and Uherské Hradiště in Moravia promoting sustainable development and cultural preservation in the area with the help of a 'Bilé Karpaty' magazine.

Through international conferences in the mid-1990s, considerable support came from abroad including some from neighbouring countries (e.g. Polish organisations in Bielsko-Biała and Katowice) but also further afield: there were visits by expert teams (e.g. two UK organisations: 'Charity Know How' and 'Ecotourism Limited') while the Landscape Stewardship approach through responsible landownership (Serafin et al. 1998) was introduced through exchange with the Quebec-Labrador Foundation at the Atlantic Centre for the Environment at Ipswich (Maine), USA. Capacity was enhanced by further domestic organisations: notably the regional association of towns and communities of the Stredné Považie region while the other groups mentioned in connection with the meadows have also extended their concern to the community dimension. Financial support came from various sources in Czech Republic, Slovakia and further afield including the Trust for Mutual Understanding, the Rockerfeller Brothers Fund, the umbrella organisation known as the Environmental Partnership for Central Europe (EPCE) and the Regional Environment Centre for CEE based in Hungary. Projects were undertaken which (along with the growth of civic and community associations) contributed momentum for the present ER. In Slovakia, concern to protect communities and control the blight of new weekend cottages inherited from the communist period led to an international conference on rural landscape renewal at Nová Bošáca in 1993. This meeting agreed on the need for practical implementation of a programme for sustainable living in areas with active local interest and a group emerged within the PLA comprising Červeny Kameň, Krivoklát, Lednica, Mikučovce and Vršatské Podhradie (Považská Bystrica District) and Dolná Súča, Horná Súča, Horné Srnie and Trenčianska Závada (Trenčín District). Work has been going on to build a development model of ecological, cultural and economic synthesis – with full local participation – including sustainable use of natural resources; tourism as a supplementary source of finance; health recovery; solar energy – as well as power based on wood, straw and wood (including bark and sawdust).

With the five Považská Bystrica District communities as a model microregion, expert teams explored 15 main issues with local activists and a local regional development association ('Resources of Biele Karpaty') emerged with the help of communities, entrepreneurs and schools – and the cooperation with Pruské Vocational Agricultural School where the association is based. Activity has included a regional conception of tourism development (with agrotourism promoted through the international ECEAT/Eurogîtes catalogue); promotion of handicrafts and folklore along with photodocumentation of regional architecture linked with housing upgrades; community renewal including a telehouse pilot project on the Austrian 'Waldviertel' model – albeit dependent on digital telecommunications – and conversion of old fire station buildings; fruit growing as a strategy for alternative agriculture; local services including education, health and waste management; energy efficiency; and the conservation of mineral springs and water courses. On the Czech side the greatest progress has been made as Hostětín through sustainable local services (a reed bed sewage system, a biomass heating system and extensive use of solar panels) and maintenance of local orchards in association with a small factory in the village producing juice, jam, alcohol and dried fruits – but noticeably without any diversification into rural tourism to extend the range of accommodation beyond the nearby spa and treatment centre of Luhačovice. Although only a single village is involved here, the scale of innovation is quite remarkable and it has a powerful demonstration role. It is largely thanks to Veronica and EPCE – that it was possible to establish 'a unique coalition of civic groups, local leaders, state officials, farmers, businessmen and foreign donors to nurture a broad range of small-scale local initiatives that provide a strong impetus for economic development in the Bílé Karpaty' (Klimkiewicz 2002 p.163). However, there remain a good many historical monuments which are afforded very little publicity, like the castles ('hrad') at Brumov-Bylnice and Lukov.

NGO activity has extended to other parts of the region and although the cases are too numerous to mention in detail it is appropriate to mention the groups affiliated to EPCE in Hungary, Poland, Romania and Slovakia; and also the work done by Carpathian Bridge ('Priashev'): an international association of ecological organisations combining the Ukrainian 'Carpathian School' (Lviv) with 'Pcola' from Slovakia (Stará Lubovna) and the Foundation of Support of Ecological Initiatives from Poland (Kraków). There is also a TACIS CBC project (1999–2001) for a Carpathian Transfrontier Environmental Network – covering the Romanian and Ukrainian parts of Bucovina – to preserve biodiversity in active participation with stakeholders, with a pilot conservation project covering landuse management and sustainable ecotourism. Finally reference should be made to the support of the World Bank Global Environmental Facility (GEF) in two important Carpathian projects. First there was a programme (also supported by the USA MacArthur Foundation) for cooperation between Poland, Slovakia and Ukraine for conservation within in a trilateral Eastern Carpathian International Biosphere Reserve of 164,000ha which includes Europe's largest stand of beech forest as well as distinctive mountain meadow country called 'poloniny'. Second, there is GEF-supported work in the Piatra Craiului Mountains of Romania on large carnivore conservation in which the Munich Wildlife Society was much involved along with the Romanian State Forestry organisation 'Romsilva'. Both projects have demonstrated the need to look at the Carpathians comprehensively as a major transfrontier ecoregion to conserve rare birds, plants and animals through a network of protected areas. At the same time the community aspects cannot be overlooked since poor people must be convinced that conservation is in their interest. Both projects have demonstrated the importance of rural tourism. Slee (1999) has assessed the potential for sustainable tourism in the Ukrainian Carpathians following a CoE mission during 1997–9. This is an area dependent on small farms, with a distinctive cultural landscape and good environment (low fertiliser and pesticide application and lack of Chernobyl-related damage) but lacking much of the industrial employment provided under communism. He comments that 'the furtherance of transfrontier tourism makes both political sense in binding countries to a shared vision of sustainable tourism and economic sense in that the combined product of the different countries provides an additional attraction to visitors' (Ibid p.3).

These initiatives were all helpful in the run-up the 'Kraków Declaration' – when conservationists backed by the Polish Academy of Sciences emphasised the opportunity to safeguard natural heritage in the ECE mountains (Nowicki 1998) – which in turn gave rise to a project for the whole Carpathian Ecoregion under WWF auspices: combining a protected area network (reconcilable with the wider EU 'Natura 2000' programme) with sustainable development negotiated in each locality between NGOs and stakeholders. The initiative was endorsed by the Carpathian Convention was approved by heads of state in Bucharest in April, 2001 and made further progress at a ministerial conference in Kiev in May, 2003. Implementation will take time, for capacity varies considerably across the region, but sustainability is relatively well understood in the Polish-Slovak border region which is considered to merit protection in its entirety (Wieckowski 1999). Joint management extends over several protected areas (most notably the Tatra National Park, which is also an ER) and rural tourism is well established on the basis of linear routes for rambling and an extension of rafting (popular on the Dunajec since the 1830s and now envisaged on the Poprad). There is also support through improved rail and bus routes (e.g. Krosno-Prešov, Barwinek-Dukla Pass-Vysny Komarnik and Zakopane-Poprad) while new frontier posts are exploiting new potentials such as pilgrimage tourism at Uście Gorlickie (a major centre of Poland's Ruthenian or Lemko population noted for its Orthodox churches and folk events) on the Novy Sącz-Bardejov route (Birek 1995).

Industry plays a relatively minor role when the CBC rationale is dominated by the environmentalists for whom sustainable development is seen primarily in terms of organic agriculture, fish farming and rural tourism, along with services and cultural-sporting programmes of the kind reported on the Bulgarian-Romanian border as a result of agreements between local authorities, especially in the Giurgiu-Ruse area. Indeed manufacturing is often seen rather narrowly as an unwelcome legacy through pollution, as in the case of the Horná Orava PLA, when conditions are complicated by inadequate provision for sewage. Here in the Beskid ER the development agenda is strongly influenced by environmental and tourism interests.

However, there are cases of modern high-tech industry as in the case of ABB at Elbląg (Poland) in ERB producing the newest generation of turbines in the firm's network and attracting more investments of this kind should be a more prominent aspect of CBC strategies. For it is unlikely that tourism will be sufficient to employ large numbers however much the local attractions are boosted by improved information and interpretation, more waymarked itineraries for cycling and walking – and better frontier roads of the kind advocated for the Polish-Slovak frontier north of Žilina where the 'Tatra' branch factory at Čadca needs to be extended by a further complement of light industry (Lauko 1999). Meanwhile local resources – especially timber – should support further manufacturing according to the community woodland model (Ioras et al. 2001). Hence the scope for FDI to seek out cheap labour for light industry higher using advanced technology, perhaps with incentive provided by major road improvements over the next decade (Grimm 1999). Certainly for the moment cross-border traffic is often light away from the main European transport axes and provision is often geared to local requirements: hence new crossing points on the Romanian-Ukrainian border for Ukrainians living in Romania at Poenile de sub Munte in Maramureş and for Romanians in Herta area of Ukraine (close to the town of Dorohoi in territory which was always Moldavian until it was demanded by the Soviets in 1940, along with Bessarabia and northern Bucovina). Local arrangements are also needed at Lendava on the Croatian-Slovenian border to allow farmers access to their land as the shifting course of the river (the frontier marker) has transferred land from one country to another (Borut 1997).

The Hungarian-Romanian Border

Growth of activity very largely follows the historic agreement between the two countries in 1996 and the policies of the centre-right government in Romania during 1996–2000 when the European agenda was belatedly embraced. This change of policy came after six years of extreme caution over when academic supporters of Romania's first post-communist governments criticised ERs motivated by xenophobia. Thus Deica and Alexandrescu (1995 p.10) dismissed the CER as serving only Hungarian self-interest through 'a dominant political substratum devoid of objective economic reasons'. However, after 1996 CBC offices were established in Arad and Békéscsaba and twinning links developed to the extent that 12 Hungarian municipalities were involved in twinning with Romania (including Debrecen with Oradea), 11 with Yugoslavia and three with both. At a higher level cross-border links have been formalised through the CER which has gradually attracted support from Romania, especially since 1996 and the ER Danube-Criş-Mureş-Tisa (DCMTER) of 1998, which includes the Serbian part of Banat as well as Hungarian and Romanian territory (Nagy 2001).

The profile is much influenced by the environmental agenda (also to some extent by economic and commercial relations) which is dominated by the rivers which comprise the Tisza/Tisa and its various Romanian tributaries including the

Szamos/Someş and the various Körös/Criş rivers. Given the danger of flooding in eastern Hungary an international monitoring system is needed while the risk of pollution (demonstrated in 2000 by the release of cyanide from a lagoon managed by gold processor 'Aurul' in Baia Mare) gave rise to an emergency programme including research to identify all areas of risk. Fortunately, the Danube River Protection Convention came into effect in 1998 and an International Commission for the Protection of the Danube River (ICPDR) was set up the following year to ensure water protection through monitoring laboratories and an accident emergency warning system. Hence the pollution crisis in 2000 produced a rapid response through a Baia Mare Task Force involving all the countries with an interest in the Tisza basin, but particularly Hungary and Romania. Studies for the ICPDR revealed 144 risk sites (including 42 high risk – mainly concerned with mining and energy). There is considerable clustering with nine high risk states close to Baia Mare while five high-risk and 10 low-risk sites are clustered around Tisaújváros and a further two high-risk and 28 low-risk sites form a loose cluster covering the Debrecen-Szolnok axis. The Carei/Nagykaroly area has emerged as a joint Hungarian-Romanian flood-retention zone (with socioeconomic dimensions) where cross-border restoration will be assisted by WWF. This arises from the designation of the Szamos as a pilot river basin for testing EU Water Framework Directive guidelines and it will give rise to a nature conservation programme close to the border in the Baia Mare area of Maramureş.

As regards business opportunities, Bihor and Hajdú-Bihar CCIs cooperated as early as 1992 for mutual exchange of information and the organisation of exhibitions and fairs which have helped some Hungarian enterprises to extend their operations into Romania. Industry parks are now becoming a commonplace with provision in Arad and Timişoara on the Romanian side and in Békéscsaba in Hungary – along with business zones in Bihar (Berettyóújfalu-Biharkeresztes) and Záhony – and another at Csenger to follow, with a border crossing at Letavertes (Baranyi et al. 1999). Local border transport gives scope for cross-border labour flows to industries in the main towns, but the volume is limited since most industry is incapable of expansion due to obsolescence and productivity as well as poor infrastructure. So to maximise pluriactivity it is recommended that communities should build on former cooperative activities in food processing (with development agencies to organise cooperation with farmers), agricultural machinery, basket making and construction work. Small FDI projects have started by people with personal contacts with the border area including Battonya, Elek, Deszk, Ruzsa (Pal and Nagy 1998, 1999). However the pace of industrial growth is slow, for spending power is low, while catering for export markets requires computers, web sites and Internet access which may not be possible due to lack of digital telephone systems. There is also some caution due to the imposition of the Schengen frontier when Hungary joins the EU, although Romanian membership is expected to follow after 2006. The outstanding success of Timişoara has prompted reference to a 'Timişoara model' which involves a cosmopolitan mentality (with peaceful cohabitation in a multinational environment) and an attractive city with low land prices. The privileged location of the city in the west of the country has been cleverly exploited by a policy aimed at stimulating regional collaboration. Foreign companies are now numerous – e.g. 'Continental' (Germany) and 'Solectron' (USA) – while Italian capital is involved in banking and in projects for a technology park and the 'Interporto' international road transport link.

The development of business is dependent on the transport system and unfortunately the improvement programme for Hungary's single carriageway Highway 4 – Budapest-Debrecen, which gives access to Oradea and Satu Mare – has been rather slow, with the Szolnok bypass followed only recently by others for Törökszentmiklós and currently Abony. Other routes to Romania have developed from the Budapest-Szeged motorway project with a partially-modernised single carriageway branch road from Kecskemét to Békéscsaba and Arad and the link from Szeged to Arad and Timişoara. The latter city was formerly accessible only through Arad – and attracted the explanation that Arad's political clout ensured that its rival would be relatively isolated (Batt 2002b) – but since 2002 it has been possible to travel direct from Szeged to Timişoara via Sânnicolaul Mare with the opening – at Cenad/Kiszombor – of one of a number of new crossing points to appear since 1996, with positive demographic consequences for the rural communities involved (Ancsin 1999). The fortunes of Sânnicolau Mare have been transformed quite remarkably by the new border crossing and the road improvements particularly on the Romanian side. Once totally isolated within closed borders 62kms from Timişoara, this small town of 13,000 has suddenly become one of the most accessible places in the country and several new foreign investments are helping to reduce the high unemployment rate. There remains the possibility that the direct railway between Szeged and Timişoara will be restored and consequently avoid the detour currently necessary via Békéscsaba and Arad and work to the advantage of another disadvantaged border town: Jimbolia. And further reconstruction of the once-unified Banat rail network could reopen links with Baziaş and Zrenjanin. In the north, a new road bridge at Sighet will complement the rail crossings at Câmpulung pe Tisa and have a positive effect on cohesion on the Romanian-Ukrainian border.

Other aspects of business concern agriculture and especially the Csenger animal husbandry project which followed an initiative by Romanian community leaders wishing to support cattle farmers on both sides of the frontier through increasing the output of high quality agricultural produce and creating more jobs in the process. Ultimately there will be a model farm with dairy and meat processing – which can provide technical and professional agricultural extension services (supported by the British Know-How Fund). Csenger agrotechnical innovation centre is also assisting the border region while agreements have been reached to refine sugarbeet from Békés at Arad's 'Zaharul' factory and to establish joint cultivation of vegetables in the cross-border Méhkerek-Salonta area. Meanwhile, tourism has links with the rivers and the control of pollution hazards and there is a pilgrimage dimension through the importance of Teremia Mare for Catholics. The Assembly of Border Localities in the Oradea area is working on ways in which tourist areas in Romania can serve the people living in Hungary and it seems that Budureasa commune in the Apuseni Mountains could develop its capacities (Ilies & Dehoorne 2002). Tourism in the Maramureş mountains is also becoming more accessible through the rebuilding of the Tisa bridge and the improvement of access to an area of great value for biodiversity and ethnography. CBC may also be discussed with regard to supermarket shopping in Hungarian towns like Békéscsaba, Debrecen and Szeged. And in education there are university links involving Szeged, Timişoara and Novi Sad while Csenger Cross-Border Vocational Training Project has arisen out of an agreement between schools in Csenger and Satu Mare.

Concluding Discussion

It is a contention of this chapter that CBC across broad frontier regions constitutes an important aspect of regional policy because the problems inherited from communism have been substantially neutralised and FDI is making a powerful impact, along with other funding geared to infrastructure and social development. However, the impact is highly variable with frontier regions providing a microcosm of ECE as a whole. It remains a matter of some doubt whether all parts of the region will find a rewarding role in the global economy after a long history of backwardness (a situation which communist central planning sought to alleviate as an autonomous system for the Eurasian periphery). While it seems likely that more sophisticated regional policy will focus on the remaining pockets of poverty and deprivation after EU enlargement, there must clearly be more awareness of the perceived constraints in the minds of investors and better promotion of the weaker regions to correct the excessively negative evaluations. By making a virtue out of peripherality, ways must be found to build on what has already been achieved to improve cohesion with a further widening of horizons to attract more FDI to border regions. Reconstituted civil societies are showing some skill in packaging the 'local' to catch the 'global' but the harvest is still awaited in much of the region. Stronger organisations are needed with wider interpretation of the sustainability principle.

A Balance Sheet for CBC

There have undoubtedly been opportunities to develop synergies (Gruchmann and Walk 1996; Hajdu 1997) in view of the handicaps of the communist era and the funding geared specifically to CBC projects, while the EU approach to regional planning has given the cross-border dimension an added significance – though it has to be said that actions in frontier regions do not depend on formal structures and programmes and (for example) much of the investment in new business arises from individual perception of opportunity in providing services along main lines of communication or exploiting cheap labour to produce goods for export. Equally however, it too simple to dismiss CBC programmes in general and ERs in particular as a 'fad' inspired by the EU and international financial institutions or to deride anomalies in the boundaries of these regions, influenced as they are by the decisions of individual municipalities (which can sometimes result in simultaneous membership of two or even three ERs – such is the complexity on Slovakia's borders in particular). The enormous variations in the size of ERs – when the CER, Bug ER and DCMTER (150.0, 82.2 and 77.4th.sq.kms respectively) are compared with just 1.53sq.kms for Těšínské-Slezsko and 1.47 for Silesia – has drawn ironic comment in the academic press although more constructive comment would value the development of organisations dedicated to extending opportunity (Deica and Alexandrescu 1999). The boundaries of cross-border regions – whether ERs with NUTS III status or more informal units for short-term CBC programmes – do not derive from a masterplan. Rather they reflect local perceptions over a sense of community that extends across a frontier, highlighted by the intimacy of inter-ethnic relations and the growth of organisations – with their lobbies and initiatives – keen to reach out to both sides of a boundary line. There is big difference between the situation on Hungary's Trianon frontiers (where no border existed pre-1918) and elsewhere in the former Habsburg Empire, compared with the frontier (say) between Bulgaria and Romania which constitutes a long-standing ethnic divide reinforced by the Danube river.

ERs in particular have given rise to soul-searching over the extent of decentralisation that is desirable. The centre may reluctant to allow municipalities to get involved in external relations, for some governments 'are afraid of an economic or cultural annexation of their own territories' (Suliborski 1995 p.34) – especially where local administrations might be taken over by ethnic minority interests – and Wecławowicz (1996 p.167) has highlighted the specific case of Poland's western border where 'the great wealth and economic disparity between Germany ... and the relatively underdeveloped former Regained Territories raises the threat of losing economic and political control to Germany'. However, this cloistered attitude now appears to be largely a thing of the past given the new Europe grounded in the principle of subsidiarity. For sovereignty is in no way affected and supporters of CBC feel that alarmist views flatter the ER concept with a coherence that goes way beyond reality. And if some border areas appear privileged through a surge in business and employment this is another inevitable characteristic of the ECE regional mosaic. Indeed the sovereignty issue imposes constraints because PHARE CBC projects tend to highlight one particular side of a border (instead of total integration) given the allocation of funds to particular countries, while in addition, Brown and Kennard (2001 p.22) point out that schemes involving cooperation between two states have to be discussed through the management structure of each of the countries involved; while there were also tensions within the EU when PHARE/TACIS was overseen by the External Relations Directorate while Interreg for the pre-2004 member states was the concern of the Regional Policy and Cohesion Directorate (although the latest phase of CBC – Interreg III – sought to minimise these problems). Further problems arise from the federal system of government in Germany (Kennard 1997, 2000b). Hence the DCMTER is not alone in finding that truly joint implementation has not been possible despite best efforts over the coordinated and synergistic use of funds.

There is uncertainty given the Schengen syndrome and the need for effective frontier administration which must compromise the spirit of openness inherent in CBC. EU border regulations are obliging candidate countries to introduce tighter controls on their eastern frontiers when their domestic political interests make for closer contact with former Soviet states. At one stage Slovakia's ambivalence towards the EU seemed to raise problems that would arise in the event of non-simultaneous entry for Czech Republic and Slovakia, but governments in Slovakia since 1999 have prioritised EU membership and both countries joined in 2004. But deep divisions now threaten the CER and other ERs involving CIS territory: concessions over visa formalities can ease these problems, but if ethnic minorities linked with EU member have privileged status this will inevitably be seen as divisive by other CIS citizens. On the other hand, Schengen will deliver improvements through elimination of internal borders, although the run-up to enlargement has seen a big increase in movement ('traffic shock') and consequent frontier delays especially for trucks, despite revised work practices. Three vehicle crossing points in Bratislava now seems insufficient at rush hours and summer holidays when there is queuing. There are insufficient crossing points on the Czech Republic-Poland border. Questions are also being asked about the costs of increased transit traffic in terms of environment and crime: motorway projects in sensitive areas are attracting opposition. Conventional roads are becoming congested while toll motorways (as in Hungary) may have little traffic because they are too expensive for many users. Motorway projects in sensitive areas are attracting opposition: notably the Prague-Dresden motorway in the Šumava where tunnelling is proposed as one option and abandonment in favour of rail transport as another. There is also a crime problem concerned mainly with drugs and refugee smuggling, including some young men arriving in Szeged to avoid military service in Romanian and Serbia (Pal 2000). Car crime is a problem while UN sanctions on Serbia gave rise to a substantial clandestine trade in petrol across the Danube which brought quick profits to the Romanian fishermen living in the Iron Gates.

Community Attitudes Towards CBC

There is certainly much variation in the level of interest in CBC and on any particular border a range of perceptions can usually be noted. A sense of community may be weak where borders constitute long-standing cultural watersheds and while business interests may dispassionately exploit new horizons, trust may be compromised by a troubled past or simply from mutual suspicions. Tension arising from a contested frontier in Spišz arises from long-standing division between Poland and Slovakia, apart from World War Two when Polish Spišz became part of Slovakia (Kaluski 1999) and confidence is also undermined by tensions on the Belarus-Poland border. In general Hungary has good reason to support CBC as fully as possible in the interest of her minorities in neighbouring countries who have come to see the Trianon frontiers as less of an outrage given the recent increase in permeability, yet self-interest can give rise to cynicism among the other national groups, while a large Roma minority may also generate prejudice and may well be perceived as a weakness by potential investors. Again, while Slovaks want to participate in Austria's prosperity, the Austrians have reservations: so for the time being 'the Austrian population [feeling marginalised in the east of the country as if it were still in the Russian Zone] utter fears, whereas their Slovakian neighbours mainly express their hopes' (Fridrich 2002 p.101). However experts on both sides are positive because they see the potential of a dynamic Vienna-Bratislava core. And there are not only mental barriers but a difficult financial situation for Slovak municipalities: they can hardly ever complete water/sewage schemes so many Austrian mayors see no point keeping in touch with Slovak neighbours: 'collaboration with a poor neighbour does not pay' (Ibid p.96). On the Austrian-Czech border entrepreneurs think that integration lags behind potential and despite some conservatism most Czechs are keen to retrain and work in Austria. Yet for Austrians the border provides protection for the local labour market 'from unpredictable overborder migration' (Havlicek 2002 p.117. No doubt EU membership will overcome the need for offers of jobs from the Austrian side.

Polish public opinion favours CBC despite little knowledge and fear of neocolonisation. There is good awareness of ER activity in Prenzlau or Pasewalk (Germany) but less so in Poland: a wide belief that the ER is good for cooperation over environment, tourism and culture ('knowing neighbours' but without emotional connections) but only a minority of Germans think that the economic situation is better on account of the ER, for Germans are sensitive that 'Aussländers' will take their jobs (and there is a particular dislike of Romanians and Ukrainians). The situation is not helped by a tendency to identify with a small 'Kreis'-type area rather than the much larger ER. Negative attitudes over employment reflect the postwar decline of textiles in Guben and the social disintegration that gave rise to witch-hunts by xenophobic right-wing youths during 1999–2000: hence a negative image for CBC with local entrepreneurs 'blaming local politicians for not supporting 'their' entrepreneurs and leaving them alone with their problems of survival' (Matthiesen and Bürkner 2001 p.47). For their part, the Polish population in Gubin – the product of postwar repopulation from the east – fears that Germans will return and take over by buying up Polish soil: reservations which reinforce social distance. The tendency to demonise can also be seen in Greek attitudes to the Albanian border, coloured by the flood of illegal immigrants (Green 1997 p.97); while closer integration on the Czech-Polish border (where an economic development organisation was set up in 1993) is allegedly restrained by a fear of competition in tourism, despite the potential efficiency gains from an integration of facilities.

The Business-Industry Question

Border regions need more investment in industry, for the population is considerable in many cases – e.g. in the DCMTER where major urban groupings exist – and there could be demographic increases of the kind advocated by Djurdjev (1999) through the resettlement of refugees from Bosnia and Kosovo. Yet the literature produced frequently fails to emphasise this point. Indeed the strength of the programme based on environmental protection, organic farming, sustainable rural tourism and infrastructure – emerging from SWOT analyses (Zahorakova 2002) – conveys an impression of unreality. Industrial opportunities may be limited for various reasons including technology: along the Austrian-Hungarian border there is a positive attitude to visitors on both sides – and Austrians crossing for shopping and pleasure are complemented by Hungarians travelling for business – but there has been little FDI from Burgenland in Györ-Moson-Sopron because of lower industrial quality standards in Hungary. Further problems arise from the poor local government-business links and inadequate promotion although this may be symptomatic of problems over competency of local authorities: those in Romania are perceived as having lower competency than Bulgarian and Hungarian counterparts and it is the country's eight development regions which seem best able to work with CCIs, NGOs and institutes to build capacity over the promotion of FDI in manufacturing. NGOs also have to work within their limited capabilities and it is probably easier to mobilise rural communities to become involved in tourism without substantial outside assistance than to form industrial promotion groups, for unless their is industry established with entrepreneurs seeking expansion there is little than can be done separate from the activities of national agencies which will tend to work on the basis of international position. Thus sustainable tourism for the Upper Orava based on the local landscape and environment values – along with sport and culture – enjoys the backing of the Polish-Slovakian Chamber of Tourism and the 'Wspolna Orawa' Association in meeting the demand for better information through the Internet and conventional publicity materials and ensuring the protection of culture in the Kysuce region. Meanwhile the legacy of environmentally-burdensome industries – traditionally concerned with engineering, clothing and wood processing – needs to be overcome through more advanced processing to the stage of final production using new environmentally-friendly technology. Yet the Slovak-Hungarian ER profile (Dubcova and Kramarekova 2002 pp.169–269) highlights the push on agriculture, tourism and environment with only rare mentions of industry and the intensity of entrepreneurial relationships is weak. On the Bulgarian-Romanian border cross-border commuting is non-existent because economic weakness in underlined by the lack of transport connections: restricted to one bridge since the ferry services proved to be unviable in the context of high harbour taxes on both sides of the river.

Most serious however is the shortage of funding for radical improvements in infrastructure which can only be achieved slowly, except in the case of major European transport corridors which carry a high priority. The CER suffers from an acute lack of funding, comprising as it does the poorest regions of the countries in question (Batt 2002a). Although virtually the entire territory was once part of the Habsburg Empire, the area is remote from capital cities, little known in the outside world and was subject to highly variable levels of support from the national governments until the changes in Romania (1996) and Slovakia (1999). Moreover, with falling levels of industrial and agricultural activity and a diverse range of legal regulations and administrative procedures, there is the further drawback of limited expertise in management and marketing, a poorly-developed business-financial infrastructure and serious transport defects: hence a low level of FDI and modest private sector involvement. Other border zones continue to be isolated like the Šumava (formerly part of the Iron Curtain) and the Locva Mountains beside the Iron Gates where the tourist potential will remain latent as long as the roads remain unmodernised and the old multi-ethnic community of Romanian shepherds and foresters, Czech cattle farmers, Serb gardeners and German shopkeepers is eroded through out-migration. Furthermore, where cross-border trade reflects changing price differentials between neighbouring countries sudden shifts can have a dramatic effect. In western Poland, Gorzelak (2001 p.325) alludes to reduced German buying at Polish bazaars with the appreciation of Polish currency and equalisation of prices: German demand 'dramatically declined in the second half of the 1990s and nowadays has ceased to be an important factor of local and regional development in western Poland'. Meanwhile in eastern Poland commerce was 'dramatically influenced by the Russian crisis of 1998 and since them has not fully recovered' (Ibid). Thus CBC is likely to remain for some time a domain of highly variable achievement.

References

AEBR/LACE (1997), *Cross-Border and Inter-Regional Cooperation on External Borders of the EU*, Euregio, Gronau.

Altzinger, W. (1997), 'Cross-Border Development at the Vienna-Bratislava Region: A Review', in U. Graute (ed), *Sustainable Development for CEE*, Springer, Berlin, 89–114.

Ancsin, G.S. (1999), 'A Demographic Analysis of Hungarian and Romanian Border Settlements on the Southern Great Plain', in C. Gruia, G. Ianoş, M. Torok-Oance and P. Urdea (eds), *Proceedings of the Regional Conference of Geography: Danube-Criş-Mureş-Tisa Euroregion – Geonomical Space of Sustainable Development*, Editura Mirton, Timişoara, 341–8.

Bachtler, J., Downes, R. and Gorzelak, G. (eds) (2000), *Transition Cohesion and Regional Policy in CEE*, Ashgate, Aldershot.

Baran, V. (ed) (1995), *Boundaries and their Impact on the Territorial Structure of Region and State*, University M. Bel Faculty of Natural Sciences, Banská Bystrica.

Baranyi, B., Balcsok, I., Dancs, L. and Mezo, B. (1999), *Borderland Situation and Peripherality in the Northeastern Part of the Great Hungarian Plain*, Centre for Regional Studies, Pécs.

Barjak, F. and Heimpold, G. (2000), 'Development Problems and Policies at the German Border with Poland: Regional Aspects of Trade and Investment', in M. van der Velde and H. van Houtum (eds), *Borders Regions and People: European Research in Social Science*, Pion, London, 13–31.

Barlow, M., Dostal, P. and Hampl, M. (eds) (1994), *Territory Society and Administration: the Czech Republic and the Industrial Region of Liberec*, University of Amsterdam/Charles University, Amsterdam/Prague.

Batt, J. (2002a), 'Transcarpathia: Peripheral Region at the "Centre of Europe"', in J.Batt and K.Wolczuk (eds), *Region State and Identity in CEE*, Frank Cass, London, 155–77.

Batt, J. (2002b), 'Reinventing Banat', in J. Batt and K. Wolczuk (eds), *Region State and Identity in CEE*, Frank Cass, London, 178–202.

Berenyi, I. (1992), 'The Socio-Economic Transformation and the Consequences of the Liberalisation of Borders in Hungary', in A. Kertesz and Z. Kovacs (eds), *New Perspectives in Hungarian Geography*, Academy of Sciences, Budapest, 143–57.

Bičik, H. and Stepanek, V. (1994), 'Long-Term and Current Tendencies in Land Use: Case Study of Prague's Environs and Czech Sudetenland', *Acta Universitatis Carolinae: Geographica*, 29, 47–66.

Birek, U. (1995), 'The Opening of a Road Crossing as a Factor in the Economic Development of Border Communities using the Example of the Uście Gorlickie Community in the Beskid Niski Mountains', in V. Baran (ed), *Boundaries and their Impact on the Territorial Structure of Region and State*, University M. Bel Faculty of Natural Sciences, Banská Bystrica, 101–4.

Boar, N. (1999), 'Turism transfrontalier maramureşan', *Analele Universităţii din Oradea: Seria Geografie*, 9, 76–81.

Bojar, E. (1996), 'Euroregions in Poland', *Tijdschrift voor Economische en Sociale Geografie*, 87, 442–7.

Borut, B. (1997), 'Croatian Real Estates in Lendava Municipality as a Constituent of the Border Problems', in M. Pak (ed), *Sociogeographical Problems*, Department of Geography University of Ljubljana, Ljubljana, 183–93 (in Slovenian with an English summary).

Brown, C. and Kennard, A. (2001), 'From East to West: Planning in Cross-Border and Transnational Regions, *European Spatial Research and Policy*, 8, 15–28.

Bufon, M. and Minghi, J.V. (2001), 'The Upper Adriatic Borderland: From Conflict to Harmony', *GeoJournal*, 52, 119–27.

Buursink, J. (2001), 'The Binational Reality of Border-Crossing Cities', *GeoJournal* 54, 7–19.

Ciechocinska, M. (1992), 'The Paradox of Reduction in Development in the ECE Fringe Areas', in M.Tikkylainen (ed), *Development Issues and Strategies in the New Europe*, Gower, Aldershot, 189–209.

Corrigan, J., Suli-Zakar, I. and Beres, C. (1997), 'The Carpathian Euroregion: An Example of Cross-Border Cooperation', *European Spatial Policy and Research*, 4, 113–24.

Danko, L. (1996), *Development of Tri-Border Regional Economic Cooperation: Establishment of the Special Entrepreneurial Zone*, Carpathian Border Region EDA, Sátoraljaújhely.

Dawson, A.H. (1997), 'Two Maps of Poland', in A. Dingsdale (ed), *Transport in Transition: Issues in the New CEE*, Nottingham Trent University Trent Geographical Papers 1, Nottingham, 30–41.

Deica, P. (ed) (1998), *Euroregiunile din Europa Centrala si de Est: Zonele Transfrontaliere de Romania*, Academia Romana Institutul de Geografie, Bucharest.

Deica, P. and Alexandrescu, V. (1995), 'Transfrontiers in Europe: the Carpathian Euroregion', *Revue Roumaine de Geographie*, 39, 3–11.

Deica, P. and Alexandrescu, V. (1999), 'Is the Danube-Tisa-Mureş a New Type of Euroregion?', in C. Gruia, G. Ianoş, M. Torok-Oance and P. Urdea (eds), *Proceedings of the Regional Conference of Geography: Danube-Criş-Mureş-Tisa Euroregion – Geonomical Space of Sustainable Development*, Editura Mirton, Timişoara, 15–20.

Djurdjev, B. (1999), 'Refugees and Village Renewal in Yugoslavia', *GeoJournal*, 45, 207–13.

Dobraca, L. and Ianoş, I. (1998), 'Gegenwartige und kunftige Veranderungen an der rumanisch-bulgarischen Grenze', in F-D Grimm (ed), *Grenzen und Grenzregionen in Sudosteuropa*, Sudosteuropa Gesellschaft Sudosteuropa Aktuell 28, Leipzig, 110–20.

Dokoupil, J. (2002), 'Approaches to the Typology of Czech Borderlands', in A.Dubcova and H.Kramarekova (eds), *State Border Reflection by Border Region Population of V4 States*, Constantine the Philosopher University Department of Geography, Nitra, 22–32.

Domański, B. (2000), 'Types of Investment and Locational Preferences of European American and Asian Manufacturing Companies in Poland', in J.J. Parysek and T. Stryjakiewicz (eds), *Polish Economy in Transition: Spatial Perspectives*, Bogucki Wydawnictwo Naukowe, Poznań, 29–39.

Dubcova, A. and Kramarekova, H. (eds) (2002), *State Border Reflection by Border Region Population of V4 States*, Constantine the Philosopher University Department of Geography, Nitra.

Fitzmaurice, J. (1995), *Damming the Danube: Gabčikovo/Nagymaros and Post-Communist Politics in Europe*, Westview, Boulder, Col.

Flockton, C. (2000), 'Multinational Investment on the Periphery of the EU', in N.Parker and B.Armstrong (eds), *Margins in European Integration*, Macmillan, Basingstoke, 105–27.

Freiherr von Malchus, V. (1997), 'Transfrontier Cooperation in Spatial Planning at the External Border of the EU', in U. Graute (ed), *Sustainable Development in CEE*, Springer, Berlin, 65–76.

Fridrich, C. (2002), 'Past, Present and Future of the Austrian-Slovakian Border Region as Seen in Perception and Action Patterns of the Population', in A. Dubcova and H. Kramarekova (eds), *State Border Reflection by Border Region Population of V4 States*, Constantine the Philosopher University Department of Geography, Nitra, 86–109.

Gałęski, M., Jankowski, A., Watterott, G. et al. (nd), *Euroregion Neisse- Nisa-Nysa: Cross Border Cooperation 1991–1999*, Euroregion Neisse- Nisa-Nysa, Zittau-Liberec-Zielona Góra.

Gorzelak. G. (2001), 'The Regional Dimension of Polish Transformation: Seven Years Later', in G. Gorzelak, E. Ehrlich, L. Faltan and M. Illner (eds), *Central Europe in Transition: Towards EU Membership*, Scholar Publishing House, Warsaw, 310–29.

Gosar, A. (1996), 'Slovenian Responses to New Regional Development Opportunities', in D. Hall and D. Danta (eds), *Restructuring the Balkans: A Geography of the New Southeast Europe*, Wiley, Chichester, 99–108.

Government of the Hungarian Republic (GHR) (1997), *National Regional Development Concept: Background Document*, GHR, Budapest.

Green, S. (1997), 'Post-Communist Neighbours: Relocating Gender in a Greek-Albanian Border Community', in S. Bridger and F. Pine (eds), *Surviving Post-Socialism: Local*

Strategies and Regional Responses in Eastern Europe and the Former Soviet Union, Routledge, London, 80–105.

Grimm, F-D. (1999), 'Strukturen Beziehungen und Perspektiven des ostmitteleuropaischen Verdichtungsbandes Sachsen-Schiesen-Sudostpolen-Westukraine', *Europa Regional*, 7(3), 23–36.

Gruchmann, B. and Walk, F. (1996), 'Transboundary Cooperation in the Polish-German Border Regions', in J. Scott (ed), *Border Regions in Functional Transition: European and North American Perspectives on Transboundary Interaction*, Institute for Regional Development and Structural Planning, Berlin, 129–38.

Hajdu, Z. (1997), 'Emerging Conflict or Deepening Cooperation?: The Case of the Hungarian Border Region', in P. Ganster, A. Sweedler, J. Scott and W-D.Eberwein (eds), *Borders and Border Regions in Europe and North America*, San Diego University Press, San Diego, Calif., 193–211.

Hartshorne, R. (1933), 'Geographic and Political Boundaries in Upper Silesia', *Annals of the Association of American Geographers*, 23, 195–228.

Havlicek, T. (2002), 'Labour Market and Occupation-Related Migration in the Czech-Austrian Borderland', in A. Dubcova and H. Kramarekova (eds), *State Border Reflection by Border Region Population of V4 States*, Constantine the Philosopher University Department of Geography, Nitra, 110–18.

Helinski, P. (1998), *Carpathian Euroregion 1993–1998: Five Years of Dialogue and Cooperation*, CER Secretariat, Krosno.

Horvat, U. (1992), 'Some Characteristics of Demographically-Endangered Settlements in Northeastern Slovenia', *Geographica Slovenica*, 23, 261–78.

Horvath, G. (ed) (1993), *Development Strategies for the Alps-Adriatic Region*, Centre for Regional Studies, Pécs.

Horvath, G. (1996), 'The Regional Policy of the Transition in Hungary', *European Spatial Research and Policy*, 3(2), 39–56.

Ilies, A. and Dehoorne, O. (2002), 'Alternatives for the Development of Tourism in Romania', in J.Wyrzykowski (ed), *Conditions of the Foreign Tourism Development in CEE: Problems of the Development of Ecotourism with Special Emphasis on Mountain Areas*, University of Wrocław Department of Regional and Tourism Geography, Wrocław, 131–41.

Ioras, F., Muica, N. and Turnock, D. (2001), 'Aproaches to Sustainable Forestry in the Piatra Craiului National Park, Romania', *GeoJournal*, 55, 579–98.

Jansky, B. (1997), 'Elbe Project: Principal Results of Czech-German Cooperation', *Acta Universitatis Carolinae: Geographica*, 32(Supp), 71–7.

Kaluski, S. (1999), 'Nature-Related and Socio-Economic Conditions of Trans-Border Cooperation between Polish and Slovak Spišz', *Acta Facultatis Rerum Naturalium Universitatis Comenianae: Geographica*, Supp. 2/1, 217–22.

Kennard, A. (1997), 'A Perspective on German-Polish Cross-Border Cooperation and European Integration', in M. Anderson and E. Bort (eds), *Schengen and EU Enlargement: Security and Cooperation at the Eastern Frontier of the EU*, University of Edinburgh International Social Sciences Institute, Edinburgh, 53–61.

Kennard, A. (2000a), 'The Role of CEE in the EU's Regional Planning Agenda for the New Millennium', *Journal of European Area Studies*, 8, 203–19.

Kennard, A. (2000b), 'Transnational Cooperation in the German-Polish Border Area', in N. Parker and W. Armstrong (eds), *Margins in European Integration*, Macmillan, London, 128–52.

Kisielowska-Lipman. M. (2002), 'Poland's Eastern Borderlands: Political Transition and the "Ethnic Question"', in J. Batt and K. Wolczuk (eds), *Region State and Identity in CEE*, Frank Cass, London, 133–54.

Klemenčić, M. and Gosar, A. (2000), 'The Problems of the Italo-Croato-Slovene Border Delimitation in the Northern Adriatic', *GeoJournal*, 52, 129–37.

Kovacs, Z. (1990), 'The Development of the Hungarian Urban Network after the First World War with Special Reference to Border Areas', *Foldrajzi Kozlemenyek*, 38(1–2), 3–17.

Kratke, S. (1996), 'Where East meets West: The German-Polish Border Region in Transformation', *European Planning Studies*, 4, 647–69.

Kratke, S. (1998), 'Problems of Cross-Border Regional Integration: The Case of the German-Polish Border Area', *European Urban and Regional Studies*, 5, 249–62.

Kratke, S. (1999), 'The German-Polish Border Region in a New Europe', *Regional Studies*, 33, 631–41.

Krumalova, V. and Ratinger, T. (2002), 'Land Abandonment as a Threat to Wildlife and Landscape in Zones I and II of the White Carpathians PLA', in F.W. Gatzweiler, R. Judis and K. Hagedorn (eds), *Sustainable Agriculture in Central and East European Countries: Environmental Effects of Transition and Needs for Change*, Shaker Verlag, Aachen, 129–36.

Kubik, J. (1994), 'The Role of Decentralisation and Cultural Revival in Post-Communist Transformation: The Case of Cieszyn, Silesia', *Communist and Post-Communist Studies*, 27, 331–56.

Langer, A. (1990), *Ecological Bricks for our Common House in Europe: Global Challenges Network*, Verlag fur Politische Ökologie, Munich.

Lauko, V. (1999), 'Influence of Location and Natural Conditions on Development and Contemporary Situation of Kysuce Region', *Acta Facultatis Rerum Naturalium Universitatis Comenianae: Geographica*, Supp. 2/1, 231–40.

Maggi, P. and Nijkamp, P. (1992), 'Missing Networks and Regional Development in Europe', in T.Vasko (ed), *Problems of Economic Transformation: Regional Development in CEE*, Avebury, Aldershot, 29–49.

Markowski, T. and Stawasz, D. (2003), 'Growth Determinants in Polish Agglomerations', in E. Wever (ed), *Recent Urban and Regional Developments in Poland and The Netherlands*, Utrecht University, Netherlands Geographical Studies 319, Utrecht, 39–50.

Matthiesen, U. and Bürkner, H-J. (2001), 'Antagonistic Structures in Border Areas: Local Milieux and Local Politics in the Polish-German Twin City Gubin/Guben', *GeoJournal*, 54, 43–50.

Mihailović, K. (1972), *Regional Development: Experience and Prospects in Eastern Europe*, Mouton, The Hague.

Mikus, W. (1986), 'Industrial Systems and Change in the Economies of Border Regions: Cross-Cultural Comparisons', in F.E.I. Hamilton (ed), *Industrialization in Developing and Peripheral Regions*, Croom Helm, London, 59–84.

Moodie, A.E., Fuller, G.J., Cole, M.M. and Butland, G.J. (1955), 'The Upper Soča Valley', *Geographical Studies* 2(2), 63–110.

Nagy, I. (2000), 'Environmental Problems in the Seven Hungarian Border Regions', in P. Ganster (ed), *Cooperation Environment and Sustainability in European Border Regions*, Institute for Regional Studies of the Californias/San Diego State University Press, San Diego, 203–22.

Nagy, I. (2001), *Cross Border Cooperation in the Border Region of the Southern Great Plain of Hungary*, Centre for Regional Studies, Pécs.

Novotna, M. (2002), 'Vimpersko: Geographical Analysis of a Border Region', in A. Dubcova and H. Kramarekova (eds), *State Border Reflection by Border Region Population of V4 States*, Constantine the Philosopher University Department of Geography, Nitra, 46–57.

Nowicki, P. (1998), 'The Cracow Declaration', in P. Nowicki (ed), *The Green Backbone of CEE*, European Center for Nature Conservation, Tilburg, 257–62.

Pal, A. (20000, 'Socioeconomic Processes in the Hungarian-Yugoslavian-Romanian Border Zone: Approaches to the Danube-Tisza-Maros-Körös Euroregion', in P. Ganster (ed), *Cooperation Environment and Sustainability in Border Regions*, Institute for Regional Studies of the Californias/San Diego State University Press, San Diego, 223–34.

Pal, A. and Nagy, I. (1998), 'Socio-Economic Processes in the Hungarian-Yugoslavian Border Zone', in H. Eskelinen, I. Liikanen and J. Oksa (eds), *Curtains of Iron and Gold: Reconstructing Borders and Scales of Interaction*, Ashgate, Aldershot, 229–41.

Pal, A. and Nagy, I. (1999), 'The Economic Relationships of the Hungarian-Romanian Border Region', in C. Gruia, G. Ianoş, M. Torok-Oance and P. Urdea (eds), *Proceedings of the Regional Conference of Geography: Danube-Criş-Mureş-Tisa Euroregion – Geonomical Space of Sustainable Development*, Editura Mirton, 369–85.

Petrakos, G.C. (1996), *The New Geography of the Balkans: Cross Border Cooperation between Albania, Bulgaria and Greece*, University of Thessali Department of Planning and Regional Development, Chania.

Podhorsky, F. (1995), 'Boundaries of Slovakia: Transport-Geographical Aspect', in V. Baran (ed), *Boundaries and their Impact on the Territorial Structure of Region and State*, University M. Bel Faculty of Natural Sciences, Banská Bystrica, 56–9.

Raj, A. and Petrikova. H. (2001), 'Krkonose-Karkonosze Bilateral Biosphere Reserve', in P. Hodham and R. Stein (eds), *Europarc Expertise Exchange Working Group: Transfrontier Protected Areas*, Europarc Federation/PHARE, Grafenau, 58–60.

Rechnitzer, J. (1997), 'The Main Elements of National Planning Strategy in Northwest Transdanubia', in J. Jensen (ed), *Transborder Cooperation between Western Hungary and Eastern Austria,* ISES, Budapest-Szombathely, 24–30.

Rechnitzer, J. (1998), 'Die Auswirkungen der zunehmenden Grenzpassierbarkheit zwischen Ungarn Osterreich und der Slowakei auf diebenachbarten Regionen', in F-D. Grimm (ed), *Grenzen und Grenzregionen in Sudosteuropa*, Sudosteuropa Gesellschaft Sudosteuropa Aktuell 28, Leipzig, 76–101.

Scharman, L. and Tietze, W. (1990), 'Further Remarks on the Urgent Modernization of the Central European Transport Networks: Some Special Cases covering Eastern Germany', *GeoJournal*, 22, 195–203.

Seffer, J., Cierna, M., Stanova, V., Lasak, R. and Galvanek, D. (1999), 'Large Scale Restoration of Floodplain Meadows', in J. Seffer and V. Stanova (eds), *Morava River*

Floodplain Meadows: Importance Restoration and Management, DAPHNE Centre for Applied Ecology, Bratislava, 129–38.

Slee, B. (1999), *A Tourism Development Plan for the Stuzhytsa-Ushanski Park, Ukraine*, University of Aberdeen Department of Agriculture, Aberdeen.

Storm-Pedersen, J. (1993), 'The Baltic Region and the New Europe', in R. Cappellin and P. Batey (eds), *Regional Networks Border Regions and European Integration*, Pion, London, 135–57.

Stryjakiewicz, T. (1998), 'The Changing Role of Border Zones in the Transforming Economies of ECE', *GeoJournal* 44, 203–13.

Stryjakiewicz, T. and Kaczmarek, T. (2000), 'Transborder Cooperation and Development in the Conditions of Great Socio-Economic Disparities: The Case of the Polish-German Border Region', in J.J. Parysek and T. Stryjakiewicz (eds), *Polish Economy in Transition: Spatial Perspectives*, Bogucki Wydawnictwo Naukowe, Poznan, 49–71.

Suliborski, A. (1995), 'Theoretical-Ideological Dilemmas of Transborder Cooperation and Unification in Europe', in V. Baran (ed.), *Boundaries and their Impact on the Territorial Structure of Region and State*, University M. Bel Faculty of Natural Sciences, Banská Bystrica, 30–6.

Suli-Zakar, I. (1992), 'A Study of State Borders as Factors Blocking Socio-Economic Progress in North-Eastern Hungary', *Geographical Review* 40, 53–64.

Szekely, V. (1995), 'Stimulation Factors of the Frontier Regions' Development: Theoretical Reflection', in V. Baran (ed.), *Boundaries and their Impact on the Territorial Structure of Region and State*, University M. Bel Faculty of Natural Sciences, Banská Bystrica, 16–9.

Szigetvari, T. (2001), 'Regional Development in Hungary', in G. Gorzelak, E. Ehrlich, L. Faltan and M. Illner (eds), *Central Europe in Transition: Towards EU Membership*, Scholar Publishing House, Warsaw, 287–309.

Szorenyine Kukorelli, I., Dancs, L., Hajdu, Z., Kugler, J. and Nagy, I. (2000), 'Hungary's Seven Border Regions', *Journal of Borderland Studies*, 15, 221–54.

Tvrdon, J. (1996), 'Cooperation Between the Slovak Republic and the CEE states', in Council of Europe and European Commission (eds), *The Regional Place of Greater Europe in Cooperation with the Countries of CEE: Proceedings of a Joint Conference Prague 1995*, Office for the Official Publications of the European Communities, Luxembourg, 51–8.

Vaclav, T., Bohumila. T. and Josef, K. (2002), 'Hodoninsko: Border Region of Intensive Relationship with Slovakia', in A. Dubcova and H. Kramarekova (eds), *State Border Reflection by Border Region Population of V4 States*, Constantine the Philosopher University Department of Geography, Nitra, 64–71.

Van der Boel, S. (1994), 'The Challenge to Develop a Border Region: German-Polish Cooperation', *European Spatial Research and Policy*, 1(1), 57–72.

Wallace, C. (1999), 'Investing in Social Capital: The Case of Small-Scale Cross-Border Traders in Post-Communist Central Europe', *International Journal of Urban and Regional Research*, 23, 751–70.

Wecławowicz, G. (1996), *Contemporary Poland: Space and Society*, UCL Press, London.

Wieckowski, M. (1999), 'Natural Conditions for the Development of the Polish-Slovak Transboundary Ties', *Acta Facultatis Rerum Naturalium Universitatis Comenianae: Geographica*, Supp. 2/1, 257–63.

Wild, M.T. and Jones, P. (1993), 'From Peripherality to a New Centrality?: Transformation of Germany's Zonenrandgebiet', *Geography*, 78, 281–94.

Zahorakova, D. (2002), 'Preconditions of Regional Development of the Danubian Euroregions of the Slovak-Hungarian Borderland', in A. Dubcova and H. Kramarekova (eds), *State Border Reflection by Border Region Population of V4 States*, Constantine the Philosopher University Department of Geography, Nitra, 265–9.

PART TWO
REGIONAL STUDIES

Chapter 9

Investment and Development in the Western Balkans

Derek Hall

This chapter has three objectives. First, it aims briefly to examine the contemporary economic, political and security environment of the Western Balkans, an area defined here as comprising Albania, Bosnia and Hercegovina (BiH), Macedonia (officially the Former Yugoslav Republic of Macedonia: FYROM) and Serbia and Montenegro (the former including Kosovo and Vojvodina) previously known as the Federal Republic of Yugoslavia (Figure 9.1 and Table 9.1). Although Croatia is often included within this sub-region, and is statistically embraced within this chapter for comparative purposes, it is not centrally addressed in the analysis here for three reasons: (a) the scale, extent, geographical configuration and consequent regional variations of Croatia distinguish it from the other countries of the sub-region; (b) Croatians do not regard themselves or their country as 'Balkan'; and (c) through the perspective of tourism, FDI in Croatia is specifically addressed in Chapter 12. The second objective is to examine the role of the western Balkan environment in attracting or repelling FDI. And, given that FDI has been limited, the third objective evaluates measures pursued at a number of levels to address the stability and development of the region and – implicitly – to enhance its attractiveness for FDI in the future.

Since the late 1980s, as ECE has experienced the upheavals of political and economic change, the western Balkans have been subject to contrasting sets of endogenous and exogenous pressures. The fragmentation of former Yugoslavia through conflict produced enormous human and economic disruption. That subsequent processes of macro-economic stabilisation and privatisation have had limited success has served to emphasise the inappropriateness of simplistic applications of 'Western' models of development to complex cultural regions. Post-communist transformation in the western Balkans has proved to be a much more complex process than initially anticipated by some (Daianu 2001). That so many commentators have actually found this surprising is in itself instructive.

Table 9.1   The Western Balkans: key economic and demographic indicators, 2001

	Total population (mln)	GNI $pc	GNI $bln	GDP $bln	Population	Labour force	Life expectancy at birth (years)	infant mortality (per 1,000 live births)	Gross domestic investment (% average annual growth)	Gross national savings/ GDP	Total exports (fob)	Total imports (cif)	Total debt
					Average annual % growth 1995–2001						($mlns)		
Albania	3.4	1,230	4.2	4.1	0.9	13	72	21	10.9	13.8	305	1,332	838
BiH	3.9	1,270	4.9	4.5	1.3	1.6	73	13	3.0	-0.9	1,003	2,670	2,226
Croatia	4.4	4,550	19.9	20.3	-0.9	0.4	73	8	13.7	19.9	4,752	8,764	10,555
Macedonia	2.0	1,690	3.5	3.4	0.6	1.1	73	14	-24.9	6.4	1,155	1,688	1,423
Serbia and Montenegro	10.6	990	10.5	10.9	0.1	0.5	72	13	5.8	8.0	2,003	4,838	11,949

Source: WBG 2002b.

Figure 9.1   The Western Balkans

The Western Balkan Environment

Until the mid-1980s, Yugoslavia was the most liberal, decentralised and open of the
ECE states dominated by communist regimes. By contrast, Albania was by far the
most introspective, centralised and impoverished. From the mid-1980s, Serbian and
then Yugoslav leader, Slobodan Milošević encouraged an aggressive Serbian
nationalism (Wintrobe 2002), which, in the wake of political change in the rest of
the region, was to stimulate the break-up of the Yugoslav federation and unleash a
series of bitter wars of succession (Hall and Danta 1996). Albania, having thrown
off the shackles of Stalinism in 1991, experienced a virtual haemorrhaging of its

young adult population as the comparative reality of Albanian impoverishment and isolation became apparent. Following a number of political and economic crises of its own, Albania was drawn into the Yugoslav conflicts by virtue of the large ethnic Albanian populations in Kosovo (nominally part of Serbia) and Macedonia. Heightened tension and subsequent conflict between Albanians and Serbs in Kosovo from 1998 eventually drew NATO into attacking Serb positions and ultimately Serbia itself (Danta 2000c; Papasotiriou 2002). Ironically, while Macedonia had achieved a relatively peaceful independence in the early 1990s, ten years later it too was wracked by inter-ethnic conflict. Such a climate of political instability, tension and conflict, coupled to an already relatively economically poor region, has been far from conducive to FDI (Büschenfeld 1999; Hunya 2000, 2002b).

Albania

Following substantial upheaval which accompanied political change during 1991–2, Albania experienced a period of deceptively strong economic growth based on overseas remittances, the early impetus of agricultural privatisation and massive inflows of aid during the mid-1990s (Hall 1996). But efforts to consolidate democracy and a viable market economy were severely damaged by the lawlessness and economic crisis which followed the collapse of pyramid 'investment' schemes in 1996–7. During the 1998–9 Kosovo crisis, Albania acted as host to over 460,000 refugees. The economy revived in 2000, when GDP grew by 7.8%, albeit still relying on precarious engines of growth (remittances, aid, smuggling, money laundering) and emerging from a low base point. The difficult regional situation, together with a divisive political scene and weak state institutions, has constrained reform and development. Progress has been made in securing government revenue through reform of the customs and tax services, Albania's trade regime has been modernised and liberalised and the country became a member of the WTO in September 2000. But during the 1990s FDI never rose above $100mln in any one year. By contrast, émigré remittances were as high as $700mln per annum (Haderi et al. 1999; Korovilas 1999). Privatisation of small and medium enterprises is now complete and that of larger companies is progressing with some hesitancy. For example, Turkish interests took a long lease on the 1970s Chinese-equipped Elbasan metallurgical complex, although labour problems have hampered development. Once privatisation is complete, FDI inflows are likely to fall dramatically in the absence of significant greenfield investments (Hunya 2002a).

Bosnia and Hercegovina

Following a declaration of independence in April 1992 Bosnia and Hercegovina (BiH) was plunged into a devastating war which continued until November 1995. Output fell to 10–30% of pre-war levels, GDP collapsed to less than $500pc – a fifth of pre-war levels, and half of the country's pre-war population of 4.5mln was displaced. The General Framework Agreement for Peace (Paris-Dayton Agreement) which brought the war to an end, set up a central BiH government and devolved wide powers to two entities: the mainly Muslim-Croat Federation and the Serb dominated Republika Srpska. These faced a massive task to build a stable social and political structure as well as a functioning economy (Danta 2000a; Taylor 2002). BiH was also placed under 'international protection', with an appointed 'High Representative' having wide supervisory powers. The responsibilities and powers of the State of BiH were strictly delimited, resulting in a weak state lacking many of the attributes associated with statehood. The international community has overseen the application of democratic principles in elections held since 1996. However, respect for the rule of law and human rights cannot yet be said to be universal. In May 1999 a new foreign investment law was enacted at state level, and some investments followed with the progress of privatisation.

Macedonia (formally FYROM)

When Yugoslavia's federal structure started to unravel in the early 1990s, Macedonia was paid scant attention since analysts assumed that it would either stay within the Yugoslav structure, or else would be engulfed by Serbia. That neither occurred was to the credit of Macedonian leadership (Danta 2000b). The country's independence in 1991 was achieved without conflict, despite the Yugoslav government's aggressive intentions, and the agitations of the significant Albanian minority from within. Nonetheless, a NATO peacekeeping force was put in place in 1992. Greece objected both to the use of the name 'Macedonia' and to the flag adopted by the new state, and was able to rally support for its position in Europe, the UN, and the USA, and to impose embargoes on Macedonia which was also to suffer from UN sanctions placed on Serbia. Landlocked Macedonia's closest outlet is through Thessaloniki, and some 90% of its trade, valued at over $80 billion, normally uses this route. In addition, loss of Serbia as a trade partner led to a 30% decline in industrial output. Once a compromise was reached to refer to the country formally as FYROM the way became clear for international recognition.

A cooperation agreement signed with the EU in 1997 came into effect on 1 January 1998. The EBRD adopted a strategy for developing the private sector (EBRD 1999a), and the privatisation process moved ahead. However, there continues to be a need for the generation of entrepreneurial and managerial skills. The ecological situation is relatively good, albeit with concentrations of high levels of degradation. A national environmental action plan was adopted in 1996 (Bužarovski and Stojmilov 2002), and considerable potential has long been recognised for the development of a tourism industry (Allcock 1991; Stojmilov et al. 2000). In 1999 FYROM accepted some 260,000 refugees from Kosovo (UNHCR 1999). Somewhat belatedly spilling over from the Kosovo conflict, inter-ethnic tension brought the country to the verge of civil war in the first half of 2001. Macedonia's GDP and industrial production declined sharply, restructuring of loss-making companies was delayed, and investments – especially FDI – slowed down as risks increased dramatically (Kekic 2001; Hunya 2002a). A 700–strong peacekeeping force provided by European NATO member states in September 2001 was mandated to at least 2003 (EIU 2002b). Full economic recovery is not anticipated before 2004.

Serbia and Montenegro (previously FRY)

As a result of several years of conflict with its former partners, a decade of aggressive nationalism and instability, and finally several months of NATO bombing in response to the Kosovo crisis, the economy declined by 40%, unemployment rose to 30%, and inflation to more than 100%, while GDP fell to less than half and industrial production to less than a third of 1989 levels (Brankovic and Nenadovic 2002). Removal of Milošević in 2000 paved the way for political and economic reform, including privatisation legislation. The sale of three cement plants in early 2002 bore the first fruits of privatisation-led FDI (WIIW 2002). This aimed to raise an initial €150mln and to establish an appropriate climate in which to sell off the high profile JAT (national airline) and Zastava (Yugo cars and armaments) companies. At the same time, a massive physical as well as economic reconstruction programme was adopted for Kosovo (e.g. Minervini 2002), as well as for the areas of Serbia subjected to NATO bombing. In March 2002 the country was renamed Serbia and Montenegro in an agreement brokered by the EU to avert the secession of Montenegro. This, it was feared, would plunge the region into renewed conflict. Serbia and Montenegro agreed to become semi-independent states sharing defence and foreign policy. The Montenegrin capital, Podgorica, has considerable autonomy, being responsible for a separate economy, currency and customs service (Castle and Peric Zimonjic 2002).

The Flow of FDI and the Role of the Regional Environment

FDI is usually viewed as an important element of sustainable development as it: represents capital inflows; can indicate the presence of – and can generate further – international cooperation; expresses confidence by foreign investors in the country's or region's economy; indicates the openness of the country or region to the world economy; can generate export-led growth; can encourage new technologies and techniques to be introduced into and diffuse through the regional and national economy, favouring the enhancement of labour force skills; can help safeguard employment or at least help stem emigration of young adults (e.g. Carter et al. 1993); and can help transfer practices and concepts which assist the construction and consolidation of a civil society. An extensive overhaul of a country's legislative and institutional framework is generally not a necessary precondition for attracting FDI from large multinational investors or from smaller entrepreneurial investors. More important factors are likely to be: the existence of perceived business opportunities; the potential for high returns; the risk of expropriation; the ability to repatriate profits; existing tax regimes; and an often superficial and subjective 'feel' about the host country (Hewko 2002).

While FDI may have created hundreds of thousands of relatively well-paid jobs in ECE in sectors ranging from automotive to retail, it may have also destroyed a comparable number. Foreign acquisitions of local enterprises have tended to be followed by substantial employment reductions and productivity-enhancing investment. Subsequent growth has created new employment, but in some cases this has replaced existing jobs elsewhere as foreign-owned companies expand their market share at the expense of local competitors. Local suppliers have often been displaced by imported components as foreign investors incorporate their acquisitions into global manufacturing networks (EIU 2002c p.15). In Macedonia, for example, most of the 1,688 formerly socially-owned enterprises have been sold off or liquidated. The level of interest from foreign or even domestic investors has been limited. Most companies have been sold through a process of insider privatisation, with shares distributed to workers in lieu of unpaid salaries, or to managers at heavily discounted prices payable over many years. Such new owners have resisted restructuring. Yet, industrial employment has been reduced by more than half (ESI 2002 pp.9–10; see also Ádám 2000).

Nonetheless, capital inflows from abroad are of potentially fundamental importance in the western Balkans, particularly because of a low level of savings and investment. This reflects such regional factors as poverty, underdevelopment, general lack of capital, and loss of confidence in banking and other financial institutions (e.g. Drummond 2000). The latter in particular was reflected in the way in which Albania, Bulgaria, Macedonia and Serbia were all ensnared in pyramid investment schemes in the second half of the 1990s. In Albania, for example, the absence of confidence in formal (and embryonic) banking structures, coupled to a collective naïve innocence, saw much domestic investment being channelled into nine such schemes. An estimated 70% of all families had committed their savings to these schemes for returns of up to 50% per month. This partly reflected the fact that the real value of their savings was often not appreciated, not least because their origins were usually three unconventional and, temporarily, bountiful sources: (a) the sale of easily gained former state-owned homes, a one-off transaction which many found to their cost could not be subsequently reversed, or which set off a chain of undervalued sales; (b) émigré remittances, which, while playing an important role in subsistance and housing construction, have done little in funding infrastructure or productive activities (Martin et al. 2002); and (c) smuggling and money laundering. The pyramid schemes collapsed towards the end of 1996 and during the first months of 1997 (Elbirt 1997; Korovilas 1999; Lawson and Saltmarshe 2000). The very existence, popular support, and aftermath of these huge frauds well expressed the continuing fragility of Balkan democratic structures and the insinuation of a 'globalised' informal economy (e.g. Gumbel 1997; Willan 1997) in the absence of FDI-led formal economies functionally linked to global markets.

Arguably, Albania has not been alone in the western Balkans in exhibiting outward signs of an inversion or at least a distortion of conventional development models. Based on experience in Russia, Burawoy (1996; also Saltmarshe 2001 p.8) has suggested that a process of involution (in the sense of being the reverse of evolution) may take place when a post-communist economy eats away at its own reserves by funnelling resources from production to exchange. In the case of Albania, and by implication, Albanians in Kosovo and Macedonia, aspects of the informal economy – organised crime, dealing in the production of narcotics and the trafficking of drugs, arms, refugees, women, fuel, luxury cars and scrap metals, and the receipt of substantial émigré remittances from expatriate Albanians scattered across Europe and North America (Muço 2001a) – have accrued 'globalised' dimensions. This has been possible at least in part because of the reinforcement of vertical and horizontal networks of reciprocity – rising to the highest political levels between 1991 and 1998 – investing time and energy in informal exchange rather than formal production.

The small FDI inflows into the western Balkans (Table 9.2) have tended to be driven by privatisation, greenfield investments in manufacturing being particularly rare. Examples of FDI since 2000, by country of source, may be summarised as follows (Schifferes 2001; EIU 2002a; Hunya 2002a; Prentice 2002):

- Albania: Greece and Norway (telecoms); Italy (chrome); Turkey (banking, copper and metallurgy.
- BiH: Croatia, Slovenia and USA (food and drink); Germany (cement and furniture); Italy (textiles); Lithuania (alumina); Slovenia (paper).
- Macedonia: Greece (banking and petroleum); Hungary (telecoms).
- Serbia: France (cement); Germany (publishing); Italy (telecoms).

Export-oriented investors have shown little interest, such that being mostly oriented to local markets, FDI is not viewed as a major vehicle for economic take-off (Hunya 2002a p.3). Several potential foreign partners have confined their interest to other transition economies until conditions appear much more favourable; so there are stark differences even between western Balkan countries and other SEECs. At a regional level, factors acting to repel potential FDI can be summarised as: security, sanctions, instability, legality, ecology and protectionism (EBRD 1999b; Gligorov 1998; Muço 2001b pp.50–1).

Security

Security, or the lack of it, remains a significant consideration repelling investment. High political risk exists due to potential conflict and economic shocks for most countries in the region even though military activity has now abated. Although the level of risk fluctuates, potential for conflict persists from four different sources: possible future instability in BiH, should the international community cease its peacekeeping efforts; disintegration between Serbia and Montenegro; continuing fears about further civil conflict in Macedonia; and unclear aims and uncertainties about the future status of Kosovo – and similar concerns regarding the Vojvodina, albeit to a far lesser degree (Civilitas Research, 2002a).

Table 9.2 FDI in SEE, 1992–2002

Country	Stock $ mn							Stock $pc				Inflow $mln			Inflow$pc
	1992	1994	1996	1998	2000	2001	2001	2001	2000	2001	2002 forecast	2001
Albania	157	205	291	384	568	800	233	143	232	200	58
BiH	nd	nd	nd	100	340	470	125	150	130	200	35
Macedonia	29	56	44	173	381	824	403	176	442	500	217
Serbia and Montenegro	nd	nd	nd	853	990	1,155	139	25	165	500	20
Western Balkans	**(186)**	**(261)**	**(335)**	**1,510**	**2,279**	**3,249**	**225**	**494**	**969**	**1,400**	**83**
Bulgaria	65	412	831	1,488	3,309	3,997	504	1,002	689	600	86
Croatia	nd	238	874	2,439	5,202	6,703	1,530	1,126	1,502	1,100	343
Romania	544	1,272	2,209	4,480	6,561	7,698	343	1,040	1,137	1,000	51
SEE	**(795)**	**(2,183)**	**(4,249)**	**9,064**	**17,351**	**21,647**	**377**	**3,662**	**4,265**	**4,100**	**79**

Source: Gligorov 1998; Hunya 2002a; WIIW 2002: 4.

Sanctions

Sanctions and embargoes have often imposed considerable economic, social and political impacts within the region. As mechanisms acting to constrain an attractive investment environment, sanctions fall into three categories: (a) those imposed by the outside world, most notably those enforced on the then Yugoslavia (i.e. Serbia and Montenegro) by the EU and the UN, some of which were lifted at the end of 1995 (after the Paris-Dayton agreement), and later in the 1990s; (b) sanctions imposed bilaterally within the region itself, such as those imposed on Macedonia by Greece from early 1995 to late 1996; and (c) undeclared sanctions between some Balkan neighbours, and a number of other similar trade barriers which vary from time to time and from country to country.

Macroeconomic Instability

Macroeconomic instability, including high levels of inflation and unemployment, has been a serious problem for all countries of the region, although in most cases it plays a secondary role to other factors. Such instability stems from a number of almost inherent weaknesses: markets are generally small, weak and disorganised; banking systems cannot respond quickly and appropriately to business needs; extensive bureaucracy creates delays and resistance to change; and resultant shortcomings in institutional capacity persist.

Low Level of Legality

Pervasive levels of illegality have also contributed to the high risk of foreign investments in the area. Problems of fraud and corruption are found in all the countries of the western Balkans, and have persisted in the FDI arena at least in part because of slow implementation of liberal laws for foreign investments and capital movements (Pendovska and Georgievski 1999).

Poor Ecological Context

Both explicitly within individual countries and implicitly for the region as a whole, actual and perceived poor ecological conditions normally act to repel potential investors (although see Xing and Kolstad 2002). Ecological issues tend to be low on the list of national priorities: budget constraints force governments to focus on more short-term goals (as they do in most countries). Publicising the significance of ecological issues thus tends only to take place if there is an emergency, such as the need to clean up drinking water. The impacts of military conflicts, both on the ground and in the atmosphere, have had profound environmental implications (Daianu and Veremis 2001 p.5). Ecological considerations are fundamental to any strategic assessment for investment purposes, and compliance with EU standards is a key requirement for long-term accession aspirations (Staddon and Turnock 2001 p.241).

Protectionism

Some countries have also followed a path of protectionism, although often for different reasons. One of the most protectionist former components of Yugoslavia is arguably also the most successful – Slovenia (e.g. Anon 2002). Ironically, a strong political lobby in Belgrade has argued that Serbia should follow the Slovenian model of economic development and disdain the perceived need for FDI: an approach that accords with Serbian populist nationalism (Civilitas Research 2002b).

Country-Specific Factors

The significance of these and other factors naturally varies from country to country. In the case of BiH, investors complain of a cumbersome and intrusive bureaucracy; the split between the two political entities hampers economic activity, and although harmonisation is proceeding, much needs to be done; transport links are poor; the labour market is rigid; tax rates are high; legitimate business suffers unfair competition from smugglers; and the judiciary has been ill-equipped to handle matters of commercial law (Taylor 2002). In the case of Serbia, failure of the second attempt to elect a president, in December 2002, perpetuated an environment of political uncertainty. While the Serbian Investment and Export Promotion Agency pointed to foreign investments increasing from $50 million in 2000 to $160 million in 2001 and to as much as $294mln in the first nine months of 2002, more than two thirds of the FDI in Serbia has been collected from a few privatisation deals. Important greenfield investments are still lacking. This reflects a number of general and specific factors: unclear criteria for tender sales and protracted auctions; an absence of investment funds, clear stock exchange procedures and a lack of guidelines for liquidation of insolvent companies; pervasive corruption, bureaucracy and protectionism; and finally the lack of economic growth has meant that the domestic market, which many expected to expand rapidly, remains underdeveloped. Generally, a poor investment climate in the western Balkans has constrained progress in market reforms (Pissarides 2001). Indeed, the two appear to have a symbiotic relationship, and necessary kick-start factors would appear to be at least: improved regional infrastructure: communications and electricity supply and distribution are particularly poor; political stability, to remove fear of conflict; economic stability through appropriately encouraging taxation regimes, removal of harassment by petty officials, improvement of access to finance, and clarification of property rights; removal of high debt levels; removal of stifling bureaucracy; and wholesale removal of fraud and corruption (Resmini 2000; Agence France Presse 2001).

Responses to the Western Balkan Environment

The mosaic of programmes – both pre-dating and responding to the Kosovo crisis of 1999 – and agencies involved in 'supporting' the governments and people of the western Balkans provide a superficial impression of overlapping activities and objectives, unclear interrelationships and divisions of labour, coupled to limited levels of ownership and control vested in local institutions (despite claimed capacity-building objectives) (Table 9.3).

The Stabilisation and Association Process (SAP)

In the 1990s, the EU's political, trade and financial relations with the western Balkans focused on crisis management and reconstruction, reflecting the emergency needs at that time, as exemplified in Montenegro (e.g. Table 9.4). As the region began to emerge from this difficult period, responding to the need for a more long-term approach to its development, the EU launched the SAP in 1999. The EU/EC has become by far the single largest assistance donor to the region, providing support across a wide range of programmes. Since 1991, more than €6bln has been provided, and some have argued for the rapid full 'euroisation' of the Balkan countries as a means of reforming and opening up their financial systems (e.g. Gros, 2002). As the framework for the EU's approach to SEE, the SAP is designed to encourage and support domestic reform processes. It is a step-by-step approach based on aid, trade preferences, dialogue, technical advice and, ultimately, contractual relations. Since December 2000, Albania, BiH, Croatia, Macedonia and Serbia and Montenegro have benefited from EU trade preferences, with the majority of their products enjoying duty-free and unlimited access to EU markets. The next step, for countries that have made sufficient progress in terms of political and economic reform and administrative capacity, is a formal contractual relationship through a tailor-made Stabilisation and Association Agreement. At the time of writing, Croatia and Macedonia had signed such agreements. Overall, objectives of the SAP are aimed to: bring each SEEC closer to EU standards and principles, and to prepare them for gradual integration into EU structures; help national authorities to consolidate democracy and implement the rule of law; assist national governments in achieving comprehensive administrative and institutional reform; lay the foundations for sustainable economic development and growth; facilitate the process of economic and social transformation towards an efficient market economy; consolidate the peace process and foster co-operation; help ethnic reconciliation and the return of refugees and displaced persons to their homes of origin (EC 2001a, 2002a).

The EC CARDS programme (Community Assistance for Reconstruction, Development and Stabilisation) is the main channel for the EU's financial and technical co-operation programme to facilitate the SAP, with €4.65bln funding over the period 2000 to 2006. This replaces both the PHARE (ECE assistance) and Obnova (Kosovo reconstruction) programmes (EC 2001b). Its priorities are: reconstruction, democratic stabilisation, reconciliation and the return of refugees; institutional and legislative development, including harmonisation with EU norms and approaches, to underpin democracy and the rule of law, human rights, civil

Table 9.3   Summary of major support programmes in the Western Balkans

Programme	Objectives	Major supporters
SAP	Support domestic reform, economic development, stability and 'Europeanisation' in SEE	EC/EU (CARDS programme)
SP	Support for regional cooperation, economic, ethnic and political stability in SEE	EC/EU, G8, World Bank, IMF, EBRD, EIB, UN, NATO, OECD, OSCE, CoE
SECI	Encourage cooperation and integration in SEE	EC/EU, US
FIAS	Analysis of and advice on the FDI environment	International Finance Corporation, World Bank
MIGA	Promoting FDI into emerging economies	World Bank
REReP	Coordinating donor activity for environmental reconstruction in SEE	EC/EU DGE, World Bank, NGOs

See preliminary pages for key to abbreviations.
Source: EC 2001a, 2001b, 2002a, 2002b, 2002c; SECI 2002; SPC 2002; WBG 2002a, 2002d.

society and the media, and the operation of a free market economy; sustainable economic and social development, including structural reform; and promotion of closer relations and regional co-operation among SAP countries and between them, the EU and the ECE candidate countries, focusing on: integrated border management, to help to tackle cross-border crime, to facilitate trade across borders and to stabilise border regions themselves, institutional capacity building to raise awareness of EU policy and laws that the region should increasingly be moving towards support for democratic stabilisation to consolidate advances in democracy and to encourage the involvement of civil society in development, and help to plan the integration of the region's transport, energy and environmental infrastructure into wider European networks. To help meet these objectives, a regional strategy (EC 2001c) and five national strategies (e.g. EC 2002b, 2002c) were prepared. These are being followed up by annual reports monitoring each country's progress in meeting their SAP goals.

The Stability Pact for Southeastern Europe (SP)

The Stability Pact, launched at the EU's initiative in 1999, as a direct reaction to the

Table 9.4 The main objectives of EC assistance for Montenegro

Assistant objectives	Assistance actions
Alleviation of government expenditures	Providing budgetary assistance to help fund: • social welfare payments to the most vulnerable households, • electricity imports, and • additional expenses linked to the hosting of displaced persons during the Kosovo crisis in 1999.
Modernisation of infrastructure	For example: • reconstruction of a bridge in central Podgorica, • rebuilding of the road to Podgorica airport, and • supply of buses and ambulances.
Help with economic and other reforms	Support for: • public administration reform (tax reform and revenue collection), • private sector development (particularly SMEs, vital for job creation), • education (focusing on the needs of 7-15 year olds), • agriculture (livestock), and • administrative assistance (e.g. the creation of an aid co-ordination and programming unit within the Government, to ensure that aid is promptly and properly disbursed).
Humanitarian support	In the areas of: basic shelter, food, non-food items, medicines, water/sanitation, and psychosocial/community services, for refugees and displaced persons from Kosovo and for the most vulnerable of the local population.
Other	For example: exceptional macro-financial assistance; media, NGO and democracy support.

Source: EC2001d.

urgent problems generated by the Kosovo crisis, is a political declaration of commitment and a framework agreement on international co-operation to develop a shared strategy among all partners for stability and growth in SEE. It is not a new international organisation nor does it have any independent financial resources or implementing structures (SCSP 2002). Increasingly, the SP's focus is on supporting greater regional cooperation, which is also a key objective of the SAP (EC 2002c). Key objectives are to: secure lasting peace, prosperity and stability for the region; foster effective regional cooperation though observance of the principles of the Helsinki Final Act (1975); create 'vibrant' market economies based on 'sound' macro policies; integrate the SEECs fully into the European and Atlantic cooperation structures, primarily the EU (WBG 2002c). Organisationally, the SP relies on the 'special co-ordinator', Erhard Busek, former vice-chancellor of Austria, and a 29–member team – representatives of governments, international organisations and institutions. Such participants include the East-West Institute's (Prague) Action Network for South Eastern Europe which seeks to encourage regional, cross-border approaches to economic and civic development in SEE (Budway 1999). The coordinator's tasks include: bringing the participants' political strategies in line with one another; co-ordinating existing and new initiatives in the region; and thereby to help avoid unnecessary duplication of work (SCSP 2002). A regional table acts as an umbrella body to review progress and provide guidance for advancing SP objectives. It is organised through three groups, or 'working tables' – democratisation and human rights (working table I), economic reconstruction, development and co-operation (II), and security issues (III).

The Southeast European Cooperative Initiative (SECI)

Embraced within the SP, SECI is not an assistance programme but encourages co-operation among its participating states and facilitates their integration into European structures. It does this by attempting to: emphasise and coordinate region-wide planning; identify needed follow-up and missing links; provide for better involvement of the private sector in regional economic and environmental efforts; help to create a regional climate that encourages the transfer of know-how and greater investment in the private sector; and assist in harmonising trade laws and policies (SECI 2002).

The Foreign Investment Advisory Service (FIAS) in the Balkans

This is a joint service of the International Finance Corporation and the World Bank, advising governments on aspects of establishing appropriate environments for inward investment. For example, during the 2000–2002 period it undertook evaluations of administrative barriers to FDI in BiH and Macedonia as well as in Croatia and Romania. For Albania, a diagnostic review of the FDI environment included an analysis of the country's policy and legal framework (WBG 1999).

The Multilateral Investment Guarantee Agency (MIGA)

Also part of the World Bank Group, MIGA (founded in 1988) promotes FDI into emerging economies through: provision of investment guarantees against certain non-commercial risks (i.e. political risk insurance) to eligible foreign investors into MIGA member countries; improving the ability of developing member countries to attract FDI by providing capacity-building services; disseminating information about investment conditions and specific investment opportunities through three online information services. These are, first, FDI Xchange (launched in April 2002): a web- and email-based service providing customised investment information; second, the Investment Promotion Network IPAnet (established in 1995): a web portal linking corporate investors and the development community with investment information and analysis on emerging markets, featuring 700 investment information resources on SEE; and third, PrivatizationLink, which focuses more narrowly on investment opportunities arising from current privatisation activity in SEE. All the SEECs are members of MIGA and are eligible for its guarantee and technical assistance services. Thus, for example, BiH had almost US $14mln investment guarantees underwritten during the year to mid-2002 (WBG 2002a).

Case Study: EC/EU Relations with Albania

Although a trade and economic cooperation agreement with Tirana came into force in December 1992, the EC concluded in 1995 that a classical Europe agreement with Albania could not be envisaged largely for economic reasons. However, the country received almost ECU400mln of PHARE assistance for 1991–1995 and became the highest per capita beneficiary of EU assistance in CEE. This situation changed dramatically in the wake of the 1998–9 Kosovo conflict. As part of the 1999 SP, Albania's requirements for external finance would be addressed by the IMF, the World Bank, the EBRD and the EIB, as well as via bilateral assistance programmes and project finance. The longer-term prospect of Albania gaining EU membership on the basis of the Treaty of Amsterdam and the 'Copenhagen criteria' (human rights, economic and political considerations) now arose. In October 1999, as part of the EU common strategy towards the western Balkans, the EC recommended recognition of the need 'to confirm the vocation for membership' of countries of former Yugoslavia and Albania, albeit under strict conditions. In addition to the Copenhagen criteria, these countries would be required to mutually recognise each other's borders, settle all issues relating to the treatment of national minorities and pursue economic integration in a regional framework as a precondition for their integration within the EU (Hall and Danta 2000). This would have the intended effect of both reassuring such countries as Albania that their aspirations were recognised, while keeping the Balkan nations at arms length for at least another two decades (Hall 2001).

Between 1991 and 2000 the EU provided €1021mln to Albania, of which an important part took the form of balance of payment support or specific budgetary assistance linked to sectoral reforms (agriculture, public administration) or to refugee related costs during the 1999 Kosovo crisis. In humanitarian assistance, the EU provided around €140mn while sectoral programmes amount €404mln. Major areas of EU involvement include: strengthening public administration and judiciary; providing strategic advice, training and equipment to the Albanian police; working with the Albanian customs service; improving energy, transport and water infrastructures, local community development; agriculture policy advice and support for fisheries and veterinary control; a cross-border cooperation programme to develop closer links with EU neighbours Greece and Italy; raising awareness of democracy and human rights with assistance to the media and NGOs; and humanitarian assistance, including water sanitation and rehabilitation of schools in remote areas (EC 2001a). However, while the initial exclusion of Serbia from stabilisation and integration programmes provided Albania with an opportunity to take a potentially key role in the region's political stability and spatial development, this opportunity appears to have been missed. Albania's position at the intersection of two planned transport axes, one east-west the other north-south (Corridors 8 and 10) had substantial potential (Hall 1993). Yet the construction of these has been slow (Skayannis and Skyrgiannis 2002) and Albania's potential is, again, unlikely to be fully realised. Communications infrastructure is also severely constrained. Albania has an average of just 6.5 telephone lines per 100 population, although the country's two mobile phone operators have three times more users than Albtelecom's 200,000 subscribers. Yet only two percent of Albania's population has a personal computer, and only 0.6% have access to the Internet (EIU 2002a p.24).

Case Study: BiH: Stability and Development?

Over €2bln of EC funds were committed for BiH between 1991 and 2000: humanitarian assistance provided by ECHO, the Humanitarian Aid Office, totalled €1.032bln, while assistance under the Obnova and PHARE programmes amounted to €890.7mln. In 2001 assistance of more than €105mln was committed under the CARDS Programme. In addition EU member states contributed over €1.2bln in assistance between 1996 and end of 2001. Both Mostar and Sarajevo have benefited from integrated reconstruction programmes covering public buildings, houses, and water, energy and transport networks. There have also been countrywide projects in the transport, telecommunications, energy, and water sectors. Extensive mine removal has been undertaken and programmes support returnees who were made refugees as a result of conflict. A priority is to create a single economic space which permits the free movement of goods, services, capital and labour, and makes way for complementarity with the EU aquis (EC 2002a). The country has begun preparing for a stabilisation and association pact with the EU, and confirmation of the country's membership of the CoE was imminent at the time of writing. The central bank's currency board regime pegged the convertible Marka to the Euro, bringing inflation down to 3–5 percent and eliminating exchange rate uncertainties for foreign investors and businesses. In February 2002 the first government roundtable on foreign investment and the creation of a single economic space within BiH was held (EC 2002b; Taylor 2002).

Summary and Conclusions

Endogamous Growth?

As in many regions subject to conflict, the myth that FDI will naturally follow major international aid has been exposed in the western Balkans (Teil and Woodward 1999). Despite the small amount of foreign investment taking place, economic development is progressing in the region. The main force behind this growth, however, is not the wider international business community, but regional business leaders and politicians, and some analysts argue that regional commerce and trade may act as the greatest single force for long-term security in the western Balkans (Civilitas Research 2002a). A number of regional companies have shown a greater willingness to explore new opportunities, irrespective of the continuing concerns expressed internationally about the high degree of political risk. This can be attributed partly to the strong comparative advantages enjoyed by businessmen from the region, such as: similarity in language, mentality and business cultures; pre-existing links and familiarity between sellers and buyers; easy availability of valuable local information about market opportunities; and an understanding of the various irregularities that exist. Notably, local companies carry less of the risk aversion mentality that characterises international companies. Certainly, the political nationalism of the early 1990s is being replaced by economic pragmatism which is deepening economic cooperation with neighbours and aiding regional stability and growth. This has been encouraged by the SP through a network of 21 bilateral free trade agreements (FTAs) among the SEECs to help create an economic area of 55mln consumers. Yet the new private sector which is emerging in the western Balkans is of predominantly small-scale, low-capital-intensity ventures in trade and construction. Most new private businesses are based around the family as the primary business unit, and traders continue to be their own bankers and insurers.

European Dreams?

As US engagement wanes (Dassú and Whyte 2001), the western Balkans has become the testing ground for a European vision of dispensing stability and prosperity. But while the EU has been taking on greater responsibility for Balkan affairs, it has also been absorbed in debates over its own enlargement and constitutional reform (Hall and Danta 2000). At present the rhetoric of Europeanisation for the states of the western Balkans remains insubstantial as they: have no early prospect of opening formal negotiations with the EU on membership; face sharply declining aid, because assistance under CARDS was heavily front-loaded towards the early years of the programme, and is now declining sharply; and are excluded from the larger European project of strengthening economic and social cohesion across the continent (Blazyca 2001; ESI 2002).

As the western Balkans' major support, the EU needs to be more explicit about its long-term commitment to economic and social cohesion after 2004. In particular, the stabilisation and association process needs to more centrally embrace economic cohesion policies consonant with the development principles that underlie the EU's structural funds, but de-linked from the progress of states towards accession, such that the Stability Pact could become an employment and cohesion pact.Three major problems are converging to present a new crisis of economic and social dislocation in the region: the substantial political and economic adjustment required in BiH and Kosovo as they attempt to cope with diminishing reconstruction assistance; a deepening unemployment and income inequality crisis (Bisogno and Chong 2002); and a growing disenchantment in the region with democratic processes which have appeared largely unresponsive to concerns over social and economic decline (ESI 2002). Without a serious commitment from the EU to develop a new set of policy instruments, there is the danger that countries of the western Balkans will find themselves increasingly isolated from European processes unfolding around them, while continuing to be locked into a dependency relationship for development and humanitarian aid.

References

Ádám, Z. (2000), 'FDI: Good or Bad?', *South East Europe Review for Labour and Social Affairs*, 3, 73–88.

Agence France Presse (2001), *The New Balkans Question: to Invest or Not to Invest*, AFP, London <http://www.balkanpeace.org/hed/archive/apr01/hed3155.shtml)>.

Allcock, J.B. (1991), 'Yugoslavia', in D. Hall (ed), *Tourism and Economic Development in Eastern Europe and the Soviet Union*, Belhaven Press, London, 236–58.

Anon (2002), 'Home Advantage', *Business Eastern Europe*, 31(32), 5.

Bisogno, M. and Chong, A. (2002), 'On the Determinants of Inequality in Bosnia and Herzegovina', *Economics of Transition*, 10(2), 311–48.

Blazyca, G. (2001), 'Will Europe ever be 'Round and Whole'? Reflections on Economic Boundaries and EU Enlargement after the Collapse of Communism', *Geopolitics*, 6(1), 6–26. Reprinted in A.H. Dawson and R. Fawn (eds), *The Changing Geopolitics of Eastern Europe*, Frank Cass, London, 6–26.

Brankovic, A. and Nenadovic, A. (2002), 'Undertaking a Difficult Transition in Yugoslavia', *Transition Newsletter*, 13(3), 21–24 <http:www.worldbank.org/transitionewsletter/mayjun02/pgs21–24.htm>.

Budway, V. (1999), *Making a Difference in the Balkans: The East-West Institute Action Network for SEE*, Action Network for South Eastern Europe, Prague <http://www.nato.int/ docu/colloq/1999/pdf/087–091.pdf>.

Burawoy, M. (1996), 'The State and Economic Involution: Russia through a China Lens', *World Development*, 24(6), 1105–17.

Büschenfeld, H. (1999), 'Aussenwirtschaftliche Entwicklungstendenzen in den Nachfolgestaaten Jugolawiens', *Osteuropa*, 49(3), 272–84.

Bužarovski, S. and Stojmilov, A. (2002), 'Macedonia', in F.W. Carter and D. Turnock (eds), *Environmental Problems of Eastern Europe*, Routledge, London, 347–65.

Carter, F.W., French, R.A. and Salt, J. (1993), 'International Migration between East and West in Europe', *Ethnic and Racial Studies*, 16, 467–91.

Castle, S. and Peric Zimonjic, V. (2002), 'Balkans Accord Spells the End for Yugoslavia', *The Independent*, 15 March.

Civilitas Research (2002a), *Regional Business Integration in the Western Balkans*, Civilitas Research Ltd., Nicosia <http://www.civilitasresearch.com/resources/view_article. cfm?article _id=7>.

Civilitas Research (2002b), *Serbia: Political Turmoil Threatens Economic and Social Stability*, Civilitas Research Ltd., Nicosia <http://www.civilitasresearch.com/ resourcesview_article.cfm ?article_id=24>.

Daianu, D. (2001), 'Transition Failures: How does Southeast Europe Fit In?', *Journal of Southeast European and Black Sea Studies*, 1(1), 88–113. Reprinted in T. Veremis and D. Daianu (eds), Balkan Reconstruction, Frank Cass, London, 88–113.

Daianu, D. and Veremis, T. (2001), 'Introduction', in T. Veremis and D. Daianu (eds), *Balkan Reconstruction*, Frank Cass, London, 1–11.

Danta, D. (2000a), 'Bosnia and Hercegovina', in D. Hall and D. Danta (eds), *Europe Goes East: EU Enlargement, Diversity and Uncertainty*, The Stationery Office, London, 289–98.

Danta, D. (2000b), 'Former Yugoslav Republic of Macedonia', in D. Hall and D. Danta (eds), *Europe Goes East: EU Enlargement, Diversity and Uncertainty*, The Stationery Office, London, 310–20.

Danta, D. (2000c), 'Serbia and Montenegro', in D. Hall and D. Danta (eds), *Europe Goes East: EU Enlargement, Diversity and Uncertainty*, The Stationery Office, London, 299–309.

Dassú, M. and Whyte, N. (2001), 'America's Balkan Disengagement', *Survival*, 43(4), 123–36.

Drummond, P. (2000), *Former Yugoslav Republic of Macedonia – Banking Soundness and Recent Lessons*, IMF Working Paper WPO/00/145, New York.

EBRD (1999a), *EBRD Activities in FYR Macedonia*, EBRD, London <http://www.ebrd.com /english/opera/COUNTRY/fyrmfact.htm>.

EBRD (1999b), *Transition Report: Ten Years of Transition*, EBRD, London.

EC (2001a), *The EU's Relations with Albania*, DGER EC, Brussels <http://europa.eu.int /comm/external_relations/see/albania/index.htm>.

EC (2001b), *The EU's Relations with SEE: Albania Country Strategy Paper 2002–2006*, DGER, EC, Brussels <http://europa.eu.int/comm/external_relations/see/albania/csp/ index.htm>.

EC (2001c), *The EU's Relations with SEE: CARDS Regional Strategy Paper 2002–2006*, DGER EC, Brussels <http://europa.eu.int/comm/external_relations/see/news/ip01_ 1464.htm>.

EC (2001d), *The EU's Relations with SEE: Republic of Montenegro*, DGER, EC, Brussels <http://europa.eu.int/comm/external_relations/see/fry/montenegro/index.htm>.

EC (2002a), *The EU's Relations with Bosnia and Herzegovina*, DGER EC, Brussels

<http://europa.eu.int/comm/external_relations/see/bosnie_herze/index.htm>.

EC (2002b), *The EU's Relations with SEE: Bosnia & Herzegovina: Country Strategy Paper 2002–2006*, DGER EC, Brussels <http://europa.eu.int/comm/external_relations/see/bosnie_herze/csp/index.htm>.

EC (2002c), *The EU's Relations with SEE: Overview*, DGER EC, Brussels <http://europa.eu.int/comm/external_relations/see/>.

EIU (2002a), *Albania: Country Report October 2002*, EIU, London.

EIU (2002b), *Macedonia: Country Profile 2002*, EIU, London.

EIU (2002c), 'Creating More Jobs in CEE', *Transition Newsletter*, 13(4–5), 12–15 <http:www.worldbank.org/transitionewsletter/julaugsept02/pgs12–15.htm>.

Elbirt, C. (1997), 'Albania under the shadow of the pyramids', *Transition*, 8(5), 8–10.

ESI (2002), *Western Balkans 2004*, ESI, Berlin <http://esiweb.org/westernbalkans/showdocument.php?document_ID=37>.

Gligorov, V. (1998), *Trade and Investment in the Balkans*, Vienna Institute for International Economic Studies, Vienna <http://www.wiiw.ac.at/balkan/files/Gligorov.pdf>.

Gros, D. (2002), 'The Euro for the Balkans?', *The Economics of Transition*, 10(2), 491–511.

Gumbel, A. (1997), 'Albania's Export Boom in Vice and Drugs', *The Independent*, 2 December.

Haderi, S., Papapanagos, H., Sanfey, P. and Talka, M. (1999), 'Inflation and Stabilisation in Albania', *Post-Communist Economies*, 11(1), 127–41.

Hall, D. (1993), 'Impacts of Economic and Political Transition on the Transport Geography of CEE', *Journal of Transport Geography*, 1(1), 20–35.

Hall, D. (1996), 'Albania: Rural Development Migration and Uncertainty', *GeoJournal*, 38(2), 185–89.

Hall, D. (2001), 'Albania in Europe: Condemned to the Periphery or Beyond?', *Geopolitics* 6(1), 107–18. Reprinted in A.H. Dawson and R. Fawn (eds), *The Changing Geopolitics of Eastern Europe*, Frank Cass, London, 107–18.

Hall, D. and Danta, D. (eds) (1996), *Reconstructing the Balkans: A Geography of the New Southeast Europe*, Wiley, Chichester, 15–32.

Hall, D. and Danta, D. (eds.) (2000), *Europe Goes East: EU Enlargement, Diversity and Uncertainty*, The Stationery Office, London.

Hewko, J. (2002), 'FDI: Does the Rule of Law Matter?', *Transition Newsletter*, 13(3), 11–13 <http:www.worldbank.org/transitionewsletter/mayjun02/pgs11–13.htm>.

Hunya, G. (2000), *Recent FDI Trends, Policies and Challenges in South-East European Countries*, WIIW Research Report 273, Vienna.

Hunya, G. (2002a), *FDI in SEE in the Early 2000s*, WIIW, Vienna <http://www.investmentcompact.org/pdf/hunyajuly2002[1].pdf>.

Hunya, G. (2002b), *Recent Impacts of FDI on Growth and Restructuring in Central European Transition Countries*, WIIW Research Report 284, Vienna.

Kekic, L. (2001), 'Former Yugoslav Republic of Macedonia (FYROM)', in T. Veremis and D. Daianu (eds), *Balkan Reconstruction*, Frank Cass, London, 186–202.

Korovilas, J.P. (1999), 'The Albanian Economy in Transition: The Role of Remittances and Pyramid Investment Schemes', *Post-Communist Economies*, 11(3), 399–415.

Lawson, C. and Saltmarshe, D. (2000), 'Security and Economic Transition: Evidence from North Albania', *Europe-Asia Studies*, 52(1), 133–48.

Martin, P., Martin, S. and Pastore, F. (2002), 'Best Practice Options: Albania', *International Migration*, 40(3), 103–18.

Minervini, C. (2002), 'Housing Reconstruction in Kosovo', *Habitat International*, 26(4), 571–90.

Muço, M. (2001a), 'Albania', in T. Veremis and D. Daianu (eds), *Balkan Reconstruction*, Frank Cass, London, 119–31.

Muço, M. (2001b), 'Low State Capacity in Southeast European Transition Countries', in T. Veremis and D. Daianu (eds), *Balkan Reconstruction*, Frank Cass, London, 41–54.

Papasotiriou, H. (2002), 'The Kosovo War: Kosovar Insurrection Serbian Retribution and NATO Intervention', *Journal of Strategic Studies*, 25(1), 39–62.

Pendovska, V. and Georgievski, S. (1999), 'Legal Aspects of Foreign Investment in Macedonia', *Review of Central and East European Law*, 25(4), 513–70.

Pissarides, F. (2001), *Financial Structures to Promote Private Sector Development in SEE*, EBRD Working Paper 64, London.

Prentice, E-A. (2002), 'Big Business Leads the Way as Germans Conquer the Balkans', *Scotland on Sunday*, 25 August.

Resmini, L. (2000), 'The determinants of FDI in the CEECs: New Evidence from Sectoral Pattern', *Economics of Transition*, 8, 665–90.

Saltmarshe, D. (2001), *Identity in a Post-communist Balkan State: An Albanian Village Study*, Ashgate, Aldershot.

Schifferes, S. (2001), 'Yugoslavia's Shattered Economy', *BBC News Online*, 2 July <http://news.bbc.co.uk/1/low/business/1410623.stm>.

SCSP(2002), *About the Stability Pact*, SPC, Brussels <http://www.stabilitypact.org/stability pactcgi/catalog/cat_descr.cgi?prod_id=1806>.

SECI (2002), *Southeast European Cooperative Initiative*, SECI <http://www.secinet.org>.

Skayannis, P.D. and Skyrgiannis, H. (2002), 'The Role of Transport in the Development of the Balkans', *Eastern European Economics*, 40(5), 33–48.

Staddon, C. and Turnock, D. (2001), 'Conclusion: Environmental Geographies of Post-Socialist Transition', in D.Turnock (ed.), *ECE and the Former Soviet Union*, Arnold, London, 231–46.

Stojmilov, A., Temjanovski, R. and Ziko, M. (2000), 'Tourismus und Tourismuspotential Makedoniens', *Südosteuropa-Studie*, 66, 157–76.

Taylor, M. (2002), 'Help! A Divided and Troubled Bosnia makes a Pitch for Foreign Investment', *Business Eastern Europe* 31(10), 1.

Teil, B. and Woodward, S.L. (1999), 'A European "new deal" for the Balkans', *Foreign Affairs*, 78(6), 95–105.

UNHCR (1999), *FYR of Macedonia*, UNHCR, Geneva <http://www.unhcr.ch/world/euro /macedon.htm>.

WBG (1999), *The Foreign Investment Advisory Service Program in the Balkan States*, WBG, New York <http://www.seerecon.org/DonorPrograms/DonorPrograms-FIAS.htm>.

WBG (2002a), *Economic Reconstruction and Development in SEE*, WBG, New York <http:// www.seerecon.org/BusinessOpportunities/BusinessOpportunities.htm>.

WBG (2002b), *Rebuilding Kosovo*, WBG, New York <http://www.worldbank.org/html /extdr/kosovo/>.

WBG (2002c), *Stability Pact for SEE*, WBG, New York <http://www.seerecon.org /RegionalInitiatives/StabilityPact/sp.htm>.

WBG (2002d), *The Regional Environmental Reconstruction Programme*, WBG, New York <http://www.seerecon.org/RegionalInitiatives/RERep.htm>.

WIIW (2002), *FDI Inflow in Transition Countries Expected to Decline*, WIIW, Vienna <http://www.wiiw-wifo_fdi_June02_summary_eng[1].pdf>.

Willan, P. (1997), 'Mafia Linked to Albania's Collapsed Pyramids', *The European*, 13 February.

Wintrobe, R. (2002), 'Slobodan Miloševič and the Fire of Nationalism', *World Economics*, 3(3), 1–26.

Xing, Y. and Kolstad, C.D. (2002), 'Do Lax Environmental Regulations Attract Foreign Investment?', *Environmental and Resource Economics*, 21(1), 1–22.

Chapter 10

Foreign Direct Investment in Bulgaria: The First Ten Years

Francis W. Carter

Introduction

FDI may be defined in its simplest form as investment by one firm in another across borders. The aim is to acquire some degree of control, either by means of creating a wholly owned subsidiary, the acquisition of a minority/majority joint venture, or a greenfield/brownfield site (Dunning 1998). The sheer variety of home country characteristics, as well as the motives and strategies of individual firms in taking advantage of opportunities available in host countries, has made any single theory of FDI difficult to develop. Most FDI theories are aimed at investigating the economic mechanism and impacts of host and home economies, and the spatial dimension is often ignored. Exceptions are Dunning's 'eclectic' approach and Vernon's product cycle method (Dunning 1997; Vernon 1966, 1979). Within this context, the consequences of FDI location at local and regional level can be quite significant. Attempts to interpret where, why and how production occurs at an international level have been critical in the theoretical development of the locational dimension of multinational firms (Dunning 1992) Even so, further empirical research needs to be done in support of these concepts. Its influence on the development of urban centres, whole regions and local communities has to some extent been overlooked, together with the whole concept of 'commodification'.

Why Bulgaria?

Many issues concerning FDI location at local and regional level in the post-communist countries need some explanation. Systematic spatial analysis of FDI activities in countries like Bulgaria can only add to our knowledge of the Balkan region as a whole. By 1990, Bulgaria had experienced a decade and a half of lost competitiveness and industrial decline. As in other post-communist societies, the economic transition brought about a shift from a socialist economic to market philosophy. The former had been influenced by ideology and strategic-political factors because under Comecon the predominant (though declining) orientation was eastward; while the West was neglected in the 'mental maps' of the central planners.

After 1990, cost effectiveness became of utmost importance in the transition to economic efficiency. Slowly Bulgaria began to realise that without FDI and freer access to EU markets, it would be virtually impossible to modernise its economy and develop a stable middle class (Pishev 1990). Yet Bulgarian production capabilities and markets were potentially very vulnerable and FDI was regarded locally with suspicion: a clear majority of Bulgarians believed foreign businesses 'exploited' them (Dobosiewicz 1992; Garnizov 1995).

At this time, Bulgaria was the least reformed country in the region after Romania (Izvorski 1993; Jones and Miller 1997). Between 1990–1993 Bulgaria received only about one percent of total FDI won by former communist countries, partly because the country's investment potential was poorly advertised abroad. Regional wars in the former Yugoslavia certainly compounded the problem of Bulgaria's limited attractiveness, but progress towards privatisation was limited (Konstantinov 1994; Popov and Todorova 1997) and 90% of the large industrial production complexes remained under state control. Although the private sector increased its share from five percent of the economy in 1989 to a third in 1993, it comprised mainly small units specialising in trade and services (Bogetic and Hillman 1994). Meanwhile, the privatisation of medium- and large-scale enterprises made little progress due to highly politicised decision-making and the lack of clear incentives at government level, combined with a weak regulatory environment, enterprise indebtedness and resistance to major privatisation transactions at the grassroots.

However, while Bulgaria was only moderately attractive as a host country during four years of declining output and income, the mid-1990s showed signs of economic improvement and by the end of the 1990s the economy was providing a shining example to the rest of the Balkans, reflected in a growth of interest in Greece as well as other EU states (Petrakos 1997; Petrakos and Christodoulakis 2000). It managed to attract the largest amount of FDI amongst post-communist SEECs in 2000 when the Bulgarian target of $1.0bln was exceeded by 20% and the top 25 investors accounted for three-quarters of this total. The Bulgarian economy was soon poised to take full advantage of the reopening of the Danube and the prospect of closer business contacts with Serbia. This has been helped by the promise of a new Danube bridge at Vidin-Calafat funded by €190mln from the Stability Pact. Internally, realistic encouragement is being offered the existing legal framework and macro-economic indicators: GDP grew by 3.3% in 2000 and was forecast to reach 5.0% in 2001. Free movement of capital was aided by a new foreign currency law, together with simplified licensing procedures for new businesses: a move deliberately taken to encourage further investment. At the time it seemed that large energy ventures with the USA might add another billion dollars to the modest cumulative total of $3.0bln while economic reform was being negotiated in order to bring a new agreement into play when the current one expired in May 2001.

Data Sources

During the early transition years, data was available only from Bulgaria's Privatisation Agency established in 1992 with responsibility for the sale of large, strategic SOEs. 72 transactions were completed by June 1999 through participation by companies from Austria, Belgium, Germany, Korea, Sweden, Switzerland and the USA Korea (Zheliazkov 2000). In April 1995, the Bulgarian Foreign Investment Agency (BFIA) was founded to help would-be foreign investors. It had close business relations with central and local government and state institutions, as well as business organisations and other NGOs. Up to 1997 the government only provided FDI data on joint ventures and greenfield sites (from 1992) and on privatised companies (from 1993). But in 1997, new legislation was passed with the aim of establishing a unified information system, including data on all foreign investments. At the end of each calendar quarter, the Ministry of Finance, the National Statistical Institute, the Central Depository and the Bulgarian National Bank submitted aggregated data on the type and volume of foreign investments, while other central and local authorities were required to provide information as and when requested by the BFIA. But in spite of this late start, sufficient data became available to allow a spatial evaluation for the whole of the 1990s and thereby assess the impact of FDI on the country's regional development. However, there is only limited back-up information because regional data on private sector employment is not published in the statistical yearbooks, although it would have been very useful for this paper in revealing possible spatial correlations with FDI.

Bulgarian Politics and FDI

The transition in Bulgaria began in November 1989 when two main political parties emerged: the former communists under the banner of the Bulgarian Socialist Party (BSP) and the opposing Union of Democratic Forces (UDF) whose supporters wanted radical change (World Bank 1991). These parties reflected the division in Bulgarian society which has persisted to the present. Instability in the early 1990s was highlighted by no less than six BSP governments between November 1989 and January 1994, broken by an 11–month period of UDF rule from November 1991 to September 1992 which enacted important legislation to encourage FDI: notably the law on the Privatisation of State and Municipal Enterprises (April 1992), albeit heavily amended by the BSP in June 1994 (Panov 1994). And although the BSP government's economic programme of February 1995 adopted privatisation as part of a reform process envisaging radical and irreversible change, market reform and restructuring actually slowed down and a three-year freeze on restitution was imposed. It was not surprising that this attempt by the BSP to reinstate a centrally planned economy provoked an unprecedented economic disaster in 1996. FDI remained low and was concerned mainly with chemical, electronics and engineering companies. It was clear that Bulgaria's political leaders still had a lot to learn about the art of securing rapid privatisation through foreign investment (Marinov and

Marinova 1997; Tchipev 1996; Whitford 1996). A significant casualty of the 1996 crash was the $7.5mln Rover car plant launched the previous year in a former military installation in Varna with an annual capacity of 10,000 Maestro cars and vans: it was forced to close when promised government orders failed to materialise.

The election of the UDF coalition in 1997 brought a remarkable economic improvement. The general acceptance of a currency board agreement created political stability and macro-economic rejuvenation leading to accelerated privatisation, which in turn boosted FDI and provided expectations of sustainable growth. Incentives grew when the Bulgarian parliament adopted a new foreign investment law in October 1997 to stabilise the crisis-hit economy (Nicholls 1999). FDI was to further encouraged by the Foreign Investment Act of October 1997. The BFIA was now to be financed from the state budget and would operate on the basis of regional divisions. Increased FDI underpinned the planned large-scale privatisation of SOEs, of which over a thousand were scheduled for early disposal. Greek companies were prominent at this juncture, seeing Bulgaria as the first step into ECE markets. For the first time, the UDF government drew up a new public sector investment programme (1998–2001) as a consequence of progress in creating the more favourable macro-economic climate which was stimulating the investment market. The government sanctioned a proactive marketing strategy with emphasis on investment opportunities in the food industries (including tobacco and wines), textiles and tourism. According BNBank data the inflow of foreign investment in 1999 exceeded $660mln.

Deals were signed by UniCredito (Italy) for Bulbank and the National Bank of Greece for the United Bulgarian Bank. Also included were the purchases of three enterprises in Devnya (Varna) by Solvey Sodi, Neckermann's investment in the Golden Sands resort, the purchase of Nova Televisiya and Radio Express by the Greek-owned Antenna company, the construction of a $25mln refrigerator plant by Liebherr (Germany) and an investment in stores by Billa (Austria) worth over $20mln. The government succeeded in privatising virtually all large industrial enterprises and all but two commercial banks, supported by a new regulation system and improved public administration. Bulgaria also undertook to work with neighbouring countries on a common regional approach to meet EU standards (CSD 2000; Lampe 2001). Yet, delays in reforming state-owned monopolies limited the opportunities for business to take advantage of the country's excellent geographical position at the contact between Europe and Asia. And at the run-up to the 2001 Spring elections considerable political problems still faced Ivan Kostov's UDF government. Public opinion was irritated by the slow rise in living standards (e.g. low pensions and university salaries) and high-level corruption: scandals during the summer of 2000 led to nine cabinet ministers being replaced (Whitford 2001). Another problem was unemployment exceeding 20%; leaving 3.2 million people in work with the burden of supporting a retired population of 2.4 million. Meanwhile, emigration of young educated Bulgarians during transition reached the half million mark.

The Regional Distribution of FDI

The location of business activity has generated a growing theoretical literature on MNEs since the late 1980s, notably through the work of J.H. Dunning (Safarian 1999). Business development at a particular place – defined as a region, urban centre, locality or community within a country – is seen as contributing a knock-on effect (in terms of prosperity and welfare) reflected in the employment, skills, IT and income as well as regional politics and the social environment. In recent years, MNEs have exerted a particularly potent influence as their importance has grown in relation to Bulgarian firms. Dunning's eclectic paradigm suggests that MNEs have advantages in total capital stock, technology and management/marketing know-how which could be brought to Bulgaria through commodification: the gradual evaluation, treatment and perception of the factors, services and goods of production as commodities used by enterprises to create profit. Exchange value becomes much more important than the use value (formerly important under communist central planning). Thus, value may be added to locations by development and marketing techniques so that their intrinsic use value is exceeded. These factors have an effect at a number of levels within a country like Bulgaria: applying to both factors of production and locations, as well as the spatial relationships between them.

   With increasing globalisation, the only type of private enterprise which acts as a catalyst for commodification usually functions as a joint venture or is bought outright by foreign investors (Carter 1999a, 1999b; Hamilton 1995, 1999). The concept has made considerable input into FDI through first advantage theories and the 'follow the leader' scenario. Under these conditions strategic competition factors have led companies to build up very large capacities in the hope of discouraging competition. Therefore through MNEs, the significance of FDI has increased more rapidly than either trade or production (Estrin 1994). It is clear that commodification is a key concept in this process. During the communist era, factors of production and merchandise were not created as commodities, but with the transition, there has been a switch from use value to exchange value. Since 1990, this shift has involved a re-evaluation in the role of place and location in post-communist countries. It has affected factors of production such as natural resources, because prices have been liberalised so that procurement now reflects real transport costs. The old socialist values of use optimisation have been replaced by more economic values such as the cost effectiveness of extraction; although some negative influence continue, such as environmental pollution (Carter 1996; Paskaleva et al. 1998).

   Locational problems also exist regarding the creation of cheap labour markets (Paunov 1993). For example, although unemployment exists in Bulgaria and labour markets may be described as depressed, this does not automatically mean there is great job competition. Lack of employment opportunities still remain, given the high rigidity found in some post-communist countries e.g. labour hoarding in former SOEs. Another problem is that deindustrialisation – and consequent job-shedding – may proceed more quickly than new employment that can be created to exploit each regional labour market (Strong et al. 1996). At the same time, land and property are also commodified through privatisation, but the process has not spread evenly across Bulgaria because it has been more successful in city centres than rural or peripheral areas (Carter and Kaneff 1999; Kopeva 1994). Labour has entered the commodification process in those firms which were privatised but must now compete with other regions of the country, with neighbouring post-communist states and indeed on an ever wider, global scale. Such factors as skills, productivity, wage levels and working conditions have all to be taken into account. Thus a mosaic pattern of deindustrialisation, paternalistic and globalising enterprise exists. Large SOEs, especially in mining and heavy industry, have not been successfully privatised. Far from being constrained by hard budgets, they continue to receive subsidies and carry out barter: especially in Bulgaria and other countries where incomplete reform prevents commodification from penetrating the economy fully.

On the basis of experience over the past decade, more research is needed into why FDI has been located in specific regions, cities and communities. FDI and location are clearly closely interrelated because in formulating an investment strategy, choice of the host country is the first priority for MNEs (while location within the country is of secondary importance). As regards Bulgaria, foreign companies appreciate the country's large consumer demand, qualified (yet low cost) personnel, low material costs and the strategic geographical location. As FDI is primarily responsible for increasing the number of globalising enterprises accompanied by commodification, it has contributed greatly to economic restructuring and locational change. Technological modernisation, economies of scale and scope, as well as flexible production specialisation have all been introduced by foreign investors attempting to exploit domestic markets or to use the region as an export platform to the EU, or both. Bulgaria's cumulative FDI since 1989 only exceeded $1.0bln in 1997, although it exceeded $600mln in both 1997 and 1998 – reflecting an economic upturn, albeit tempered by slow privatisation and the impact of the Kosovo conflict. However, increased greenfield investment, re-equipment of privatised firms and infrastructural work should ensure some growth of FDI in coming years (Hall 2000). FDI in Bulgaria certainly suggests spatial discrimination – indeed Buckwalter (1995) demonstrates resurgent regional inequality – as foreign investors have chosen to locate in areas that can effectively serve their target markets, because FDI inflow motives are strictly connected to the success (or otherwise) of the overall effect on trade (Figure 10.1, based on BFIA data relating to a total investment of $2.34bln). Usually metropolitan areas and EU border zones are the largest recipients of FDI. 'Secondary cities' and nodal areas such as ports also receive their fair share. Areas near EU markets also attract FDI, in Bulgaria's case the Struma valley: the gateway to northern Greece (Petrakos 1997). The Black Sea coast's major urban centres attracted more FDI for tourism than for either commerce or port activity. In the early 1990s, very little FDI went to the relatively poor central region, while over two-thirds (72%) of the northern region's economy depended on manufacturing.

Figure 10.1 Distribution of FDI in Bulgaria, 1992–1999, by districts

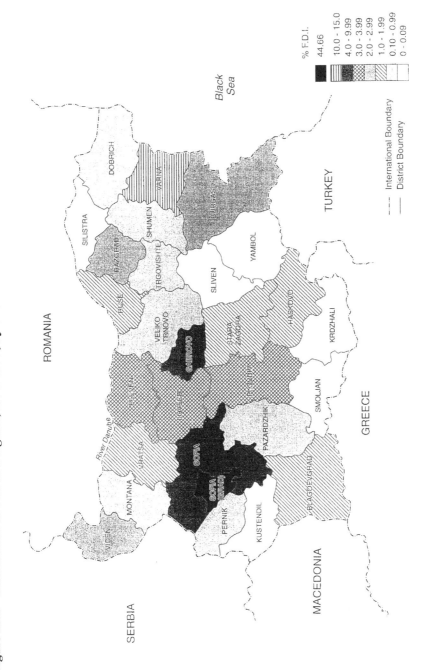

Sectoral Characteristics

Bulgaria's leading FDI countries between 1992–1998 were Belgium and Germany with 38.1%. Ten of the top 16 states were EU members. A sectoral breakdown showed that well over half (54.1%) went into industry, followed by trade (19.0%); less than a tenth was allocated to finance, tourism and transport, while construction and telecommunications had 2.6% and agriculture only 0.3% (Table 10.1). The largest privatisation investments were for the 'Sodi' chemical works at Devnya (by Solvay of Belgium: $160mln), the MDK copper smelter at Pirdop (by Union Miniere of Belgium: $103mln) – followed by the SOMAT road haulage firm (by Willi Betz of Germany: $68mln) and the Vitosha (now Kempinski) Hotel in Sofia (by Ivan Zografski of Germany: $53mln). The degree of foreign penetration by manufacturing industries is dependent on industry-specific features and the quality of the privatisation policy. FDI in Bulgaria follows a global pattern of industrial corporate integration. For example, FDI in textiles, clothing and leather is less globalised than for most other commodities, while there is only a small foreign presence in industrial branches having large, cumbersome structural difficulties and oversized capacities, like the steel industry. Technology-intensive electrical machinery and car production are often popular FDI targets.

Engineering has consistently proved to be the most attractive FDI manufacturing branch. Mechanical engineering has a significant position in the national economy with proven export capabilities. Sales reach 60 countries in both the developed and emerging markets. Much of this is based on a low dependence on imported raw materials, quality goods production and well-established trade relations. The major trading partners include other Balkan countries, the Middle East, Western Europe and the CIS. During the transition period, sales to new markets have compensated for the loss of others mainly in ECE. Foreign interest has been particularly notable

Table 10.1 Sectoral breakdown of FDI, 1992–1998

Sector	FDI $mln	Percent	Companies	Percent
Industry	1,034.2	54.1	720	6.9
Trade	362.3	19.0	8,270	79.2
Finance	205.3	10.7	160	1.5
Tourism	101.5	5.3	97	0.9
Transport	89.0	4.7	199	1.9
Telecoms	32.6	1.7	25	0.2
Construction	17.9	0.9	113	1.1
Agriculture	6.1	0.3	72	0.7
Others	60.5	3.2	787	7.5
Total	1,909.7	100.0	10,443	100.0

Source: BFIA.

in the electronics/electrical engineering sector. This is because, as a vital sector of the Bulgarian economy, it has received priority in long-term development programmes, based on export potential and minimal dependence on imported raw material. Production is concentrated largely in medium-sized (50–250 employees) and large-sized (over 250 employees) firms respectively. Most foreign investors have acquired part or total ownership of firms producing instruments and automation parts, mainly for the Middle East, India, Pakistan and South America, as well as Balkan and CIS markets. The major FDI participants in electronics are 'Pramet Bulgaria' (Czech Republic) and 'DRS Ahead Technology' (USA). The Bulgarian chemical industry is well structured and has a strong position in several major export markets. Its location has pronounced advantages especially for trade with the Middle East and Mediterranean countries. The largest plants are located near ports on the Black Sea and Danube river. These sites are favourably located for the import of raw materials, especially oil and gas via pipelines from the CIS. Fortunately Bulgaria has fewer environmental problems than in some of the other transition countries in the region, and these are being helped by a further reduction in levels of dangerous pollutants under current restructuring programmes. Foreign investment has centred around Devnya (Agropolyhim, Polymeri and Sodi) and Ruse (Orgahim and Petar Karaminchev). The sale of Sodi to the MNE Solvay Group (Belgium) for $160mln in September 1997 (already noted) was the largest foreign transaction. There are also 10 joint ventures in the chemical industry most involved with producing plastics and synthetic resins. Meanwhile the strength of state-owned fertiliser producers – still delivering profits to the national budget – should be mentioned: 'Neochim' of Dimitrovgrad and 'Chimko' of Vratsa.

Turning to light industry, agriculture and the tertiary sector, it is interesting to see foreign investors specialising according to nationality. Italian firms show interest in design and textiles, while French firms are often linked with hypermarkets and those from The Netherlands with agribusiness. However, many like to extend their activities into relatively stable domestic markets of the food, beverage and tobacco industries which constitute one of Bulgaria's major economic sectors. Most foreign interest has been in wine, beer and tobacco – albeit hampered by problems associated with the unclear ownership of agricultural land. However, the Bulgarian government has committed itself to completing the agricultural land restitution process and is making serious attempts to create a real land market which should provide more favourable investment conditions for the agricultural sector (Carter 1998; Carter and Kaneff 1999). Most FDI interest has come from Interbrew (Belgium), Kraft General Foods Ltd and Nestlé (Switzerland), TKM Fruits and Juices (Greece) and Helian Commodities (Netherlands).

With tourism designated a strategic economic sector, a comprehensive development programme is under way, covering human resources and marketing. It is being supported by the EU which, it is hoped, will help make Bulgaria more significant at an international level (Bachvarov 1999). Hotels in Nesebar, Sofia and Varna attracted investors from Cyprus, Hungary, Italy and Turkey in 1998. Nevertheless, while foreign investors may be attracted to a traditionally profitable sector, they must be prepared to spend more money particularly on upgrading hotel star ratings, as well as a greater range of packages and services offered. Transport has also been commodified. In the socialist era it was viewed at best as a social cost or discounted altogether as merely a fraction of time and distance. Now such issues as nodality and accessibility need to be accounted for financially. Since 1990 transport costs have also reflected distance and the quality/quantity of the infrastructure. Transport therefore plays an important role in the location of a firm as the cost of supplying consumer and producer markets now have to be considered.

Although Bulgaria's transport infrastructure is relatively well developed, it has suffered in recent years from insufficient financial support and poor maintenance. Even so, Bulgaria is situated on the traditional overland route connecting Western Europe with Turkey and the Middle East, enhancing the country's transit transport role significantly. A northern axis along the Danube has suffered considerably as a result of events elsewhere. The country has no other rivers that are navigable by sizeable ships. Blockading of the Danube has kept the river closed to shipping, and NATO's bombing campaign has badly hurt both Bulgaria and Romania which depend on this transport corridor. However, Bulgaria should now benefit from peace in Yugoslavia and the tourism sector (which shrank by a third in the wake of the 1999 bombing campaign against Serbia) is growing again. Road and river traffic is again passing through former Yugoslavia, lowering the costs of Bulgarian trade. Bulgarian building firms are well placed to win contracts in former Yugoslavia, while Bulgaria's oil refinery in Burgas on the Black Sea could become a major supplier. The southern axis of major urban centres from Sofia along the Maritsa and Struma valleys also offers lower risks for foreign investors particularly from Greece and Turkey. Another strong axis emanates from Sofia eastwards towards the Black Sea coast and the significant port of Varna. In broad terms, the western half of Bulgaria has attracted more FDI, especially for SMEs which need to maintain close contact (in terms of management and input/output linkages) with their home base, compared with larger firms which can be more mobile.

In terms of surface transport, four of the proposed and long-anticipated European rail and road transport corridors pass through Bulgaria. These include: Corridor IV from Calafat in Romania via Sofia to Promachin in Greece and then to Svilengrad on the Bulgarian-Turkish border; Corridor VIII from Gushevo on the Bulgarian-Macedonian border, via Sofia to the Black Sea ports of Varna and Burgas; Corridor IX which crosses from Giurgiu (Romania) to Alexandroupolis in Greece; and finally Corridor X entering from the Serbian border en route for Edirne in Turkey. Bulgaria has also begun segregating its railways according to service function, while the Black Sea and Danube ports offer direct links with the CIS, the Mediterranean region and Central/Western Europe. The infrastructures of both Sofia airport and Burgas maritime port are undergoing modernisation. Under communist central planning, Bulgaria was the major supplier of telecommunication equipment for the Comecon trading bloc, an advantage which has proved useful since 1990 transition. Connections and line quality have rapidly improved thanks to digital overlay and international installation supported by the EBRD, as well as GSM and analogue cellular services, all facilities necessary to attract potential foreign investors.

The Role of FDI in the Urban Pattern

The cumulative FDI regional pattern by district between 1992–1999 confirmed the capital, Sofia (population 1.2mln) as the top attraction with $196mln: 57.97% of the investment in 1998, a disproportionate concentration, especially because of the demands for highly-educated and professional employees. Examples include Somat/Willi Betz (Germany) in transport and Hotel Vitosha/Ivan Zografski (Germany) in the hotel sector as well as ShZI/Nestlé (Switzerland) in the food industry and Eskos Dograma/Gibu (Italy) in wood processing. Turkish capital has gone into banking while Mobicom (UK) and Mobiltelad (Bahamas) have invested in telecommunications. Along with leading provincial cities, Sofia will continue to attract FDI and will therefore induce migration from the smaller urban centres. This follows a general pattern whereby FDI shows a preference for the capital city since it is often the city best known by the investor, not to mention the market size, the financial and other services, the transport connections and the communications infrastructure. The city is also recognised as having potential as an air transport hub and there are plans to build a new airport there. Major metropolitan areas make for relatively low risk (despite the costs of agglomeration); so investors gain confidence and gradually diversify by taking advantage of services most easily found in the capital and larger urban centres. Bulgaria's capital and primate city is also the seat of national economic development and a modern service centre. All this was highlighted during 1989–1993 when three-fifths of all FDI in producer services was concentrated in the capital, with a further fifth in its surrounding region.

At the end of the 1990s Sofia's economic environment was highly dynamic, thanks to the increasing role of private businesses. The central core is the key area for business headquarters, legal and financial services and retailing – causing rents to spiral (Cheshire and Hamilton 2000) – while the leading sectors of the former socialist economy are relatively neglected. Public sector firms declined during 1996–1998 from 977 to 632, compared with an increase from 26,540 to 32,143 for private firms: a clear indication of changing economic structure and evidence of firms' capacity to enter the sought-after register of the National Statistical Institute. While unemployment is still a major problem in Bulgaria (16% in 1998 and 18% in 1999), due to privatisation and the disappearance of SOEs, Sofia has the lowest recorded level and accounts for 9.3% of all available jobs. Unemployment reached a peak of 15% in 1993 but has declined steadily ever since. Meanwhile, the wider capital city region has become much more highly valued given the guaranteed market, the concentration of skills and services. In 1998 the structure of Sofia's economy shows that almost half of all registered firms in the city are concerned with trade and repairs.

Some FDI is also located in other cities and nodal areas including the ports

(Burgas and Varna) – thanks to a rising demand for Bulgarian goods in both Germany and the CIS – and in other areas near to EU markets; while Turkey has invested in Bulgaria as a neighbouring country that was once part of the Ottoman Empire. Devnya attracted 16.1% of FDI to 1998 ($54.5mln), Ruse 7.2% ($24.5mln) and Plovdiv 5.9% ($20.1mln), while all the other 20 leading centres of FDI (fairly evenly spread throughout the country) had less than one percent apart from Peshtera (4.1%) and Sevlievo (4.1%). But it is impressive that the top ten centres accounted for 97.6% of the total. Like Sofia, these centres have experienced a real estate boom with the inevitable rise in house prices and rents. Even demand for building insulation products grew following the freeing of energy prices (Levinson 1995). As private business activity has flourished, ground floor flats have become small retail outlets, especially boutiques and small cafes. This phenomenon is particularly evident amongst many pre-communist property owners or their heirs who, after reclamation, changed their property into shops and restaurants, while foreign trade companies introduced a variety of Western consumer goods.

Overall there appears to be a clear west-east division. Investors are interested not only in Sofia but also Plovdiv and increasingly Haskovo and Lovech while Blagoevgrad and the Struma Valley has been strongly influenced by Greek entrepreneurs, particularly from Thessaloniki: e.g. Sandanski has tobacco processing supported by Liib Tobaco Mihailidis of Greece. Ruse attracted foreign capital for its chemical and engineering industries; also Cherven Briag (textiles), Pavlikeni (paper), Sevlievo, Gabrovo and Gorna Malina (instruments) in the north – as well as Lovech and Veliko Trnovo in the Danubian Plain. In the Middle Maritsa valley Plovdiv is dominant (glass and footwear), along with Kalojanovo (engineering), Liubimets (instruments), Peshtera (footwear), Stamboliiski (paper) and Yambol (textiles). Ruse on the Danube is prominent but the border towns near Greece (Kardzhali and Smolyan), Serbia (Kustendil and Pernik) and Turkey (Yambol) have not done very well, while the Macedonian frontier is generally difficult because of the emphasis on agriculture.

Many FDI projects in Bulgarian towns have involved brownfield sites through decisions to invest in a specific inherited stock of equipment, know-how and production capacity where the location is already fixed from the communist period but may well be acceptable in a new technological, logistical and market context. The result may be a wholly-owned subsidiary or a minority/majority joint venture. Timing if often crucial because opportunities in particular industries are likely to arise in all transition states and unless the investor wishes to acquire capacity in every state (a very unlikely outcome given the likely interest in a global market) much will depend on when individual plants become available and when foreign companies decide they want to get involved in the region, although if choices were possible one country's plant might be preferable to another because of its regional context as well as its capacity, technological level and labour relations. However, according to the BFIA the future lies very much in greenfield investments, especially when privatisation in completed: indeed they were already in a majority cumulatively by the end of 1998. In 2000, $500mln in greenfield investment marginally outweighed $480mln in privatised enterprises as, for example, Austria's ÖMV spent $40mln on a chain of filling stations and Liebherr built a new $20mln production facility in the Plovdiv region. It is predicted that future investment will be made in SMEs which local government will need to attract. However, the BFIA has complained about the lack of interest so far shown by local governments in attracting FDI and they will need to improve their capacities.

Conclusion

FDI had made a significant impact but most development has occurred in the capital, its surrounding regions and the coastal areas. Links with the EU through the border region with Greece have been less influential, except in the Struma valley. Pre-1990 traditional industrial centres, close to the CIS have experienced many negative trends, such as unemployment and poverty, due to the high level of deindustrialisation. All this vividly illustrates the need to appreciate the changing perception of space in contemporary Bulgaria. For the collapse of communism has seen a major change in the way space and location are conceived since the forces of commodification have impacted in varying degrees on the transition states and their constituent regions according to the level of privatisation and economic reform as well as their potential for market penetration. Thus, FDI is the major engine for economic restructuring. However, FDI is still in its early stages: following Stern (1997) there has been 'initialisation' to gain a foothold when government stability may still be in doubt and economic reforms are incomplete as well as 'internationalisation' through higher technology and human capital investment to exploit domestic and foreign markets but the stage of 'mature investment' has not yet been reached.

References

Bachvarov, M. (1999), 'Troubled Sustainability: Bulgarian Seaside Resorts', *Tourism Geographies*, 1, 192–203.

Bogatic, Z. and Hillman, A.L. (1994), 'The Tax Base in Transition: the Case of Bulgaria', *Communist Economies and Economic Transformation*, 6, 537–52.

Buckwalter, D.W. (1995), 'Spatial Inequality Foreign Investment and Economic Transition in Bulgaria', *Professional Geographer*, 47, 288–98.

Carter, F.W. (1996), 'Bulgaria', in F.W. Carter and D. Turnock (eds), *Environmental Problems in Eastern Europe*, Routledge, London, 38–62.

Carter, F.W. (1998), 'Bulgaria', in D. Turnock (ed), *Privatisation in Rural Eastern Europe: The Process of Restitution and Restructuring*, Edward Elgar, Cheltenham, 69–92.

Carter, F.W. (1999a), 'The Geography of FDI in Central-East Europe during the 1990's', *Wirtschafts-Geographische Studien*, 24–25, 40–70.

Carter, F.W. (1999b), 'The Role of FDI in the Regional Development of Central and

Southeast Europe', in Anon (ed), *Regional Prosperity and Sustainability: Proceedings of the Third Moravian Geographical Conference, Slavkov u Brna*, n.p., Brno, 16–23.

Carter, F.W. and Kaneff, D. (1999), 'Rural Diversification in Bulgaria', *GeoJournal*, 46, 183–191.

Cheshire, P. and Hamilton, F.E.I. (2000), 'Urban Change in an Integrating Europe', in G. Petrakos, G. Maier and G. Gorzelak (eds), *Integration and Transition in Europe: The Economic Geography of Interaction*, Routledge, London, 100–30.

CSD (2000), *Bulgaria's Capital Markets in the Context of EU Accession: A Status Report*, Center for the Study of Democracy Report 5, Sofia.

Dobosiewicz, Z. (1992), *Foreign Investment in Eastern Europe*, Routledge, London.

Dunning J.H. (1992), 'The Geographical Sources of the Competitiveness of Firms: Some Results of a New Survey', *Transnational Corporations*, 5(3), 1–29.

Dunning, J.H. (1997), 'Trade Location of Economic Activity and the MNE: A Search for an Eclectic Approach', in B. Ohlin (ed), *International Allocation of Economic Activity: Proceedings of a Nobel Symposium held in Stockholm*, Macmillan, London, 395–418.

Dunning, J.H. (1998), 'The Changing Geography of FDI: Explanations and Implications', in N. Kumar (ed), *Globalization FDI and Technology Transfers: Impacts on and Prospects for Developing Countries*, Routledge, London, 11–42.

Estrin, S. (1994), *Privatisation in CEE*, Longman, Harlow.

Garnizov, V. (1995), 'Politics Reform and Everyday Life: The Changing Map of Popular Attitudes to the Reform Process in Bulgaria', in E. Dainov and V. Garnizov (eds), *Politics Reform and Daily Life: Evolution of Popular Attitudes to Key Issues in the Reform Process in Bulgaria 1993–4*, Centre of Social Practices, New Bulgarian University, Sofia.

Hall, D.R. (2000), 'Bulgaria', in D.R. Hall and D. Danta (eds), *Europe goes East: EU Enlargement Diversity and Uncertainty*, The Stationery Office, London, 215–29.

Hamilton, F.E.I. (1995), 'Re-evaluating Space: Locational Change and Adjustment in CEE', *Geographische Zeitschrift*, 83(2), 67–86.

Hamilton, F.E.I. (1999), 'Transformation and Space in CEE', *Geographical Journal*, 165, 135–44.

Izvorski, I. (1993), 'Economic Reform in Bulgaria 1989–1993', *Communist Economies and Economic Transition*, 5, 519–31.

Jones, D.C. and Miller, J. (1997), *The Bulgarian Economy: Lessons from Reform during Early Transition*, Avebury, Aldershot.

Konstantinov, E. (1994), 'Rahmenbedingungen der Privatisierung in Bulgarien', in H. Roggemann and E. Konstantinov (eds), *Wege zur Privatisierung in Bulgarien: Rechtsvergleich und Rechtspraxis*, Verlag Arno Spitz, Berlin, 14–25.

Kopeva, D. (1994), 'Land Reform and Liquidation of Collective Farm Assets in Bulgarian Agriculture: Progress and Prospects', *Communist Economies and Economic Transformation*, 6, 203–18.

Lampe, J.R. (2001), 'Bosnia and Bulgaria: Crossroads for Two Economic Transitions', *Woodrow Wilson Centre East European Studies*, Jan-Feb, 10.

Levinson, A. (1995), 'The Impact of Privatization on Settlement Patterns in Southwestern Bulgaria', in M. Tykkyläinen (ed), *Local and Regional Development during the 1990s: Transition in Eastern Europe*, Avebury, Aldershot, 137–48.

Marinov, M.A. and Marinova, S.T. (1997), 'Privatisation and FDI in Bulgaria: Present Characteristics and Future Trends', *Communist Economies and Economic Transformation*, 9, 101–16.

Nicholls, A. (1999), 'A Survey of Bulgaria: The Limits of Stability', *Business Central Europe*, 6(63), 39–50.

Panov, O. (1994), 'Delayed Privatization and Financial Development in Bulgaria', in M. Jackson and V. Bilsen (eds), *Company Management and Capital Market Development in the Transition*, Avebury, Aldershot, 185–225.

Paskaleva. K., Shapira, P., Pickles, J. and Koulov, B. (1998), *Bulgaria in Transition: Environmental Consequences of Political and Economic Transformation*, Ashgate, Aldershot.

Paunov, M. (1993), 'Labour Market Transformation in Bulgaria', *Communist Economies and Economic Transformation*, 5, 213–28.

Petrakos, G.C. (1997), 'The Regional Structures of Albania, Bulgaria and Greece: Implications of Cross-Border Cooperation and Development', *European Urban and Regional Studies*, 4, 195–210.

Petrakos, G. and Christodoulakis, N. (2000), 'Greece and the Balkans: The Challenge of Integration', in G. Petrakos, G. Maier and G. Gorelak (eds), *Intregration and Transition in Europe: The Economic Geography of Interaction*, Routledge, London, 269–94.

Pishev, O. (1990), *The Bulgarian Economy: Transition or Turmoil*, Stockholm Institute of Soviet amd East Europem Economics Working Paper 6, Stockholm.

Popov, M.K. and Todorova, E.N. (1997), 'Privatisation Democratisation ou Oligarchisation de la Bulgarie Post-Communist', *Balkanologie*, 1(2), 70–84.

Safarian, A.E. (1999), 'Host Country Policies towards Inward FDI in the 1950s and 1990s, *Transnational Corporations*, 8(3), 93–112.

Stern, R.E. (1997), 'FDI Exports and East-West Integration: Theory and Practice', in J. Cooper and J. Gacs (eds), *Trade Growth in Transition Economies: Export Impediments for CEE*, Edward Elgar, Cheltenham, 329–57.

Strong, A., Reiner, T. and Symer, J. (1996), *Transition in Land and Housing: Bulgaria Czech Republic and Poland*, Macmillan, Basingstoke.

Tchipev, P. (1996), 'Financial Institutions' Role in Bulgarian Privatisation', *Communist Economies and Economic Transformation*, 8, 93–108.

Vernon, R. (1979), 'The Product Cycle Hypothesis in a New International Environment', *Oxford Bulletin of International Statistics*, 41, 255–67.

Whitford, R. (1996), 'Investment in Bulgaria: Damage Control', *Business Central Europe*, 5(31), 18–9.

Whitford, R. (2001), 'Cardinal Sin', *Business Central Europe*, 7(77), 47.

World Bank (1991), *Bulgaria: Crisis and Transition to a Market Economy*, World Bank, Washington, D.C.

Zheliazkov, Z. (2000), 'Privatisation Agency', in Anon (ed), *Economic Book Bulgaria-Europe 2000*, n.p., Sofia, 111–16.

Chapter 11

Foreign Direct Investment in Bulgaria's Wood Products Sectors: Working with the Grain?

Caedmon Staddon

Introduction

While FDI has played a relatively small part in the story of post-communist transformation in Bulgaria compared with countries such as Czech Republic, Hungary and Poland, it grew significantly in volume in the latter 1990s (Jordanova 2001). By 1995 total FDI into Bulgaria amounted to $470 million (Begg and Pickles 1998) and Buckwalter (1995) went so far as to argue that the relatively miniscule amount of FDI even made robust analysis of its impact difficult. Begg and Pickles (1998) however reveal two critical characteristics of the Bulgarian experience of industrial transformation: economic restructuring to the middle 1990s took place largely through disinvestment rather than investment (e.g. plant closure and asset stripping) and much of this deconstructive economic activity was directed by domestic Bulgarian capitalists. It appears then that the story in Bulgaria at least has been a much more home-grown affair, with rapidly formed cartels engaging in the systematic liquidation or strategic capture of Bulgarian industrial capacity, occasionally with at least the tacit collusion of government. There has not been a coherent process of Schumpterian 'creative destruction'. The agricultural sector for example suffered widespread and often very destructive liquidation of productive assets at all levels during the 1990s (Meurs and Begg 1998). Similarly there have been periodic acrimonious debates about the privatisation of economic 'crown jewels' such as Balkantourist (which controls much of the winter and summer tourism infrastructure), Neftochim (which is the Balkan penisula's largest petrochemical plant) and Bulgartabak (which controls virtually all tobacco distribution in the country). However, out of this period of primitive capital accumulation have emerged a number of domestic capitalists whose attention is now turning away from liquidation and speculation and towards longer term investment. Simultaneously the country is now attracting much more international attention from investors (the top five are Germany, Greece, Italy, Belgium and Austria) as a result of the upturn in the global economy and the more aggressively pro-market policies implemented by the new government of Premier Simeon Saxe-Coburg-Gotha, elected in June 2002.[1] Since the election the country has enjoyed low inflation (around three percent) and good growth figures (around 4.4% in 2002).

Notwithstanding this much improved economic picture, little of the already small pot of FDI has gone into the wood and wood products sectors. Even in the relatively good investment years of the late 1990s and early 2000s less than five percent of total inward investment went to the wood and wood products sectors. Linked to this there has been relatively little assistance for the sector from international donor organisations aimed at improving investment attractiveness within the sector (unlike the situation in Russia and Poland, for example, where large amounts of World Bank and EU monies have supported sectoral change in forestry and forest products). As this paper shows, the story in the wood sector has generally been one of an essentially domestically brokered restructuring, with occasional high profile interventions by foreign capital interests, particularly with respect to pulp and paper production. In the process, production has fallen dramatically across the forestry and wood products sectors (Prins and Korotkov 1994). The past decade has seen a reduction of 30–40% in the production of most wood products, including higher value added products such as joinery and pulp and paper production. There are several reasons for this sharp decrease, including: continued sustained decreases in global export prices for wood products (as with most other unprocessed and semi-processed commodities worldwide); the breakup of the centrally planned distribution systems within the former Comecon bloc; a lack of investment in the development of wood processing enterprises and new technologies; consequent deterioration of facilities, machinery and equipment in the timber, wood processing and pulp and paper industries; relatively low quality of certain products which prevents them from being competitive on the world market; and difficulties in penetrating new markets and facing the protectionist policies of some western countries.

Even so, as of the end of 1997 there were over 2,500 officially registered enterprises (state, municipal and private) devoted to 'Forestry, Logging and Related Services' in Bulgaria and 3,818 enterprises devoted to 'Wood and Products of Wood and Cork, Except Furniture', 4,337 devoted to 'Furniture and other Wood Products', and 394 devoted to the 'Pulp, Paper and Paper Products' (National Statistics Institute 1998 p.360). The total workforce is somewhere in the neighbourhood of 131,000 although the vast majority of these are employed in quite small enterprises with less than 49 employees. Currently the sector as a whole is operating at only 25–30% capacity, according to the Bulgarian Chamber of Woodworking and Furniture Industry. During the 1990s there has also been much activity within the 'black' part of the timber economy, with illegal logging, processing and export of timber taking place, particularly in the border areas (Staddon 1999, 2001). In late 2002 the Bulgarian government launched a systematic crackdown that resulted in 4,185 woodworking workshops being inspected in five months, with over 1,700 statements of violations served, mostly for tax violations. Additionally over 400 Forestry Department officials were cautioned and 33 were summarily dismissed.

As Table 11.1 shows, though the forestry and woodworking sectors have grown considerably in size and complexity since the early 1960s, overall levels of production, especially of higher value-added materials, fell dramatically after 1989 and remain relatively low. Particularly interesting – when compared with the increase in numbers of enterprises in the 1990s – is the recorded 78% decrease in total sawnwood[2] volumes between 1989 and 2001, implying a proliferation of relatively inefficient SMEs throughout the sector, an implication confirmed by detailed fieldwork undertaken in 1999 and 2000 (Staddon 2001). Overall, production of more complex wood products, such as printing and writing paper and veneer sheets, has declined more rapidly relative to technologically simpler products more congenial to small shops with limited capital plant. Consequently export volumes have increased annually by some 10% since 1989 for simpler industrial commodities, while imports of higher value added wood-based products (e.g. fine papers) have increased (Statisticheski Godishnik 1998 p.257). This seems to be a classic case of systematic disinvestment in a post-communist industrial sector, leaving a poorly capitalised and largely atavistic, inward-looking network of producers with few linkages to broader national or pan-European capitalist networks. As Smith and Swain (1998) point out, such 'isolated institutional networks' may actually inhibit the development of the sort of leaner, hyper-efficient wood products producers coveted by domestic policymakers and multilateral development planners.

This chapter is divided into six sections. The next outlines an analytic model for interpreting patterns of change within the wood processing sector based on regulation theory. Section three gives a statistical overview of FDI in Bulgaria with particular attention to the forestry and wood processing related sectors. This is followed in section four with a discussion of a number of specific FDI deals in the Bulgarian wood sector in the 1990s. Here attention is also given to the lack of foreign donor interest in sectoral restructuring projects and the implications of this for lower FDI in the sector relative to other transition economies. The fifth section presents a number of case studies of different types of FDI including foreign buyouts of wood products firms, joint ventures of various sorts, and more indirect investment through longer term production contracts with foreign firms such as Ikea. Implications of a regulationist analysis of these patterns and suggestions for future research are presented in the sixth and concluding section.

Models of Development: Just What is Transition?

I have suggested elsewhere (Staddon 1999) that a 'transition model of development' has emerged that is effecting the (re)peripheralisation of rural localities. A critical element of this process involves the mobilisation of labour power and the natural environment as key sources of primitive capital accumulation (O'Connor 1993; O'Riordan 1989). For Bulgaria's rural areas this has meant a return to an economically and politically subservient role as 'hewers of wood and drawers of

Table 11.1 Production in Bulgaria's forestry sector, 1961–2001 (cu.m but tonnes for pulp-paperboard items)

	1961	1971	1981	1989	1996	2001
Roundwood (Total Felling)	5,071,000	4,881,000	4,946,000	4,203,000	3,205,000	3,991,890
Sawnwood	1,659,000	1,709,000	1,473,000	1,402,000	253,000	312,000
Wood-based Panels	143,000	370,000	558,000	493,000	233,000	530,234
Veneer Sheets	n/a	36,200	55,000	41,000	20,000	3,264
Plywood	66,000	67,000	53,000	51,000	23,000	65,000
Particle Board	77,000	199,000	324,000	300,000	124,000	180,000
Fibreboard/MDF/etc.	n/a	68,000	126,000	101,000	66,000	281,970
Pulp, Paper and Paperboard	118,400	432,100	621,000	424,000	384,500	297,000
Mechanical Wood Pulp	13,000	13,200	19,000	8,000	2,000	n/a
Chemical/Semi-Chem Pulp	21,300	103,000	249,000	130,000	95,000	85,000
Other Fibre Pulp	n/a	30,000	20,000	10,000	10,000	10,000
Recovered Paper	n/a	85,000	128,000	128,000	128,000	80,000
Printing and Writing Paper	14,900	42,900	39,000	47,000	4,500	6,000
Other Paper and Paperboard	69,200	209,100	388,000	211,000	145,000	130,000

Source: FAOSTAT 2003.

water'. Whether it be through asset stripping, plant closure, privatisation of state functions, agricultural lands restitution or tax reform, a variety of actors and institutions are helping to divert capital into a new, largely urban-based, oligarchy. Rural places have been (re)constituted as the locales within which a productive mix of short term capital investment (and certain forms of disinvestment), flexible, semi-skilled labour and raw resources are maintained. Subsequent research (Staddon 2001) has also shown that there are at least five different structural types of firm currently operating in the wood products sector.[3] Here I have suggested that these firms could be arrayed along a spectrum with purely exogenous actors at one end (foreign buyouts) and purely endogenous actors at the other (locally rooted development). Diversity of enterprise types was theorised to be a function of at least three conditions: high unemployment – estimated to be as high as 50% in some rural areas; ease of access (legal and illegal) to well-managed and relatively plentiful stands of mature timber; and complex (pre-existing and new) industrial and financial networks that combine to create fields of possibility and constraint to enterprise managers. In this chapter I will reconsider this typology from the perspective of the capital flows into the sector (that have accelerated since the later 1990s) and at the scale of the Bulgarian sector as a whole. It should become clear in the course of the discussion that the key differentiating factor involves the integration, or lack thereof, of particular enterprise types into emergent industrial networks (Smith and Swain 1998). Interactions between 'transformative' versus 'devolutionist' industrial networks will also be examined. This tapestry of interacting networks, social conditions and differential resource access at the local scale subsequently greatly influences the development of a specific form of 'inclusive peripheralisation' empirically manifest in the Razlog Basin at the end of the 20th century (Black 1989). The question of privatisation in the forestry, timber and wood processing industries has been highly contested and, as a result, repeatedly delayed. Among the related issues are questions regarding: the transformation of forms of legal ownership; the valuation of assets; the devising of equitable and efficient means for transferring ownership (e.g. sale or distribution, coupon or voucher systems, existing employees preferences and access for foreigner investors); and the development of necessary producer services and financial sectors (Prins and Korotkov 1994).

FDI: Statistical Review

Overall FDI flows into Bulgaria have been much smaller than those in most other post-communist transition countries, though they have risen significantly during the 1990s.[4] In 1992 Bulgaria received $34.4 million, which rose to a high of $1001.5 million in 2000, before falling back during the global economic downturn of 2001 and 2002 (Table 11.2). According to preliminary data for the period January–September 2003, FDI flows have amounted to $864.2mln (some 4.6% of GDP) indicating significant improvement. Overall then there has been an acceleration of FDI into

**Table 11.2   FDI flows into the Bulgarian wood products industry, 1998–2002
($hundred million)**

Sector	1998	1999	2000	2001	2002
Total	**537.3**	**818.7**	**1001.5**	**812.9**	**458**
Forestry, logging and related service activities	---	---	0.1	---	---
Manufacture of furniture	1.4	1.0	2.1	---	---
Manufacture of pulp, paper and paper products	14.9	6.4	10.1	---	---
Manufacture of wood and of products of wood and cork, except furniture; manufacture of articles of straw and plaiting materials	8.8	10.2	15.0	---	---

Source: United Nations (2003) p.3.

Bulgaria during the transition period: less than $1.0bln during 1989–95 but over $4.0bln for 1996–2002. Even so, this is miniscule compared with the more than $50bln invested in Poland during 1989–2002. In per capita terms the differences are again marked: $345/person in Bulgaria compared with $812 in Poland and $1,274 in Czech Republic. Moreover, only 15–25% of net FDI finds its way into manufacturing and FDI stocks in the wood products sectors accounted for only 2.5% of the 1999 total. According to the Bulgarian Foreign Investment Agency (2003) wood products ranked only 9th of 14 sectors in FDI received during 1992–2001 with some four percent of all investment ($2,750mln). Moreover, the proportion of FDI into manufacturing appears currently to be falling, especially after 1999 (Jordanova 1999).

Over time the type of FDI has also changed, from an early reliance on straightforward equity purchase to other forms, including non-equity forms such as loans, revolving credit arrangements between foreign firms and Bulgarian suppliers and complex investment instruments such as 'Zunk Bonds' (which effectively securitise debt). In 1996 equity capital type FDI accounted for fully 95% of all FDI, while by 2002 this had fallen to 67% (Table 11.3). Only two percent of all FDI results in minority stakes, though other forms of FDI relationship are beginning to emerge. Those companies included in a July 2000 KPMG survey – referred to by Jordanova (1999) – went on to say that their main motives for investing in Bulgaria were: established relations with regular customers from the region (28%); market potential; Bulgaria's favourable geographical location; and the existence of a skilled labour force (47%) and low labour costs (35%). Indeed, this latter point is hard to dispute as average wages for the economy as a whole are less than $2.00/hour. Interestingly, Table 11.3 suggests that notwithstanding the criticisms levelled at Bulgaria's investment climate in the 1990s (by European governments and major multilateral institutions such as the IMF), levels of purely equity capital investment have remained relatively robust, suggesting that investors saw at least longer term potential. Other forms of FDI, especially reinvestment of corporate profits are what appear to suffer immediately as a result of shocks such as the Russian/East Asian economic crisis of 1997/8 and the global slowdown of 2001–2.

Table 11.3   FDI flows into Bulgaria by type of investment, 1996–2002 (%th)

	1996	1997	1998	1999	2000	2001	2002
Total	109.0	504.8	537.3	818.7	1001.6	812.9	873.7
Equity capital	104.0	491.9	504.6	500.3	754.8	566.7	588.2
Other capital	5.0	12.5	−17.0	350.9	188.5	240.0	232.7
Reinvested earnings	0.0	0.4	49.8	−32.5	58.3	6.3	52.8

Source: Bulgarian National Bank.

Forms of FDI in Wood Processing

As noted above, little of the relatively small trickle of FDI into Bulgaria is finding its way into forestry and related industries. Most restructuring in this sector can therefore be related to domestic processes, including voucher privatisations and the reorganisation of intra- and inter-sectoral linkages (Staddon 2001). For example, a Bulgarian paper trading company, Gerhard, has recently bought a 55% majority stake in Trakia Paper of Pazardjik – producing corrugated board and testliner – and plans to invest $6.1mln in the company over the next five years. There have been numerous similar purchases of state-owned wood products companies by Bulgarian holding companies in the 1990s, often motivated by purely speculative interests as well as a more proactive desire to broker production contracts with foreign firms. Meanwhile a growing number of partnerships between foreign firms and Bulgarian holding companies are effectively enabling foreign access to cheaper Bulgarian production capacity without the necessity of taking a direct financial stake through privatisation or joint venture deals. Naturally this kind of 'soft' FDI is rather difficult to accurately track or assess, but it is also likely to become more common as macroeconomic stabilisation in Bulgaria makes it possible for foreign firms to form reliable and profitable non-equity based relationships with Bulgarian firms (note the recent shift away from simple equity purchase as the primary vehicle of FDI).

In any case, most of the largest foreign investors in this sector have been in pulp and paper and include: Adut-Adox Skalica of Slovakia; Iskilar Packaging of Turkey; Norekom GmbH – a German distribution firm; Profitech Enterprises of Nicosia, Cyprus; and Trace Paper Mill of Greece. The Privatisation Agency has signed contracts for the sale of large stakes in three Bulgarian paper companies. Profitech Enterprises have invested $545,000 in a 22% stake in Unipak of Pavlikeni which produces paper and board for packaging: some 67% of the company is being prepared for stock market flotation with only 6.6% reserved for Unipak employees. The Agency completed a deal to sell a 26.4% share in Stradjiza-located Velpa-91 to Norecom GmbH for $1.1mln in 2001 and the company went on to increase its stake to 81% in June 2003. Much larger was the outright purchase of Celhart Paper – with a 300 strong labour force producing craft paper, fluted cardboard and packaging paper – by Iskilar: one of the largest paper packaging manufacturers in Europe with markets all over the world. More recently the German pulp, paper and packaging conglomerate Mahr-Melnhof has purchased a kraftliner plant at Nikopol in the north of the country, following an earlier acquisition in Slovenia.

Table 11.4 shows some of the other major deals brokered by the Privatisation Agency during the period 1992–8, including several outside the pulp and paper area. At the end of 1997 Swedish Match paid $2.35mln for a 58% stake in PLAM Bulgarski Kibrit JSCo, Bulgaria's leading match manufacturer with some 400 employees in its factory and head office at Kostenets, 70 km southeast of the capital. From the point of view of the domestic market it was a strategic purchase as the company controls virtually all the Bulgarian match market and has capacity to export to neighbouring countries. Kronospan of Austria bought lumber producer Bules of Burgas in mid-1997, thus further adding to its growing European network of wood products manufacturings involving 17 locations in seven countries (including Bulgaria, Lithuania, Poland, Romania and Ukraine). With few exceptions these deals were for majority stakes in the Bulgarian companies involved, though as noted above in the later 1990s these seems to have been a greater willingness to enter into other kinds of privatisation or partnership deals. For example, Bulgarian Furniture Company paid 30% in cash and the rest in Zunk Bonds[5] for 'Ludogorie 91', while other privatisations to benefit Bulgarian and foreign firms have involved the liquidation of residual holdings from the mass privatisation programme, with consistent undervaluation of the companies concerned. Again, these financial complexities suggest that the true impact of FDI on the wood sector is not reducible to tables of simple FDI figures. There has been some debate in Bulgaria about the precise forms of FDI and privatisation with some commentators noting that current Bulgarian law does not require much disclosure by foreign firms interested in investing in the country.

Table 11.4   **Wood sector privatisations registered with the Bulgarian Privatisation Agency, 1992–1998**

Company	Date	Stake (%)	Price ($)	Investment Contracted ($)	Purchaser
Escos Dogramma (Sofia)	9/94	80	390,000	2.2mln	Gibu Ltd (Italy
Pirinska Mura (Bansko)	11/95	67	2.8mln	3.5mln	Evrotech Ltd Bulgaria (USA)
Bules JSCo (Burgas)	5/97	51	3.2mln	---	Kaindl/Kronospan (Austria)
KMH (Belovo)	11/97	58	6.4mln	10.7mln	Trace Papermilin (Greece)
Plam Bulgarski Kibrit (Kostenets)	12/97	58	2.4mln	1.9mln	Swedish Match Group (Netherlands)
ZKO Hadji Dimitur (Svoge)	---	75	1mln	3.05mln	Adut Adox (Slovakia)
Lesoplast JSCo (Troyan)	7/98	3	43,000	---	Welde (Austria)
Tselhardt (Stamboliiski)	8/98	23	2.1mln	---	Iskilar Holdings (Turkey)
ZMK Nikopol	10/98	31	1.2mln	---	Enerholding (Cyprus)
Rulon (Iskar)	12/98	75	2.0mln	2.3mln	Europak (Austria)

Source: Bulgarian Privatisation Agency.

One factor strongly related to the structure and pace of wood sector FDI is the relative absence of multilateral development agencies within the Bulgarian sector, affecting the types of plant and sectoral restructuring that can facilitate subsequent privatisations of foreign investments. For example, starting in 1994 the EC launched the programme 'Assisting the Privatisation of the Russian Wood Processing Industry' which assessed over 300 wood sector companies in Russia and eventually selected six for more detailed assistance in restructuring: Sevzampel (furniture, including upholstered) and UIFK from the Leningrad region, Podomoskovije from Moscow region and three from Novosibirsk: Bolshevik (doors, windows and mouldings), Shatura (particle board and panel-based furniture) and Tasharanski Leskombinat. Shatura in particular has profited from these EC investments in the rationalisation of its (furniture) production process (claiming a 20–30% increase in productivity) and expanding its export markets. The Bulgarian wood sector too has benefited from multilateral assistance, though to a lesser extent. In 2002 the World Bank, through its affiliate the International Finance Corporation (IFC), provided $25mln to help Tselhart modernise production on top of two previous grants from this source. Other recipients include Kronospan (Burgas) to upgrade pressed board production. More generally, the World Bank has committed $30mln to restructure the forestry management system (including fibre supply) and in 2003 it loaned $3.6mln to a wood residue power generation plant (biomass boiler) in Sofia to help Bulgaria satisfy its Kyoto commitments. Interestingly the UNECE/FAO conference in March 2003 in Romania concentrated on knowledge transfer and partnership building around the sound use of wood.

Case Studies

This section reviews two common types of FDI activity: branch plant buy-outs, as in Lesoplast and Bules, and joint ventures like the Velpa-91–Norecom tie up in 1998. A third, softer, form of FDI is also treated through the development by the furniture MNE Ikea of longstanding contractual relations with various Bulgarian wood products manufacturers. This could be considered a kind of 'flexible' FDI, particularly once account is taken of the many contractual stipulations (around economic, social, and environmental issues particularly) that Ikea uses to bind its suppliers to its overarching corporate philosophy.[6] In some cases the names of companies has been altered to protect confidentiality of information collected through personal interviews with the author in 2000 and 2001. It is important to note that only first tier suppliers are examined as there may be fairly dense networks of lower tier suppliers of everything from preprocessed wood stocks (e.g. match blocks for flooring) to specific component parts like fastenings or colourings.

Majority Foreign Ownership

JK Furniture is a wholly owned branch plant of a Belgian company which has been producing tables, chairs and bedroom suites for export to Western Europe since 1992. As is often the case with such deals initial entry into the region was provided by an emigré Bulgarian businessman now living in Western Europe (Koparanova 1998; Mihailova 1997). The plant has recently been extensively modernised, with much investment of new German and Italian equipment, some of which utilises computer controlled cutting systems designed to reduce reject rates and wood waste. During 1999 the plant underwent another expansion, this time involving improvement of stock handling and drying facilities. Both modernisations were designed to reduce wood wastage, estimated to be as much as 20% of raw materials, and improve the overall quality of wood stocks. The company produces a fairly wide range of furniture items including chairs, cabinets, tables and even some doors and windows. All products are straight-cut, with little apparent use of wood-bending or laminate furniture technologies.

The workforce of approximately 50 people – average for the sector[7] – is well-trained and, by local standards, well paid at an average wage of approximately $100/month. A bonus scheme is also in place to encourage higher production levels. As is common in the wood sector male employees outnumber female employees, with women primarily occupied in the sanding/finishing, assembly and packing parts of the operation. The managing director (MD) stresses the need for his workers to be multiply-skilled as the variability of orders (both volume and type) from the parent company means that workers are frequently redeployed within the plant. Interestingly the majority of employees do not come to the company with either experience or specific training in the woodworking trades. Unlike many plants in the region JK Furniture bears no filial relations to former state enterprises like Bules, Lesoplast, Piriniska Mura or Velpa-91. The MD, one of the few employees with a background in the wood processing sector, has suggested that getting specifically trained personnel was less important than having hardworking flexible workers (indeed here as elsewhere in the sector the local Wood Industry Technicum came in for much criticism for its 'outdated' teaching and its apparent inability to inculcate a 'new' work culture of flexibility and diligence). One exception to this pattern was in the lacquering room, where the need to conserve lacquer and to coat items to uniform thickness without runs or missed spots meant that skilled men formerly employed at a nearby state furniture company were employed. The MD also noted that this part of the production process entailed specific health and safety issues requiring proper management by skilled technicians.[8]

The plant utilises primarily beech (Fagus sylvaticus), with lesser amounts of white pine (Pinus peuce, leucodermis), as basic construction materials as this wood is particularly popular in Western Europe. Currently the plant sources wood materials largely from the southeastern Strandzha region at a cost, for beech, of approximately $50/cu.m for 'wet' beech and $75 for air-dried. Recently an onsite kiln was installed to bring all wood stocks down to approximately 10% moisture content, pretty much the standard level for furniture production. Like the MD for RPP (see below), the MD of JK Furniture expressed considerable concern about the potentials for cartelisation of wood supply as a result of the current restructuring of industrial forestry.[9] Supply of fibre and chipboard panels has been a particular problem in recent years as Bulgarian production has declined and imported supplies are relatively expensive. Currently the MD is developing plans to incorporate further stock conserving technologies into the production process, including 'finger-jointing' machinery to join short end pieces into useable lengths.

JK Furniture is a classic example of a foreign-owned branch plant sited in a location with a highly profitable combination of skilled labour, cheap raw resources and low levels of government interference. Unlike foreign branch plants in other parts of the world such as Mexico and Malaysia however it does not show any sign of being especially footloose. On the contrary, JK Furniture has an air of permanence, or at least relative security, which is unusual in a sector dominated by insecurities at both the input and output ends of the production process. This is indicated by the large investment in fixed capital stocks (like buildings), but also by the fact that the company has operated in Bulgaria since 1992, weathering the tumultuous economic storms that drove many footloose plants from the country. Its wholly export orientated production has also undoubtedly helped it weather domestic economic crises such as the 1997 financial meltdown. In time its production may well come to be at least partially absorbed by a resurgent domestic market.

Joint Ventures

Located in the southwest, Rila Pulp and Paper (RPP) is a kraft pulp and paper plant which at its apogee employed approximately 1,000 people in the production of packing, corrugated and other types of rough brown papers. It is a classic expression of the central planning system for though the region has good supplies of softwood, RPP's size was such that it always required softwood stocks from other regions of the country in order to operate efficiently. Under communism, this supply network extended as far as the USSR's Komi Autonomous Region which was for several decades effectively annexed to the Bulgarian national forest fund. In world terms however RPP is relatively small, with an installed capacity of 100,000t/yr of kraft pulp and 70,000t/yr of brown liner papers, the latter using 70% post-consumer recycled paper (Holmes and Hayter 1993). Capacity was increased to 120,000t/yr through a plant teardown and rebuild in 2000, but significant work remains to be done at other plant bottlenecks before this increased capacity can be realised.

In 1995 RPP's management was restructured as part of a Bulgarian-Swiss-Italian joint venture, in which the Swiss-Italian partners took a 51.8% share in June 1997. The same Swiss-Italian group also bought a 75% share in another pulp and paper mill in Silistra and is integrating its Bulgarian acquisitions into a regional-scale production complex for packaging products. The new management team included an English MD grafted onto the old management team to represent the interests of the Italian partner, a common pattern in Bulgarian JVs (Kopanarova 1998). He has developed a master plan for further investment in the modernisation of the plant itself, some of which is currently in very poor condition and has also argued that the business must greatly enhance its orientation to foreign – especially Turkish and Greek – markets. As of summer 2000 RPP was producing approximately 30,000t of kraft liner annually, constituting approximately a third of Bulgaria's total production of this commodity. Smaller amounts of related products are also produced and where possible marketed, including turpentine (400t/yr) and lignin (a potential source of power).[10] Even with the recovery of regional markets in 1999 and 2000 the MD estimates that the mill is realising only a quarter of its economic potential.

He suggests that the plant faces three key challenges to its successful restructuring. First, there is a serious problem with lack of access to investment capital. Initial attempts to raise even modest amounts of investment capital within Bulgaria after the 1997 agreement were unsuccessful; investors were non-existent and Bulgarian banks were unwilling to loan money to the former state enterprise on any terms.[11] Currently the MD is vigorously pursuing investment capital from outside Bulgaria, especially from international development institutions such as the EBRD and PHARE, which have shown a willingness in the past to provide funding to specific industrial plants: unfortunately, to date only Tselhart in central Bulgaria has managed to win EBRD or other funds for plant modernisation. Much of the planned $15 million investment is earmarked for a systematic overhaul of all sections of the plant, while specific additional allocations are to be made for the complete renovation of recovery boilers, offline since a breakdown in the mid-1990s. Lack of access to normal operating credit arrangements has an especially deleterious effect as it forces the plant to operate on a 'stop-go' basis. When the plant receives payment from a customer, on delivery of goods, it is able to pay for normal operating costs. During interim periods it is forced to go into a sort of stasis, unable to pay wages or even utilities. At one point in 1999 the plant's electricity was cut off for bill non-payment! Both Mihailova (1997) and Dobrinsky (2000) point out that the lack of normal rotating credit facilities and the financial (over)leveraging of former state industries combined to create such 'credit sinks' throughout the economy in the 1990s – leading to the securitisation of SOE debt through the Zunk Bonds referred to above. Unlike the cases of FDI discussed by Koparanova (1998) however, RPP's Swiss-Italian partners do not appear to be providing any more than a minimal amount of operating capital.

The second key challenge involves the need to introduce labour flexibility into the plant. Partly as a consequence of the specific terms of the joint venture deal, the plant remains mired with an unsustainably large workforce. Of the remaining 650 workers, the MD has suggested that up to two-thirds should ultimately be made redundant, potentially affecting up to 2,000 jobs indirectly in the local economy. Somewhat unusually for a former state enterprise, the plant benefited from significant technological refitting in the late 1980s, but unfortunately the labour process was not significantly restructured to realise the benefits of this large investment. At present RPP manifests the all-too-common situation in which state firms appear to be maintaining a social holding function, propping up regional employment whilst other firms restructure more aggressively. The third key challenge is raw lumber supply, which is ironic for a plant surrounded by 'ripe' supplies of spruce, fir and pine (Staddon 2000, 2001). The MD claims that because the restructured plant will be producing well below originally planned volumes it should be possible to survive only on local timber resources. The key problems here are that timber marketing arrangements remain unclear, that both prices and supply (even from the state forest directorates) are volatile, and that clearly separate and monopolistic supply networks seem to be emerging within the restructured state forestry apparatus itself (Krassimirova 2000; Staddon 2000). Optimal softwood stock, with diameters of greater than 40cm can cost up to \$120/cu.m and is purchased through a wide network of middlemen representing timber producers from all over the country. Thus raw material costs are difficult to maintain with the result, for example, that the sister plant in Silistra produces fine papers at a cost of DM150/unit in a market supporting only DM80–90/unit. In this respect RPP may be a good example of what Smith and Swain (1998 p.44) call a 'cathedral in the desert'; that is to say a reasonably viable enterprise existing in a state of 'blocked' integration into international industrial networks, preventing both upstream and downstream enterprises from truly reaping the synergies inherent in closer coordination of activities.

Contracting Out: Ikea's Experience

Ikea has 50,000 workers and 159 stores in 29 countries making and selling over 10,000 articles around the world and is one of the largest furniture companies in the world, having pioneered the idea of mass-market flat pack furniture distribution. In general, Ikea does not manufacture its own products, but works through a complex network of over 2,000 suppliers in 56 countries. However in the past decade it has acquired a number of its own factories, some of which function as training units and set standards for other suppliers for production economy, quality, and environmental awareness. To secure timber supplies Ikea is also developing joint ventures or financing arrangements with forest owners and wood processors in a number of countries including Poland, Slovakia, Russia, Romania and Bulgaria. Already Ikea-linked forest companies such as Swedwood and Sveaskog (both based in Sweden) have purchased stakes in Polish and Latvian forest lands. For these reasons its FDI impacts, whilst sometimes indirect, could be significant and deserve serious consideration, particularly in terms of its linkages with upstream parts of the value chain: forest management, logging and lumber processing.

While Poland is currently the third largest supplier to Ikea, after Sweden and China (de Haan and Oldenzeil 2003), the company started sourcing timber and finished products in Russia almost 20 years ago, opening trading offices in St Petersburg 12 years ago and Novosibirsk in February 2003 and its first retail outlet in Moscow three years ago (Ikea 2003). Currently the chain has over 50 suppliers in the country and Russia in particular is identified by the company as a growth region although it insists that like all product supply regions it must comply with the company's four-step certification scheme for ensuring adequate social and environmental standards. Ikea plans to build a \$20mln timber processing plant in the Vladimir region as part of its estimated $140mln expansion programme in Russia. The company has also entered into a partnership with the Worldwide Fund for Nature (WWF) aimed at boosting conservation of so-called 'high conservation value forests' in Bulgaria and Romania as well as China and Russia (WWF 2003). The company sources a number of items from Bulgarian producers as part of its hyper-efficient JIT global production system. Small toys, kitchen items (such as cutting boards), tables, chairs and other items are produced by at least a half dozen SMEs in Bulgaria (de Haan and Oldenzeil 2003). Many wood processing and furniture making enterprise MDs interviewed in 1999 and 2000 either held contracts with Ikea, had held contracts (which were not renewed for some reason) or had discussed the possibilities in previous years. Two examples of companies with current contracts include Traina Plast which manufactures trays from wood veneers and Ludogorie-91 (located in Kubrat) which produces chairs. Both of these enterprises found they needed outside help to re-engineer their production lines to boost product quality and input efficiency.

In some Bulgarian Ikea plants wages are paid hourly, but for most it is piecework, with bonuses for high quality and quantity. Workers earn between BGL120–200/month when they start with the company compared with BGL120–150 (€60–75) for those employed without contract and BGL180–220 (€90–110) for the others – Ikea's contracted plants in Poland pay around €500/month. Women tend to do manual work rather than operate machines and thus tend to earn less. Moreover, four out of five women interviewed started at no more than BGL150 (120 in the case of two recently-employed Moslem women) compared with BGL150–200 for the five men interviewed (de Haan and Oldenzeil 2003). There are small extra emoluments for specific kinds of hazardous work, especially in painting, lacquering and varnishing. Strangely, local customs about the prophylactic effects of yogurt endure and some managers make the soured whey available to workers in more hazardous parts of their operations. Interestingly, international labour organisations have shown some interest in Ikea's CEE operations; recently Canadian Teamsters Local 938 participated in seminars with Polish and other labour activists leading to unionisation of nine out of 13 Polish Swedwood plants.

Overall these direct contracting relationships are typified by a high degree of enterprise dependence upon Ikea, whose contracts are guaranteed up to a year in advance, but which can account for up to 100% of factory output. Some MDs interviewed expressed bitterness that initial contracts or expressions of interest by Ikea (and other Western firms) had not continued and many were mystified as to why apparently successful completion of contracts was often not followed up with more contracts. The stakes are of course quite high for such dependent enterprises. Actually, general understanding of Ikea's global flexible sourcing strategy is relatively poor, with the result that MDs often seemed to think that the lack of follow-on contracts was 'personal', a feeling no doubt exacerbated by the view that – at a stroke – an Ikea production contract could turn the fortunes of a struggling furniture plant around. There is no consensus as yet as to what it might mean for

Ikea or other MNEs to enter into ownership of Bulgarian forests, partly because this is still technically against the law (although there is pressure to change this as a function of EU accession requirements for a free market in land resources). But there is nothing to prevent an MNE from entering into JVs with Bulgarian enterprises that could control either timberlands or logging companies. An indirect form of control is exercised through Ikea's insistence on certain standards of environmental performance among its suppliers. Thus, in Poland, which is its third largest supplier nation, nearly 4.0mln.ha are Forest Stewardship Council (FSC)-certified compared with very little in Bulgaria.

Conclusions

To date there has been relatively little FDI into the Bulgarian forestry and wood products sectors compared with countries such as Czech Republic and Poland. Even so, there are distinct patterns to those flows which have been established. First, they are growing, especially since the year 2000, and the share of simple equity ownership is decreasing relative to more complex financial arrangements. The number of joint ventures and foreign takeovers effected as a result of the privatisation agenda driven by multilateral development agencies and successive Bulgarian governments appears to have levelled off. Already there have been several cases of earlier 'speculative' privatisations being turned over to other foreign concerns (Mahr-Melnhof in the case of ZMK Nikopol), which may have more proactive business development strategies. However, Bulgaria's accession to the EU will almost certainly fuel renewed interest by foreign investors, as it did during and after earlier waves of accession. In Spain investment increased after accession in 1986 from $1.8bln/yr (1981–5) to $7.5bln (1986–90) while the comparable increase in Portugal was from $0.2 to $1.2bln (Jordanova 1999). For the forestry and woodworking sectors accession may act to consolidate their emerging status as semi-peripheral producers; far cheaper than Western European producers, and competitive with Asia as function of the combination of order lead times, wages and (especially) transport costs. One positive impact of this future would be the maintenance of a fully developed wood sector product chain from timberlands to final product, as opposed to the export-led raw and semi-processed situation that has come about in other transition countries, particularly in Africa and Latin America. Owusu (2001) says that similar transition-type conditions in Ghana and other African countries created a situation whereby overall production grew rapidly, but largely in raw resource and semi-processed goods manufacture into Western Europe and USA.

Overall it seems that FDI into the wood sectors is having the effect of consolidating large- and medium-sized enterprises, with the likely result that the immense number of micro-enterprises observed by Staddon (2001) will either restructure through mergers into larger more sustainable firms or fail entirely. Somewhat ironically this will perforce reduce rather than increase the level of entrepreneurialism in the sector as a structure of fewer entrepreneurs and more workers (re)asserts itself. This will however be based on a real (rather than speculative) reintegration of wood processing enterprises into Bulgarian and European economic networks – true 'transformative' industrial networks in Smith and Swain's terms (1998). So far these processes, and the parallel process of developing wood sector enterprises through forward and backward linkages within the structure of domestic demand, have had the effect of improving levels of pay, management and health and safety in the workplaces, although at the cost of drastically reducing the number of jobs in the sector and redefining them as work in the Western European sense. Happily, a number of wood enterprises have made a point of retaining some aspects of social provision more redolent of the communist period, including on-site health care and holiday houses in the mountains and at the seaside. In terms of spatial patterns, these processes are clearly changing the logics of location as the previously dense and diffuse network of producers is reorganised into identifiable poles of development in the southwest, along the east-west E4 motorway corridor, in the northeastern Balkan mountains (particularly between Troyan and Nikopol) and in the central Rhodope Mountains adjacent the Turkish border. There are also environmental improvements as firms strive to achieve the world standards required for competitiveness generally (ISO9000), or those favoured by Western European markets like FSC or Ikea's simplified form of ISO14001–certified EMAS.

Notes

[1] The new premier's economic team is young, Western educated and enthusiastic, but one of the biggest challenges will be the management of the high expectations of the population. The new government's foreign policy stance is strongly pro-NATO and pro-EU.

[2] Sawnwood, unplaned, planed, grooved, tongued etc., sawn lengthwise or produced by a profile-chipping process (e.g. planks, beams, joists, rafters, scantlings, laths, boxboards, 'lumber' and sleepers) and planed wood which may also be finger jointed, tongued or grooved, chamfered, rabbeted, V-jointed, beaded etc. Wood flooring is excluded. With few exceptions sawnwood exceeds 5mm in thickness.

[3] The five enterprises discussed here are in fact composites of actually existing enterprises in the Razlog Basin contacted by the author in 1999–2000.

[4] There are three main types of FDI accounted in macroeconomic statistics: equity capital (including inter-company loans, net liabilities, suppliers credits and debt securities) and reinvested earnings. FDI flows denote the amount of new investment in a given period (e.g. quarterly or annually). FDI stock is the value of the share of affiliate enterprises and reserves (including retained profits) attributable to the parent (or investing) enterprise, plus the net indebtedness of affiliates to the parent enterprise. It is important to consider both FDI flows and FDI stocks to obtain a true picture of FDI impact in the host economy.

[5] 'Zunk Bonds' are freely tradeable financial instruments that essentially securitise the debt of SOEs. They are denominated in either $ or BGL and pay interest semi-annually until their maturity date, generally 25 years after issue. They are a way of converting debt into capital for reinvestment, generally on international markets and are therefore a form of 'Brady Bond' as originally invented during the Reagan administration for Latin American countries.

[6] It could be argued that this is a much more subtle sort of FDI relationship since Ikea is clearly using its size and the carrot of ongoing production contracts to 'invest' socially and environmentally in its suppliers while distancing itself from the financial risks of the more obvious cash equity investment route.

[7] A 2002 Bulgarian Chamber of Woodworking and Furniture Industry study suggested the average number of employees for furniture makers is about 58.

[8] The finishing of wood products is almost always an area of special concern as is sawdust management: see later discussion related to Ikea.

[9] The author found some evidence of this in research on the timber auctioning process. There were cases where it seemed that agreements had been reached ex ante about how bidding for sections of timber (by law a sealed bid process) would go.

[10] The nearby municipality of Razlog and the company have jointly developed an initiative for installing a biomass (wood residue) boiler to provide heat for the plant and the local district heating system.

[11] Or they were completely unrealistic about terms: the MD tells of one bank that demanded the plant's title deeds in return for a relatively modest loan.

References

Begg, R. and Pickles, J. (1998), 'Institutions Social Networks and Ethnicity in the Cultures Transition: Industrial Change Mass Unemployment and Regional Change in Bulgaria', in J. Pickles and A. Smith (eds), *Theorising Transition: The Political Economy of Post-Communist Transition*, Routledge, London, 115–46.

Black, R. (1989), 'Regional Political Ecology in Theory and Practice: A Case Study from Northern Portugal', *Transactions of the Institute of British Geographers*, 15, 35–47.

Buckwalter, D.W. (1995), 'Spatial Inequality Foreign Investment and Economic Transition in Bulgaria', *Professional Geographer*, 47(3), 288–298.

De Haan, E. and Odenzeil, J. (2003), *Labour Conditions in Ikea's Supply Chain: Case Studies in India Bulgaria and Vietnam*, Centre for Research into Multinational Corporations, Amsterdam.

Dobrinsky, R. (2000), 'The Transition Crisis in Bulgaria', *Cambridge Journal of Economics*, 24, 581–602.

FAOSTAT (2003), FAOStat Forestry Data, FAO, Geneva <http://www.fao.org>.

Holmes, L. and Hayter, R. (1993), *Recent Restructuring in the Canadian Pulp and Paper Industry*, Simon Fraser University Department of Geography Discussion Paper 26, Vancouver.

Ikea (2003) <www.ikea.com>.

Jordanova, Z.T. (1999), 'FDI in Bulgaria: The Basis for the Formation of Strategic Alliances of the Type "East-West" in the Process of Preparation for the Joining the EU', *Factus Universtitatis*, 1(7), 57–62.

Koparanova, M.S. (1998), 'Overview of FDI in Bulgaria in the Middle of the 1990's', *East European Economics*, 36(4), 5–74.

Krassimirova, K. (2000), 'Izpravyat Konsultativni Cveta Sreshto Gorskata Mafia', *Tema*, March 31, no pagination.

Meurs, M. and Begg, R. (1998), 'Path Dependence in Bulgarian Agriculture', in J. Pickles and A. Smith (eds), *Theorising Transition: The Political Economy of Post-Communist Transition*, Routledge, London, 243–261.

Michailova, S. (1997), 'The Bulgarian Experience in the Privatisation Process', *Eastern European Economics*, 35(3), 75–92.

National Statistics Institute (1998), *Statistical Yearbook*, National Statistics Institute, Sofia.

O'Connor, J. (1988), 'Capitalism Nature Socialism: A Theoretical Introduction', *Capitalism Nature Socialism*, 1, photocopy with no pagination.

O'Riordan, T.F. (1989), 'The Challenge of Environmentalism', in R. Peet and N. Thrift (eds), *New Models in Geography*, Unwin Hyman, London, 77–104.

Owusu, J.H. (2001). 'Determinants of Export-Oriented Industrial Output in Ghana: The Case of Formal Wood Processing in an Era of Economic Recovery', *Journal of Modern Africa Studies*, 39, 51–80.

Prins, K. and Korotkov, A. (1994), 'The Forest Sector of Economies in Transition in CEE', *Unasylva*, 45, accessed electronically at <http://www.fao.org/docrep/t4620E/t4620E00.htm #Contents>.

Smith, A. and Swain, A. (1998), 'Regulating and Institutionalising Capitalisms: The Micro-Foundations of Transformation in Eastern and Central Europe', in J. Pickles and A. Smith (eds), *Theorising Transition: The Political Economy of Post-Communist Transition*, Routledge, London, 25–53.

Staddon, C. (1999), 'Economic Marginalisation and Natural Resource Management in Eastern Europe', *Geographical Journal*, 165, 200–8.

Staddon, C. (2000), 'Restitution of Forest Property in Post-Communist Bulgaria', *Natural Resources Forum*, 22(3), 237–246.

Staddon, C. (2001), 'Restructuring the Bulgarian Wood Processing Sector: Linkages between Resource Exploitation Capital Accumulation and Transition in a Post-Communist Locality', *Environment and Planning A*, 33, 607–28.

WWF (2003), *WWF and Ikea Cooperation on Forest Products* <http://www.wwfchina.org/english/downloads/Forest/ikea-wwf1.pdf>.

Chapter 12

The Impact of Foreign and Indigenous Capital in Rebuilding Croatia's Tourism Industry

Peter Jordan

Introduction

By her 40 million foreign tourist overnights (out of a total of 45mln) in 2002, Croatia ranks eighth among European countries (closely behind Germany and Greece having 42 million and 47 million respectively) and by her 753,000 bed-places (August 2002) the country takes the same position: behind Austria and in front of The Netherlands. Among countries in transformation Croatia is clearly the leader in tourism:[1] a situation evident in the communist era, for in the second half of the 1980s (1986 and 1987) Croatia – as a part of former Yugoslavia – reached her climax as a destination of mass and seaside tourism with almost 60 million foreign tourist nights a year. In the course of Yugoslavia's dissolution, however, and due to the wars accompanying this process from 1990 to 1995, Croatia's tourism experienced almost a total collapse followed by a period of tiresome recovery. This makes Croatia definitely a special case among countries in transformation and one which is very significant in terms of investment and especially foreign investment. At the start of transformation the tourist industry was characterised by a dual system of private and public ownership (albeit with a clear dominance of the public sector) contrary to the situation in most communist countries. Small tourism enterprises like family pensions, smaller restaurants and inns were privately owned[2] in addition to the widespread system of private room renting. But this private sector was limited by law and in spite of its respectable size it was essentially complementary to the local and regional 'flagships of tourism', i.e. hotels and larger catering facilities which were always publicly owned and – typically for the former Yugoslavia – organised as self-managing enterprises. These work organisations ('Radna organizacija')[3] were owned by their staff, who were entitled to elect the managers, although in practice the managers depended on bank credits and were exposed to political influence to the extent that they might be merely 'puppets on a string'. Especially on the northern coast almost all hotels and other major tourism facilities of a micro-region or commune[4] were in the hands of such an enterprise, which in turn exerted considerable influence on the overall economic, social and political situation and development of the region, at least on the coast and on the islands, where tourism was the dominant economic factor. It seems worthwhile to investigate the effects of these two distinctive features on investment and more specifically FDI in Croatian tourism and in turn to ask how this investment can help to rebuild Croatia's tourism industry. Analysis is based on statistical data (mainly Croatian Central Bureau of Statistics, Ministry of Tourism, Croatian National Bank, Institute of Tourism Zagreb), Croatian scientific literature, interviews with Croatian experts and personal participating observation of Croatian tourism since 1980.

Development of Croatian Tourism after 1990

It is only a slight exaggeration to say that Croatia's tourism industry emerged after an almost total collapse with the same profile and structure as before (Jordan 2000, 2002). This is true for the product as well as the organisational structure. Anyone who assumed that the destruction of the industry and the almost total collapse of demand by war would be grasped as an opportunity to construct a new industry complying with new travel trends and the principles of a market economy was definitely wrong. Rather, the wars in and around Croatia – accompanied by a collapse of the demand as well as a consequential political stand-still – served to discourage investment, impede reforms and conserve existing structures, even more so in tourism than in other branches of the economy. After the political crisis had come to an end, some of the tourists 'returned', but they met the same highly seasonal, spatially-concentrated, low-quality seaside tourism business and the same dual structure of small private undertakings and large, regionally monopolist tourism enterprises: the labels had changed but the quality of service remained the same (Jordan 1997, 2002). Foreign investors were discouraged by almost a decade of weak business, the political insecurity of the Balkans, to which Croatia (as a part of former Yugoslavia) was still seen to be affiliated, the political and legal insecurity in the country itself and economic-strategic inactivity by the government as well as the slow pace of Croatia's European integration. The main difference between current tourism and tourism in the communist era is the much higher price level, caused by the hard-currency policy in connection with the Kuna (strict parity with the German Mark and now the Euro), high tax rates, high overhead costs in many enterprises and the need to cover high costs in the course of a very short season. Nevertheless Croatia still has the image of a low price and mass destination.

The collapse of tourism in Croatia did not coincide exactly with the dissolution of Yugoslavia and the outbreak of war in 1990, but was preceded by a slowdown already evident in the late 1980s due to the growing economic and political crisis in Yugoslavia as well as new trends in European tourism – cheap packages from distant destinations – unfavourable for recreational and seaside tourism (Figure 12.1). In August 1990, the first clashes in the hinterland of Zadar – correctly interpreted by the international media as the prelude to war – immediately gave rise

Figure 12.1   Tourist nights in Croatia, 1977–2002

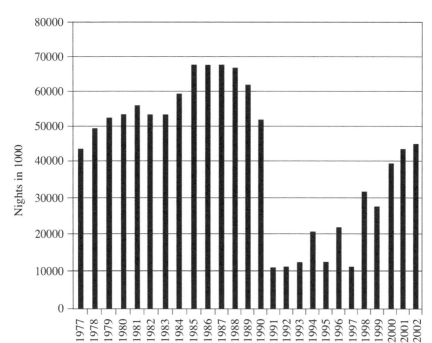

Source: CBS 2003a; Ministry of Tourism 2003a.

to a sharp decline in tourist overnights. The 'Ten Day War' in Slovenia in late spring 1991 and the full outbreak of war in Croatia the following summer pushed figures down to just 15% of the all-time peak in 1986. While the coast south of the Maslenica bridge, northeast of Zadar – i.e. most of Dalmatia – was practically empty of tourists in 1991 and 1992 and did not host many more till 1995, the north (Istria and the Kvarner) enjoyed a location relatively remote from areas of warfare and close to the travel markets of Slovenia, Austria and northern Italy which continued to deliver flows of excursionists, longer weekend vacationers and owners of weekend and vacation homes. This dual situation lasted until 1995, when the Croatian army re-established Croatian control over the territories of the 'Serbian Republic' (May and August) and the end of war in Bosnia and Hercegovina made the southern coast safe. The military operations of 1995 discouraged tourism but although the southern coast recovered as a tourist destination during 1996–9 it could not match the north with its better location in relation to the key markets. The Kosovo crisis and the consequent NATO air strikes on Serbia and Montenegro in 1999 also had an impact on Croatian tourism. But since 1999 tourist overnights have gone up continuously in all coastal areas and reached two thirds of 1986 in 2002. 2003 is considered to have been an even better season.

Figure 12.2   Foreign tourist nights in Croatia, 1977–2002

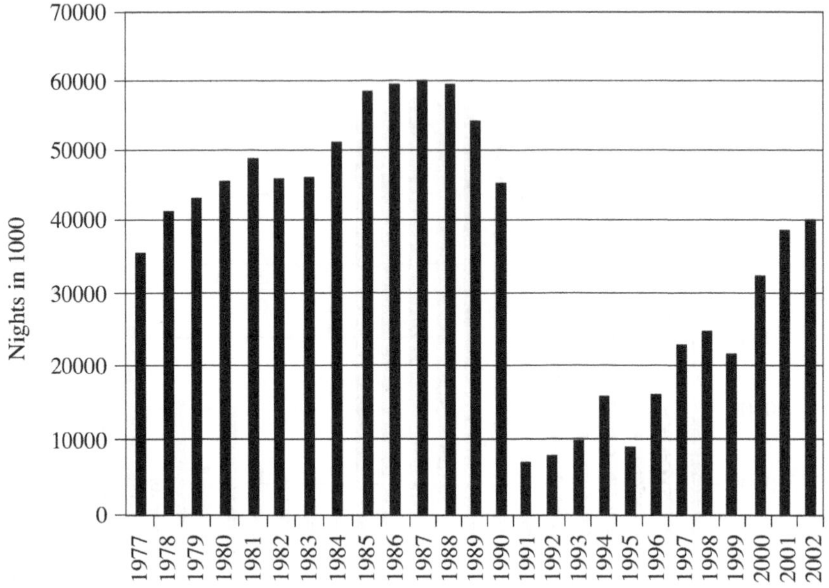

Source: as for Figure 12.1.

Long-term development of foreign tourist overnights (Figure 12.2) is not so different and looks even more favourable, although the many visitors from other former Yugoslav republics (mainly Slovenia) were only classified as foreigners after 1991. Decline of tourist nights in the early 1990s was accompanied by a collapse of tourist revenues and employment (Figures 12.3 and 12.4). However, gross revenues regained their 1990 level as early as 1996 because of price increases. Employment declined dramatically after the outbreak of wars and is even in 2001 was significantly lower than before. In Figure 12.4 it is represented according to two diverging sources. Actual damage to tourism facilities by warfare was rather exceptional, at least on the coast. Even hotels at the Plitvice Lakes – located right in the ephemeral Serbian Republic and near to zones of intensive warfare – were not only undamaged but carefully treated as army headquarters. By contrast, inland from the coast and also in Slavonia, tourism facilities – albeit of rather local importance – were often destroyed or heavily damaged. Much more important was: first, the closure of many hotels due to a lack of demand, especially along the southern coast; second, the use of many hotels (also on the northern coast and in Zagreb) as camps for refugees and displaced persons for several years; third, the lack of investment and maintenance due to the economically-desperate situation during the war; and fourth, the lack of income normally generated by tourism which

Figure 12.3   Revenues in Croatian tourism, 1990–1996

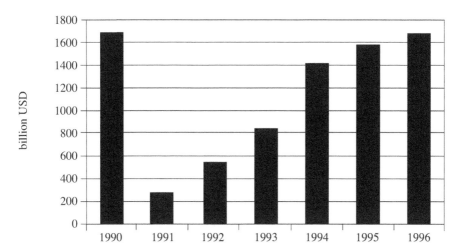

Source: Čižmar and Polijanec-Borić 1997, p. 292.

provided a living for a substantial part of the coast and island population. However investment and maintenance had already slowed down in the late 1980s because of the economic crisis and so most larger hotels experienced an investment and maintenance break of well over a decade after the high levels of demand in the 1970s and 1980s when they were constructed or previously renovated for the last time. The situation was better in the former small-scale private sector (at least after the war years) since it profited from gaps in the large-scale provision and adapted more flexibly to demands in the industry. Nevertheless, during the peak crisis seasons of the early 1990s, private room renting went down nearly to zero, which had a strong impact on the income of the local population as well as on urban Croatians, who used to offer rooms in their vacation homes at the coast.

Current Structures of Croatian Tourism

Types of Accommodation and their Spatial Distribution

The main type of tourism in Croatia is seaside recreation connected with summer sports. A survey conducted by the Institute of Tourism in Zagreb and the Croatian National Bank in 2000 at border stations (including air and sea) comprising 29,480 visitors (Hendija 2001 p.275) found that 40% came for relaxation and entertainment, while 22% were in transit, 16% visited relatives and friends, 13% came for business and the rest for shopping, health treatment and education etc. This produced a concentration of tourism on the coast reflected in the spatial distribution of tourist nights (indicated by columns) and tourism intensity (tourist nights per

Figure 12.4   Employment in Croatian tourism, 1990–2001

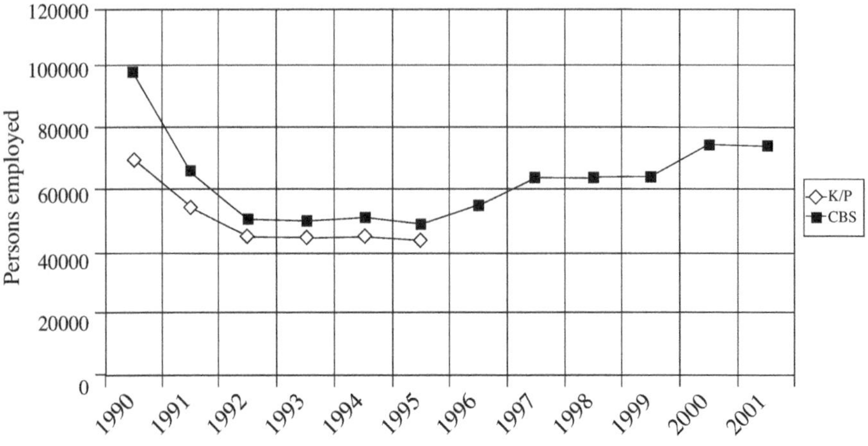

Source: Kovačević and Plevko 1997, p. 240; Central Bureau of Statistics (http://www.dzs.hr), Ljetopis 2002, table 23–1.

inhabitant) shown for 2002 in Figure 12.5 by counties. The seven coastal counties accounted for 96% of all tourist nights (leaving just 4% for the 14 interior counties), a situation reflected by the spatial distribution of accommodation shown in Figure 12.6 which credits the coast with 97% of all beds – and 92% of all hotel beds – even though most towns and cities are located in the interior. Within the coastal zone, there is a clear dominance for the northern coast comprising Istria and the Kvarner. After a balance in the 1960s and early 1970,[5] this dominance became apparent during the late 1970s and in the 1980s when motoring along the panoramic coastal highway down to Montenegro lost its attraction and the popularity of shorter holidays and long weekends favoured destinations closer to the generating markets. Taking into account that about 95% of foreign tourists in Croatia enter the country by road, technical and traffic conditions on the major international routes are also critical. In this respect too, the southern coast is handicapped, since some might still regard the transit routes across Bosnia and Hercegovina as 'adventurous' while access routes through Croatia are ill-equipped and overcrowded during the season. The situation will improve when the motorway from Zagreb via Karlovac to Zadar is completed, possibly in 2005 (Weber 2002). Despite remarkable cultural attractions, along with festivals, spas and health centres (Österreichisches Ost- und Südosteuropa-Institut 2003), these could not in the communist period – as they still cannot today – distract visitors from their main activity of seaside recreation or extend the season at any degree (indeed the seasonal peak has sharpened since the wars). And while there are opportunities in the interior it is only the capital city of Zagreb – as a business and cultural centre – and the national park of Plitvice with its natural cascade of lakes that attract a large numbers of tourists.

Figure 12.5   Tourist nights and tourism intensity in 2002 by county

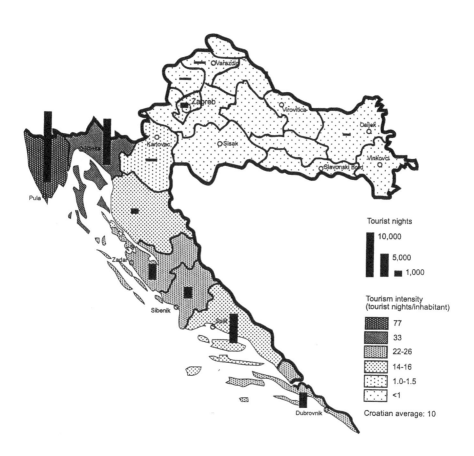

Travel Markets

Foreign tourism is paramount for domestic tourism is limited (5.0 million persons or 11% of all tourist overnights in 2002)[6] and is in no position to compensate for the decline in visitors from abroad. In 2002, tourist overnights from EU countries had the largest share, especially Germany 27%, Italy 12% and Austria 9% (Figure 12.7), followed by ECE transition states (Slovenia 12%, Czech Republic 11% and Poland 6%). This structure is the result of remarkable shifts during the 1990s. While before 1990 the West (in political terms) had more than 80% in all foreign tourist overnights, and countries as remote as Netherlands and UK were among the strongest markets, the wars practically stopped this inflow, except for some

Figure 12.6   Tourist beds and hotel beds in 2002 by county

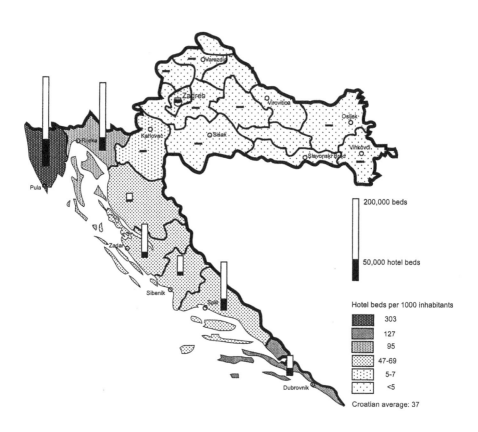

Austrians and Italians well acquainted with the local situation. Tourists from transition countries looking for cheap offers partly filled this gap (Figure 12.7, 1994 column) for during 1991–5 Czech tourists were sometimes in a majority among a very reduced total number of overnights; while Slovenes (especially on the northern coast), Hungarians and Poles also helped to fill the gap. In 1994, less than 40% of all foreign tourist nights were spent by tourists from the West and in 1997, when the conflicts in and around Croatia had come to an end, a structure similar to the current one became evident with half the foreign tourists coming from the EU and half from the ECECs. Meanwhile, Germans, Italians and Austrians had 'returned', but not the British (Fox and Fox 1998) or the Dutch. This shift of markets meant, however, a decline in possible per capita revenues (albeit compensated for by higher prices) since spending by ECE visitors is low compared with Germans or Italians: in

Figure 12.7   Structure of foreign tourist overnights by country of origin, 1990–2002

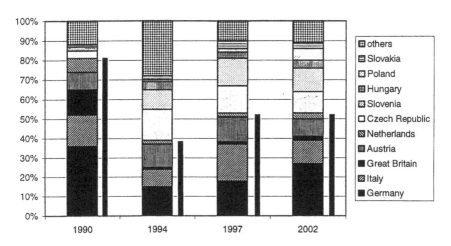

Source: as for Figure 12.1.

particular, the Czechs bring food and drink with them to ensure 'cheap holidays' on the coast.

Figure 12.8 shows dominant travel markets by counties in 2002. Germans have the majority in the two most tourism-intensive counties on the northern coast (Istria and Primorje-Gorski kotar), along with Dubrovnik-Neretva county in the far south, where before the wars the British used to be dominant. Slovenes, although well-represented along the whole northern coast, form the majority in the Lika-Senj county while Czechs dominate the central coast: Šibenik-Knin, Split-Dalmatia and Zadar counties. This coincides with a high share of non-hotel tourist capacities (camp sites and private rooms) and the lowest share of quality hotel beds at the coast. Italians have high shares along the whole coast, including the south where they arrive by ferryboat from the opposite side of the Adriatic. Yet while they are in second place at best in all coastal counties, they constitute a majority in the eight interior comprising Slavonia and the Drava and Sava valleys: explained by their inclination for hunting, which affects the whole of Croatia (including the coast and the islands out of season) but is statistically significant only in those parts of the country where the volume of tourism is very low. Austrians are well represented along the whole coast with a focus in the north, though they have a statistical majority only in the county of Krapina.

Accommodation and Catering

The short season with more than two thirds of all tourist nights realised during just

Figure 12.8   Dominant travel markets in 2002 by county

two months of full occupancy in July and August makes for an accommodation structure with a relatively low share of capital-intensive capacities: hotels, apartment hotels and hotel villages (Figure 12.9). Their share in the overall number of beds is just around a quarter, although it has grown from 22 to 29% during 1988–2002, while the total number of hotel beds increased from 204,000 to 219,000 when the marinas are included into this category. The number of hotels in the narrower sense amounts to 420, comprising around 49,390 rooms with 96,625 beds. The average size of a Croatian hotel is 231 beds and they score a record average annual occupancy rate of 37.0% (Hendija 2002 p.410). But in the coastal tourist regions outside the cities even these capital-intensive capacities are rarely open throughout the year: most operate from Easter until All Saints Day (November 1). On the larger islands there is usually just one hotel open throughout the winter. The

Figure 12.9   Accommodation structure, 1988–2002

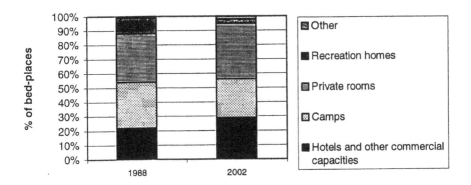

Source: as for Figure 12.1.

situation is different only in resorts with alternative attractions like Opatija which tries to relive its glorious past as a winter resort by offering health facilities (Jordan and Peršić 1998). Complementary capacities like camp sites and private rooms always had and still have the largest share of the total accommodation on offer. They mainly help to accommodate tourists in the peak season and their occupancy rate is consequently low: only 9.3% in the case of private rooms (Hendija 2002 p.409). Private room renting is nevertheless important as an additional source of income for the local population and contributes to the stability of rural areas and the islands despite an often-critical economic situation. Many of those who offer private rooms are, however, people from the cities, who rent their weekend-homes during summer. Meanwhile campsites have declined in importance compared to 1988 (from 32 to 27%) in accordance with international trends, while private room renting has gained relatively (from 34 to 39%) – though not in absolute figures – while recreation homes of all kinds have declined sharply (from 11 to 3% and from 105,000 to 25,000 beds) in line with trends in socialist and trade union tourism (though to a reduced extent compared with other former communist countries).

When the quality of the August 2001 accommodation is considered, almost exactly half the beds in hotels and all-suite hotels (47,724 out of 95,428) involved establishments with at least three stars (CBS 2003a; Ministry of Tourism 2003a). Only 2,231 beds (2.3%) were offered by four-star hotels and 2,981 (3.1%) by five-star hotels. The spatial distribution (Figure 12.10) shows concentrations of quality beds in Istria and Dubrovnik-Neretva counties, while the gap in quality on the central coast is striking. Apart from Dubrovnik-Neretva county – with the 'Excelsior' (362 beds) and the 'Villa Orsula' (28 beds) in Dubrovnik as well as the

Figure 12.10   Beds in Croatian three-, four- and five-star hotels, 2001

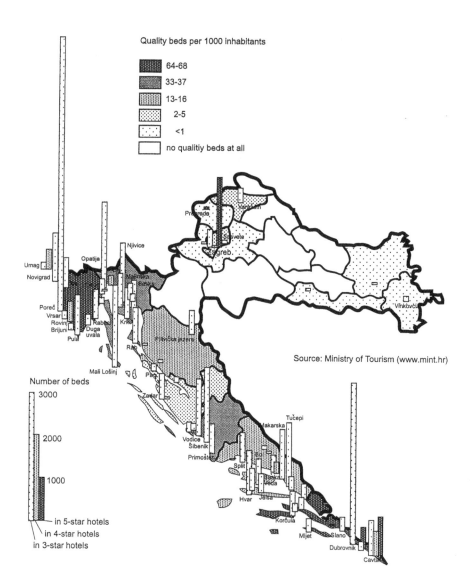

Source: Ministry of Tourism (www.mint.hr)

'Croatia' (888 beds) in nearby Cavtat – five-star hotels are confined to Zagreb (the traditional 'Esplanade' with 268 beds, 'Opera' with 792 beds and 'Sheraton Zagreb' with 588 beds) and the rural environment of the Hrvatsko zagorje where the 'Dvorac Bezanec' in Pregrada has 55 beds.

The accommodation and catering industry may be classified into six organisational categories (Kunst 1998 p.133): (a) large, multi-section companies ('tourism conglomerates') – typical of the northern coast (Istria and Kvarner) – which usually cooperate with tour operators and use their negotiation power as regional quasi-monopolists on the supply side; (b) smaller companies – though still with diversified activities, typical of smaller towns where mass tourism has not penetrated to the fullest extent – which also cooperate with tour operators while also serving individual tourists; (c) independent hotel businesses, typical of large towns and cities; (d) catering facilities of the urban type which do not offer accommodation; (e) tourism agencies and companies; and (f) marinas which companies offering accommodation and services of all kind for nautical tourists.

A survey by an international tourism consultant (Horwath Consulting 1999) compared the competitive position of the Croatian hotel industry with several Mediterranean competitors (France, Greece, Portugal and Spain) as well as Austria and Hungary. The fieldwork was carried out in 1998 with data referring to the business year 1997 and the results are presented here according to Čižmar and Šerić (1999). The Croatian sample comprised 58 of the best hotels according to market and financial performance and technical conditions. The capacity of 13,193 rooms was equivalent to 70% of the total Croatian capacity in three- to five-star hotels (although this international classification system had not been adopted in Croatia at the time). Their regional structure reflected the overall dispersion of hotels in Croatia. The key results of the survey indicated: (a) a hotel accommodation share only half the level of the other countries; (b) a much bigger average size; (c) a low average occupancy rate of 41.4% – compared with a range for the other countries extending from 73.3% in Greece to 58.2% in Hungary – and one below the common business viability threshold; (d) a low average room rate of $40.1: higher than Hungary ($35.1) and close to Greece ($42.4), but well below the rates in the most of the other countries (c.$75–80) and especially so in the case of France ($112.0); (e) reflecting low occupancy and room rates, the lowest average annual room revenue per available room of ($10,335): close to Greece ($11,872) and Hungary ($13,960) but well below the realisations in the other countries which ranged from $31,422 in Portugal to $43,564 in France; (f) limited use of a direct marketing approach (direct mailing and loyalty card programmes) and technologically-superior marketing media (telemarketing and websites); (g) the lowest position for gross operative profit (the surplus of revenue over current costs): 16.6% compared to an average of 29.2%; but (h) an average performance over the efficiency of operational departments (e.g. accommodation and food and beverages) because of high labour productivity given the low salaries and the high share of seasonal work after a dramatic reduction of personal in the early 1990s (see Figure 12.4). The gap between the very low overall efficiency and the much more acceptable efficiency for operational departments is explained through high costs in the area of undistributed expenses (Cižmar and Šerić 1999 p.311) i.e. for the two administrative levels (hotel and company management), for maintenance efforts (given the substantial back-log) and for water and energy which are relatively highly priced in Croatia. Taking all these features into account, the interim conclusion emerges that the Croatian tourism industry is by its inherited and current structure not very well prepared for privatisation and – overall – is not attractive for new investment, particularly FDI, because of: (a) low occupancy due to a short season concentrating on sun and sea tourism; (b) low tourist expenditures generated by such tourism; (c) large hotels with small rooms (and lacking additional features) offering a uniform product required by such mass tourism; and (d) the partly-hyperthropic administrative structures which increase costs.

Privatisation and Investment in Tourism

Aside from these structural characteristics, privatisation faces policy-generated problems that are partly common to all countries in transformation and partly specific. First, a basic problem the Croatian tourism industry shares with other transition countries is that investment during the communist period was not really guided by entrepreneurial criteria, but by political goals concerned with increased quantitative output, employment and job security and foreign currency revenues; and also regional development goals. There was a similarity with industrial policy which also created structures with a low investment efficiency – an unfavourable ratio between invested capital and output – so that production units were neither profitable nor viable under the conditions of an unprotected market. Second, specific problems arise from war in and around the country during 1990–5; political crisis and conflict in the Balkans, with the potential crisis regions of Kosovo, Macedonia, Montenegro and Bosnia and Hercegovina (at least up to 1999) affecting Croatia insofar as the country was perceived to be part of the Balkans by many Europeans. Third, there was only limited international acceptance of the Croatian government until January 2000 (comparable with Slovakia up to the end of the Mečiar government in 1998) and Croatia was excluded from European integration processes until 2000. Fourth, the tourist industry was considered as a sector of national interest (due to its size and economic role in Croatian economy and society) comparable to the energy or military sectors in other countries.[7] When initiating reforms in the industry, Croatian governments and political actors have to deal not only with the economic aspect, but must regard tourism as a kind of 'social institution' embedded into local society, especially in peripheral regions (like the islands and rural areas along the coast) where it frequently forming the focus of both the economy and the social life.

To understand the difficulty and slow progress of privatisation in this sector of the Croatian economy – regarded in Europe and also at home as a major asset – reference should be made to the rapid decline in investment efficiency (the ratio between invested capital and output in GDP units) since the first investment boom of 1967–71 when Yugoslav policy makers saw foreign tourism as a means to earn foreign currency and directed a larger share of capital into this sector. This was quite unique among communist regimes which normally regarded tourism – the institutionalised form of serving the better-off and mainly Westerners in the Yugoslav case – as a bastard in a working class society. While it took an average of 1.55 input units to produce one unit of national product in Croatian tourism during 1960–2 (Blažević 1997 p.219), this ratio worsened to 4.23:1 during 1972–5 period. If the 1.55:1 ratio might be seen as a basis for advance to higher revenues in the future, the situation in the early 1970s was clearly signalling significant over-investment according to all economic criteria. The ratio was even worse in tourism than for the Croatian economy as a whole (1.97:1) (Ibid). Although another tourism investment boom of this size failed to materialise, the index continued to deteriorate to 5.64:1 by 1989.

The main reason for this unfortunate situation was the very short season and the consequent low occupancy rate. Attractive hotels were sitting amidst beautiful scenery for two thirds of the year, operating at full costs despite the small number of visitors during much of this time.[8] They were closed in winter but depreciation continued and some current costs had to be covered. Thus the Croatian insiders Dragičević et al. (1998 p.244) concede that the tourist industry to a large extent has no market value. This assessment, of course, applies to the average holiday hotel and not the typical city hotel or for hotels in spas and at traffic nodes which had indeed much better occupancy rates. But it is true for the vast majority of seaside resorts along the coast and on the islands where the demand is such as to be best catered for through the flexible accommodation potentials of private room renting and camp sites. In comparable situations in Austria, private rooms and others rented by commercial village inns usually provide the only accommodation available apart from campsites. In the case of Austrian winter tourism, with its considerably higher prices (though generally appropriate given the quality that is on offer), the threshold for mono-seasonal hotel viability is a more or less full occupancy for about four months from December to Easter. Over-investment in Croatia was possible because banks had to comply with party political decisions seeking increasing numbers of beds, tourist arrivals and overnights and not least high employment. In a way, tourism played the same role as heavy industry in other communist countries and it is no surprise that Croatian tourism enterprises grew increasingly indebted to the banks.

Soon after Croatian independence in 1991 there was provision for privatisation under the first national tourism law (based on the Austrian model) and the Law on Transformation of Socially-Owned Enterprises (April 1991). During 1992 there were management and employee buyouts, public offerings of shares and public auctions of shares at the Zagreb stock exchange. In an environment of war, external pressure and national defence, the tourism industry with its 1992 employment level standing at 50,768 (CBS 2003b Table 23–1) or 46,160 (Kovačević and Plevko 1997 p.240) – albeit a sharp decline from the 108,206 employed in hotels and restaurants in 1989 (CBS 2003b Table 23–1) – was regarded as a sector of national interest that 'should not be sold off to foreigners, who would just pick out the pearls and close-down the rest'. Shares of the tourism enterprises were given first to the managers and (former) employees, frequently in exchange for salaries and pensions that had been withheld, and secondly to the Croatian banks to which the tourism enterprises were indebted. Their engagement was a way of protecting the capital placed in the industry in communist times: a kind of economic necessity, rather than (if at all) a strategic decision of engagement (Čižmar and Poljanec-Borić 1997 p.296). At the time, the banks anticipated high revenues when the war ended and they therefore hoped to sell their assets at the right moment – until it became clear that the opportunity would not arise unless the economic situation in the industry improved and a basic restructuring was undertaken. So for both managers/employees and banks, the acquisition of shares was (mostly) just a means of converting claims into shares for no new capital or know-how was introduced.

In addition, the state subsidised tourism enterprises to save them from bankruptcy and these subsidies were later converted into shares. The Privatisation Law as of March 1996 introduced the mass privatisation method for a certain category of individuals. Thus, in 1997 the large tourism enterprises (not small hotels and restaurants, which were private from the beginning) had mostly a trilateral ownership (Čižmar and Poljanec-Borić 1997 p.295) one third was held by the state or other public shareholders (the Croatian Privatisation Fund and the Retirement Fund), one third by the banks (privatised, apart from rare exceptions), and a third by small shareholders (mainly former and current employees and managers). None of these parties had a strategic perspective for the further development of the enterprises or a readiness to intervene into their internal structures and economic goals. Due to their divergent personal interests and lack of organisation, small shareholders have never had the capacity for strategic intervention; while the banks regarded their engagement as a temporary one (without an entrepreneurial perspective) and the state did not intervene strategically either, despite what should have been its obligation to do so. Although a Development Strategy of Croatian Tourism ('Strategija razvoja hrvatskog turizma')[9] and a Master Plan ('Glavni plan razvoja hrvatskog turizma') were elaborated by experts and adopted by the Ministry of Tourism by 1993 and 1994 respectively, the government hesitated to act accordingly given first, the dilemma between a policy of privatisation according to economic principles and one which approached a national asset not just from the economic viewpoint but second (and by no means least), commitment to regional development and socio-economic spatial disparity equalisation which called for the retention of employment. While the first option would have meant the closure of many, if not the majority of existing hotels (Dragičević et al. 1998 p.244) the latter's extra-economic rationale required not merely passive subsidy but active support in financial and strategic terms involving both state investment and the application of fiscal policy instruments such exemption from income, profit and property tax for certain regions and socio-economic situations or the application of different VAT rates.[10] Arguably for reasons of political opportunism it was not possible to make a clear decision either: to apply economic principles, which would have aroused opposition among the local population – especially in economically-disadvantaged parts of the country, and also among local political functionaries (the so-called 'basis' of political parties) who usually aim to defend existing structures[11] – or: to treat the tourist industry as a social category and an instrument of regional development, which would have required the allocation of a major share of the state budget or the renunciation of tax revenues (both difficult to accept in a context of budget restrictions and pressing needs in other fields).[12]

The actual strategy of subsidising many (at least the larger) enterprises to save them from bankruptcy came close to the latter solution, but it lacked the active component of strategic investment and planning. Yet the costs were still high, so the government was urged to get rid of its shares as soon as possible by 'finalising the privatisation process'. The ownership – as previously described – meant a standstill and a lack of investment. So the industry stagnated even in comparison to all other sectors of the Croatian economy. Branko Blažević (1997 p.224 especially) documents this situation for the war years 1990–5 when depreciation of existing assets was much higher than investment into substitution, expansion, reconstruction and modernisation: in 1990 10.23 times higher, in 1991 3.67 times, in 1992 7.28 times, in 1993 9.39 times, in 1994 4.67 times and in 1995 3.97 times. This process, in which depreciation is not balanced by adequate investment, is called 'disinvestment'. It was hardly surprising in a war situation when the travel markets had collapsed, yet such a moment could have been used to restructure the business fundamentally, given the political will to apply strategic concepts was inject investment capital. Unfortunately, the Croatian government remained undecided. Capital was not available: neither private domestic capital (very scarce in Croatia) nor foreign capital: indeed until 1999 no foreign investor participated in the ownership of Croatian hotels (Dragičević et al. 1998 p.246). Moreover, after the wars (1995) investment in tourism remained rather limited because in addition the general development of the market:[13] (a) the political and macro-economic conditions were adversely affected by the strict currency policy, which further reduced the competitiveness of Croatian tourism enterprises; (b) a high (22%) and uniform VAT rate was introduced abruptly in 1998 without exemptions;[14] (c) high interest rates were combined with a lack of public support for long-term, low interest loans rates or government financial guarantees for international borrowing; along with (d) 'the not very encouraging process of privatisation' and the country's 'modest overall – and particularly political – image' (Radnić and Ivandić 1999 p.53).

Some increase in interest by foreign investors in the Croatian economy was evident in 2000[15] with a remarkable 44% growth in tourist overnights over 1999, the installation of a new government led by the Social Democratic Party[16] (favourably received by the international community, with visible signs of new European integration dynamics) and amendment of the legal framework for foreign investment through the Investment Promotion Law of July 2000. A study by the Vienna Institute of International Economic Studies and the Vienna Institute of

Economic Research (Hunya and Stankovsky 2002 Table 3) shows a growth of the total FDI stock in Croatia from $3,150.9mln in December 1999 to $3,863.3mln in December 2000 (with the tourism sector increasing from $68.0 to $121.8mln). The most important investments into the economy as a whole came from Austria, followed by Luxembourg, Germany and the USA. Austria was also the dominant source of foreign investments up to the first quarter of 2003. In a ranking of the FDI stock in Croatia from 1993 to the first quarter of 2003 Austria is first with a share of 23.6% followed by Germany with 22.4% (Croatian National Bank 2003), although Austrian investment has gone largely into banks with only slight involvement in tourism so far.[17]

The new Investment Promotion Law of July 2000 set incentive measures and granted tax benefits for certain kinds of investment. Incentive measures comprised funds for initial capital and for real estate usage, assistance in the creation of new jobs and vocational training as well as re-training. Tax concessions were related to the profits tax, which is normally 20%. However, if an investment exceeds HRK10mln and creates more than 30 new jobs, profits tax is reduced to 7.0% for ten years. If the investment exceeds HRK20mln and creates more than 50 new jobs, the tax is reduced to 3.0% for ten years. And there is no profit tax to be paid for ten years, when the investment is more than HRK60mln and creates more than 75 new jobs. In addition, tax breaks are granted for investment in certain areas specified by law. Foreign investors are guaranteed equal treatment to domestic investors, free transfer of profits, ownership on real estate except in certain territories designated by law and in agricultural and forest land. But companies registered in Croatia are treated as domestic, even when they are owned by foreigners (Ministry of Economics 2003).

In the case of tourism, the government – represented by the Ministry of Tourism and by its Privatisation Fund (CPF) – made a special effort to 'finalise' privatisation by offering state shares in tourism enterprises. This effort was accompanied by promoting measures, e.g. by the instalment of a website (Ministry of Tourism Zagreb 2003b) where 'first insight' information for investors is provided. It includes a document database on companies eligible for privatisation where state ownership exceeds 25%, descriptions and images of assets/facilities, ownership structure, contacts and bids announced. The list of facilities as of June 2003 comprises: (a) hotel, motel and hotel village facilities at a size of in total 30,208 beds (32% of all beds listed in this category in August 2001): included are three-star hotels with a total 9,750 beds (23% of all beds in three-star hotels) but no four- or five-star hotels (with the exception of some hotels in Split all are typical holiday hotels); (b) 'aparthotels', apartment villages and apartments with an overall capacity of 1,512 beds; (c) hotel apartment villages with an overall capacity of 2,907 beds; (d) boarding houses, guest rooms with an overall capacity of 151 beds; (e) campsites and hotel villages with an overall capacity of 1,148 persons; (f) campsites with an overall capacity of 11,230 persons; and (g) one marina. A list of tourism companies eligible for privatisation and with state ownership exceeding 50% (as of August 2002) comprised 43 companies. 32 of them – and mostly those with the largest state-owned shares – are based in the southernmost coastal counties of Dubrovnik-Neretva and Split-Dalmatia. Tellingly enough Istrian companies are not represented on this list, while the county of Primorje-Gorski kotar (the Kvarner region) is represented only by five albeit large and prominent companies including the Liburnia Riviera Hotel.

In the course of this new and 'final' effort to privatise, potential investors were invited to invest into the Croatian tourism sector in two ways: (a) by purchasing company shares, (b) by purchasing real estate (such as hotels) in those companies that decided to sell facilities. In contrast to buying real estate, the rule governing the purchase of company shares is that the investor takes over the debts of the company, although because of the rather heavy debts in some cases the investor may not have to pay the nominal value for the actual purchase price of the company shares is reached through three steps: the first ('Nominal Value Sale') involves sale at an initial price equal to the nominal value; the second ('Discount Price Sale') concerns a lower initial price set by CPF when shares cannot be sold at the nominal value; and the third ('Special Conditions Sale') deals with special conditions of sale set by CPF – that may go as far as a symbolic price – if the shares cannot be sold at the lower initial price. In addition to the CPF, which manages the state-owned shares, the Privatisation Investment Funds (PIF) manage the shares of individual coupon holders. PIF, by law, must sell these shares within five years from their inception. Also the shares held by banks and some other larger holders as well as by former owners are not managed by CPF, but by the respective owners themselves. A potential investor has to contact these owners directly and they are independent in their decisions.

By early 2003, 11 major privatisations of hotel enterprises and hotel facilities took place based on these regulations (though the list is based on casual evidence and may be incomplete and incorrect in details): seven by Croatian investors and four by foreign investors (Ministry of Tourism 2003). And already in 2000 Casino Austria had bought shares in the five-star Hotel Esplanade – the most traditional (built in 1925) and exquisite hotel in Zagreb – which it began to refurbish in November 2002 (Lagler 2003). In 2001, shares in four companies were sold and two real estate transactions were executed (Ministry of Tourism 2003): most remarkably the acquisition of 71% shares in Hoteli Cavtat by the Spanish enterprise Endicott Compania de Turismo, partly owned by German shareholders (Thomas Cook AG). Hoteli Cavtat owns the four-star 'Albatros' (566 beds) and the three-star 'Epidaurus' (619 beds) in Cavtat near Dubrovnik (Ministry of Tourism 2003b; Lagler 2003). The other three share transactions involved 62% of the shares in Croatia Hotels Cavtat – owner of the five-star 'Croatia' in Cavtat (888 beds) – and two small companies: Hotel Alan in Starigrad-Paklenica and Turisthotel Crvena Luka in Biograd na moru, both on the central coast. The large Opatija-based company Liburnia Riviera Hoteli sold its 'Hotel Paris' and the company Hoteli Podgora sold its 'Hotel Aurora' in Podgora near Makarska (Croatian National Bank 2003). In 2002 67% of the shares of the Primošten company with facilities on the central coast were sold, in addition to real estate by Liburnia Riviera Hotel: its 'Hotel Park' and

'Villa Frappart'. In 2003 Liburnia Riviera Hoteli sold further assets: the 'Agava', 'Astoria', 'Miran' and 'Palme' hotels, all located on the Opatija riviera and later in August 2003 it was reported on Austrian television (ORF2, Magazin Eco) that the Austrian entrepreneur Haselsteiner, shareholder of the Austrian-based international construction company Strabag, had acquired shares in the Grand Hotel Park, owner of the (currently) three-star, but very traditional, Grand Hotel Park in Dubrovnik and was renovating it.

But in contrast to these success stories some privatisation efforts failed. In early 2003 the purchase of the company Sunčani Hvar by a consortium in which a Slovenian bank was participating was stopped due to the opposition of the Croatian Peasant Party, a member of the ruling coalition. According to newspaper reports the Peasant Party was afraid of local animosities against the Slovenian engagement. Besides political interventions into transactions, corruption, non-business promoting behaviour and bureaucratic neglect are accusations frequently heard. A considerable part of the enterprises privatised declined in value after they had been sold, or did not receive the investments announced, or lost jobs. Reports from Istria (Hotel Island Katarina, Rovinj) prompt the conclusion that legal security for foreign investors is not too well developed and ranks in some cases second to the power of important local players (Die Presse 2003 pp.4,7).

Outlook and Conclusion

Up to the year 2000 privatisation of Croatian tourism enterprises was only to a very limited extent connected with real investment. A new and 'final' round of privatisation started in 2000 under somewhat better political, market and legal conditions and has scored some successes so far. But even if markets continue to revive it may be assumed that this final stage of privatisation in tourism will not lead to a sell-off of all hotels in which the state currently holds shares. The government will very likely be forced to find other solutions for the majority of true holiday hotels in regions south of Istria and the Kvarner, when occupancy rates are low and where opportunities to push the enterprises across the viability threshold even after considerable investment and reconstruction are modest. In modification of a forecast given by Dragičević et al. (1998 p.247), it is likely that for some time after the current privatisation round an ownership structure of the Croatian tourism industry will comprise in general terms: (a) small and medium city and holiday hotels as well as catering facilities (an extended version of the 'private sector' in communist times) with management relying on family business owned mainly by individual Croatian citizens; (b) prominent and traditional urban hotels in cities all over the country and in larger towns on the coast (and also, exceptionally, holiday hotels at prominent locations), owned by foreign and international investors, but operated (at least after an initial phase) by Croatian managers; (c) smaller and larger companies with diversified activities on the northern coast (Istria, Kvarner) and partly on the southern coast (especially in the Dubrovnik area) owned partly by domestic institutional owners (banks and funds) and partly by international financial consortia, investment funds and chain hotels, with the latter engaging domestic managers at the operational level of the individual hotel; (d) marinas owned partly by domestic institutional owners (banks and funds) and partly by international financial consortia and investment funds, with the latter engaging domestic managers at the operational level; and (e) smaller and larger companies with diversified activities as well as individual larger hotels in typical summer holiday areas on the central and southern coast, especially on the islands with the state as the majority owner and strategic investor.

To put it more directly, foreign investors and also domestic private investors will indeed 'pick out the pearls'. Moreover, a large number of hotels will not find a private investor and even if the Croatian state decides to invest strategically in order to reconstruct and reprofile many of them, there will remain a good many that will have to be closed-down over the longer term because the burden for the state budget will be too high. There is the hope that timely and substantial strategic investment – both domestic (private and public) and foreign – will generate a new type of quality and economically sustainable tourism, in line with current market trends, will exert overspill effects on other actors in tourism and other sectors of the economy. But there is little doubt that the concentration process will be enforced in spatial terms and spatial disparities – with implications for the socio-economic situation in general – will become more pronounced.

Notes

[1] Sources: (Croatian) Central Bureau of Statistics: CBS (2003a, 2003b); Eurostat (n.d.); Institute of Tourism Zagreb (2003); World Tourism Organisation: Ministry of Tourism (2003a, 2003b); and WTO (n.d.). The four maps in the chapter are all based on data from CBS (2003a) and Ministry of Tourism (2003a). Erroneously in some statistics (including WTO statistics) Hungary is shown as having the larger number of tourist arrivals. But tourist arrivals in Hungary are counted at border stations and include all transit visitors without an overnight in Hungary, contrary to the usual definition.

[2] Active and former guest workers in Western countries played an important role in small-scale private tourism, investing their money and applying their language skills.

[3] An example for a Working Organisation in Tourism and Trade in the Kvarner Region is presented by Jordan (1989).

[4] Croatian communes ('opčina') in the communist era were large and corresponded more to counties.

[5] In pre-communist times, starting with the nineteenth century, tourism was focused on the northern coast, except for some exclaves in the south like Dubrovnik and Hvar.

[6] Many Croatians possess weekend houses ('vikendica') on the coast or in other beautiful parts of the country where they do not count as tourists.

[7] And also small-scale agriculture in some Alpine counties where the Alpine cultural landscape is regarded as a cultural heritage (also as a fundamental factor for Alpine tourism) and farmers are considerably subsidised for landscape gardening.

[8] However, a larger share of seasonal labour and the fact that a seasonal labour force was

engaged in accordance with declining bed occupancy before and after the main season helped to reduce current expenses (Jordan 1989).

[9] The Development Strategy of Croatian Tourism – as well as other official documents on Croatian tourism – comprise the following main issues (Dragičević et al. 1998 p.243): the country possesses tourism potentials of global value; the industry must be thoroughly restructured and modernised due to the low technological, management and marketing standards inherited from communism; the national tourism product has to accord with concepts of quality management and sustainable development; commercial tourism must have access to all types of real estate ownership; the tourist industry must be open for management competition, management concentration and majority ownership; privatisation should be completed according to the principle that the risk is with the dominant owner; and an independent Croatia sets the basis of an internationally-competitive Croatia tourism.

[10] In fact Croatia has introduced (January 1998) VAT at a uniform rate of 22%. Before that accommodation in hotels was not taxable (Randić and Ivandić 1999 p.53).

[11] This phenomenon however, as well as the political handling of it, is neither typical for Croatia nor for post-communist countries in general, but it can be observed in many Western countries when it comes to privatising public enterprises.

[12] It is estimated that investment necessary to improve existing structures in tourism up to an economically-sustainable level amounts to about E3.0bln, which almost equals the actual market value of Croatia's tourist industry (Dragičević et al. 1998 p.245).

[13] The pre-war market structure – characterised by a dominance of Western markets – did not fully revive, but remained partly replaced by ECE markets with a lower purchasing power (Figure 12.7). International market trends further detracted from Croatia's tourism product.

[14] VAT for tourism enterprises in France, Greece, Italy and Spain varies from 5.5 to 19.0% (Petrić 1998 p.163).

[15] But already in 1999 some major FDI transactions had taken place, such as the purchase of a 35% share of Croatian Telecom by the German Telecom and the purchase of a two-thirds share of Privredna banka by the Banco d'Italia (Ministry of Economics 2003).

[16] The government comprised a wide centre-left coalition involving the Social Democrat Party, the Croatian Social-Liberal Party, the Croatian Peasant Party, the Istrian Democratic Party, the Liberal Party and the Croatian People's Party; replacing the national–conservative government of the Croatian Democratic Union which had governed the country from independence and had been shaped by the 'pater patriae' and president Franjo Tudjman.

[17] However, this may be regarded as indirect tourism investment since banks are important investors in tourism enterprises. Austrian banks that act on the Croatian market or have invested in Croatian banks are: Raiffeisen Zentralbank (since 1994); Bank Austria Creditanstalt, which bought Splitska banka (the third largest Croatian bank) in 2002 and HVB Croatia; and Erste Bank, which runs a Croatian branch company in addition Steiermärkische Sparkasse and has bought Riječka banka. Hypo Alpe Adria Bank and Volksbank are represented by branch companies in Croatia. Next to Austrian engagement in the Croatian bank sector is the Italian interest represented – besides Banco d'Italia – by UniCredito and Banca Commerciale Italiana (Die Presse 2003).

References

Blaževi'c, B. (1997), 'Efficiency of the Investment and Disinvestment Process in the Croatian Hotel Industry', *Turizam*, 45, 215–32.

Central Bureau of Statistics (2003a), <http://www.dzs.hr>.

Central Bureau of Statistics (ed.) (2003b), *Ljetopis 2002* <http://www.dzs.hr/ljetopis 2002>.

Čižmar, S. and Poljanec-Bori'c, S. (1997), 'The Privatisation of the Tourist Sector in Croatia During the Transitional and War Conditions', *Turizam*, 45, 289–300.

Čižmar, S. and Šeri'c, M. (1999), 'Market Effectiveness and Internal Efficiency of Croatian Hotel Industry', *Turizam*, 47, 300–15.

Croatian National Bank (2003), <http://www.hnb.hr>.

Die Presse (2003), *Internationale Sonderbeilage*, March 12.

Dragičevi'c, M., Čižmar, S. and Poljanec-Bori'c, S. (1998), 'Contribution to the Development Strategy of Croatian Tourism', *Turizam*, 46, 243–53.

Eurostat (n.d.), <http://europa.eu.int/comm/eurostat>.

Fox, J. and Fox, R. (1998), 'In Search of the Lost British Tourist', *Turizam*, 46, 203–19.

Hendija, Z. (2001), 'Motives of Foreign Visitors Travelling to Croatia and Domestic Visitors Travelling Abroad 2000', *Turizam*, 49, 275–6.

Hendija, Z. (2002), 'Accommodation Capacities in Croatia 2002', *Tourism*, 50, 409–10.

Horwath Consulting (1999), *Analiza stanja i rezultata turističkog sektora u Hrvatskoj*, Horwath Consulting, Zagreb.

Hunya, G. and Stankovsky, J. (2002), *WIIW-WIFO Database: FDI in CEECs and the FSU*, n.p., Vienna.

Institute of Tourism Zagreb (2003), <http.//www.iztzg.hr>.

Jordan, P. (1989), 'Gastarbeiter im eigenen Land: das Problem der saisonalen Arbeitskräfte im Fremdenverkehr der jugoslawischen Küste am Beispiel des Touristikunternehmens "Jadranka", Mali Lošinj', *Österreichische Osthefte*, 31, 683–714.

Jordan, P. (1997), *Beiträge zur Fremdenverkehrsgeographie der nördlichen kroatischen Küste*, Klagenfurter Geographische Schriften 15, Klagenfurt.

Jordan, P. (2000), 'Restructuring Croatia's Coastal Resorts: Change Sustainable Development and the Incorporation of Rural Hinterlands', *Journal of Sustainable Tourism*, 8, 525–39.

Jordan, P. (2002), 'Croatia', in Carter, F.W. and Turnock, D. (eds), *Environmental Problems of ECE*, Routledge, London, 330–46.

Jordan, P. and Perši'c, M. (eds) (1998), '*Österreich und der Tourismus von Opatija (Abbazia) vor dem Ersten Weltkrieg und zur Mitte der 1990er Jahre*', Wiener Osteuropa Studien, 8, Vienna.

Kovačevi'c, Z. and Plevko, S. (1997), 'The Macroeconomic Implications of the War and its Impact on Tourism in the Republic of Croatia', *Turizam*, 45, 233–47.

Kunst, I. (1998), 'Market Structure of Croatian Tourism Sector', *Turizam*, 46, 123–39.

Lagler, M. (2003), 'Privatisierung: einige Probleme, aber auch viele interessante Projekte', in Anon (ed), *Die Presse Internationale Sonderbeilage*, Die Presse, Vienna, p.3.

Ministry of Economics Zagreb (2003), <http://www.mingo.hr>.

Ministry of Tourism Zagreb (2003a), <http://www.mint.hr>.

Ministry of Tourism Zagreb (2003b), <http://mint.hr/investment_opportunities.htm> Österreichisches Ost- und Südosteuropa-Institut (2003), *aos-web* <www.aos.ac.at>.

Petri'c, L. (1998), 'Tourism Policy: Goals and Instruments', *Turizam*, 46, 140–68.

Radni'c, A. and Ivandi'c, N. (1999), 'War and Tourism in Croatia: Consequences and the Road to Recovery', *Turizam*, 47, 43–54.

Republički zavod za statistiku (ed) (1989), *Promet turista u primoskim op'cinam 1988*, RZS, Zagreb.

Weber, J. (2002), *Kroatien: Regionalentwicklung und Transformationsprozesse*, Mitteilungen
   der Geographischen Gesellschaft in Hamburg 92, Stuttgart.
World Tourism Organisation (n.d.), <http://www.world-tourism.org>.

Chapter 13

Strategies of International Investors in Hungary's Emerging Retail Market

Erika Nagy

Introduction: Reconstructing the Geography of Retailing

While this chapter examines FDI in Hungarian retailing, a context must also be provided in terms of a transformed retail sector that is playing an important role in the economic restructuring of emerging market economies – but one that has not yet been properly interpreted by the geographers of ECE. In this chapter the spatial aspects of retail restructuring are analysed with the help of case studies from Hungary, an emerging economy that is considered exemplary in terms of the agents and mechanisms driving the transition process from the late 1980s (Jones Lang Wootton 1998; Lupton 2002; Nagy 2002). In the CPEs, retailing was considered simply as a set of mechanisms channelling goods to 'final' consumers; and in the official statistical system it was considered one of the 'non-productive sectors'. Yet the need to provide households with a wider range of convenience and durable goods was appreciated, so that later waves of expansion in floorspace involved some particular modes of retailing (e.g. department stores) as well as organisational changes[1] in the sector as a whole. Developments in retailing and their impact on patterns of consumption were analysed in the framework of 'mainstream' retail studies rooted (paradoxically) in neo-classical economic principles (Beluszky 1966; Kovács 1987), and had a strong applied (planning) orientation – although the alternative forms of retailing existing in socialism's 'second economy' led to analysis by economists and geographers, as did the introduction of new regulations on private enterprises in 1982. The role of private enterprise in the accumulation of financial, social and knowledge capital was considered, but the limits of this process (through stable prices and a centralised distribution system) were also recognised and emphasised (Belyó 1995; Douglas 1995; Nemes Nagy and Ruttkay 1987).

In the early years of the transition (1988–1993), rapid liberalisation made the retail sector a highly contested and structured sphere of the economy. Yet despite the profound changes that have occurred since 1989 – a series of innovations, a substantial growth in retail employment and the restructuring of retailing and consumption spaces – the role and significance of the sector in the transition process has not yet been adequately interpreted and theorised. So this chapter will attempt to do this through a political economic perspective, paying attention to transition phenomena relevant to corporate strategies, such as the surge of enterprise, the very rapid spatial and organisational centralisation and the failures of the regulation system – all discussed in the context of the circulation of capital (Ducatel and Blomley 1990; Harvey 1989). Inspired by the 'new economic geographies of retailing' that interpret the sector as a form of capital with a key role in economic restructuring (Wrigley and Lowe 1996), analysis will focus on corporate strategies and their spatial aspects through which particular features of emerging markets can be revealed, the impact of international retailers on local 'culture' – as a set of attitudes determining the decisions made by retailers, suppliers and consumers (Hughes 1996; Sklair 2001) – understood and the consequences of the inconsistency and deficiencies of state regulation explored. Emphasis will be placed on the food retailing sub-sector that first experienced the proliferation of small businesses, the adaptation of new strategies by domestic corporations and the aggressive expansion of MNEs that stimulated a significant shift in the structure and the geography of consumption.

It is hypothesised that the retail sector has had a significant role in the accumulation of capital in the emerging market economy, but through various 'models' of accumulation. Most small-scale domestic enterprises took a short-term view on their business with only limited scope for investment; a strategy countered by major international agents who introduced long-term strategies that seemed to have a stabilising effect (e.g. by reintegrating food producers into their supply system) as well as a key role in retailing innovations in the emerging market. International retailers also had a decisive impact on shaping and transforming consumer groupings through the introduction of new forms of retailing and the spread of the ideology of consumerism (Sklair 2001). However, they were forced to adjust their strategies in the light of rapidly changing market conditions, particularly, the emerging regulation system and the entry of new competitors. Corporate strategies were much influenced by spatial disparities at the level of ECE as whole as well as the national and local scales as they were alternately reinforced and reshaped by retail developments. The chapter will examine such issues as the fragmentation of the market in terms of organisation, capital and retail space, the deficiencies in retail services and the blooming of 'black' and 'grey' economies rooted in the inconsistencies and inadequacies of market regulations and also the peculiarities of consumer attitudes. The corporate strategies of MNEs will be discussed in the context of domestic market structure (with particular regard to their competitors and suppliers) and both national and local policies. Also, the impact of the expansion of international agents on the national market will be summarised; along with the responses of domestic retailers and consumer attitudes towards shopping: expectations and disappointments, an emerging consciousness, adaptations to new forms and techniques of shopping and the restructuring of demand. But these matters have to be considered in relation to the corporate strategies in which foreign capital plays a major role.

Analysis of the structural and organisational transition of the retail market rests on retail statistical databases, news published about major domestic firms and MNEs in the media, and on the reports of leading consulting firms involved in real estate and retail development. In addition to corporate websites, reference is also made to interviews with chief executives of the top ten food retailers (2002–2003) and with chief executives and/or owners of SMEs in the mid-1990s and 1998–9 that illuminated the process of organisational transition, the changing corporate strategies of domestic and international retailers, and the impact of national (and local) policies on the decisions of retailers. Consumer reactions to the transformation of shopping facilities may be studied through the Debrecen Shopping Survey of 1999 that focused on the attitude of shoppers towards the new forms of retailing, the impact of new facilities on the structure of consumption and also on the changes in shopping routes. Analysis of the spatial aspects of the transformation is also supported by the interviews already above and by a series of empirical studies on the transformation of retail space in major Hungarian county towns of Szeged and Győr in 1994 and 1998–9. This diversity of source material revealed much about the nature and impact of corporate strategies employed in an emerging market and largely overcame the persistent problems of scarce information on the spatial processes in retailing and the inconsistency of national statistics concerning economic activities (especially pre-1997).

Structural Characteristics of the Emerging Retail Market

This section considers fragmentation in retailing in Hungary in he early 1990s when deficiencies in retail services and the regulation system set the scene for the entry of retail capital, followed by rapid expansion in an increasingly prosperous business environment.

Fragmentation of the Market and Retail Space

Deregulation of economic activities and particularly the establishment of the legal framework for free enterprise and association were considered as cornerstones of the dissolution of the CPEs. Furthermore, national policies aimed at channelling domestic social and financial capital – as well as international resources – into the national economy to stimulate restructuring and manage the transition crisis (Sárközy 1989). Therefore, the entry of new agents into domestic markets was encouraged and the conditions for their activities were highly liberalised. Retailing was the primary scene for starting new businesses and accumulating capital due to its (relatively) low capital intensity, the rapid circulation of resources and also to the deficiencies in the retail services inherited from the socialist 'economy of shortage' (Kornai 1980). But the initial low level of investment in retailing reflected both the acute shortage of capital and the role of the sector in the accumulation of 'small capital': taking of only five percent of all domestic investments during 1989–95 while 9–11% of the GDP was generated annually during this same period. However, despite the unstable and unfavourable conditions of the structural crisis, retailing was a highly contested sector in the early transition years to 1996 for the overriding trend of organisational fragmentation of the market (e.g. with the disintegration of state corporations) was complemented by the expansion of international retailers – e.g. Delhaize Group, Julius Meinl, Metro, Tengelmann and Tesco in food retailing, and Baumax, Metro/Praktiker and Tengelmann/OBI in the DIY sector – making for renewed centralisation.

The dynamism of enterprise in retailing and wholesaling – included together with repairs in a single category in the Hungarian statistical system since 1991 – outpaced all the other economic sectors until 1994: there was a tenfold increase in the number of individual entrepreneurs (from 29,296 to 233,775) and a 2.5–fold increase in the number of retail/wholesale companies. At the peak of the rush in retail enterprise (1993–1994), one-third of all domestic economic enterprises were concerned with retailing. But the founding capital of domestic enterprises was the lowest in the retail sector – only one-tenth of the national average – and due to the lack of resources and increasing wage costs, the enterprise rush involved the rise of self-employment: two-thirds of companies with legal entity[2] employed less than 11 persons and 95% of individual entrepreneurs had no registered employees at all. Although net number of enterprises rose until 2000, the majority of newly-opened shops were run by self-employed individual entrepreneurs with very limited capital investment and a floorspace below 20sq.m; while the bankruptcy rate was extremely high. The rise in the number of food shops was particularly rapid in the first four years of transition, although their lifespan was only half to two-thirds of the sector average and food retailers had the most pessimistic view of the future of their enterprise (according to interviews with executives of local large and medium-sized retail companies in Győr and Szeged in 1994): although they believed that small independent food shops improved the access to convenience goods, lack of capital prevented improvement in the quality of service (in terms of the choice of goods, the shopping environment and time-saving). As a result concentration accelerated in all respects so that in 1996 five percent of the stores accounted for 43% of the food retail turnover (Mohácsi 1998). However, small food stores remained a significant element in household survival strategies, particularly in the backward eastern regions of Hungary where the number of individual enterprises still exceeded the national average in 2001.

The rise in the number of retail enterprises was particularly high in Budapest, the northwestern counties and the Lake Balaton area, where growth was stimulated by domestic, as well as German and Austrian investors. Many retail enterprises were founded in the south (particularly Csongrád and Bács-Kiskun counties) but the majority – even international investments, predominantly of Yugoslav origin – were very small in scale. In general terms, the retail sector in the eastern regions of Hungary – small in scale and particularly fragmented in structure – was maintained by cross-border shopping tourism, the flight of small capital from the war-torn Balkan countries and also by the high unemployment rate. However, although the opening rush resulted in a significant increase in major cities as well as in rural regions, the spread of new shops had a hierarchical character. Furthermore, from 1994 onwards the network was increasingly centralised to the benefit of the largest county towns (primarily cities inhabited by more than 100,000 people). The concentration was particularly rapid and reached a higher level in the backward regions. Since major county towns were uniformly favoured by retail developers with only minor regional differences,[3] spatial inequalities in the development of the store network are best revealed at the level of small towns and villages.

The fragmentation process was supported also by the transformation of SOEs. The management of retail firms had a decisive role in the re-organisation and also in first stage of privatisation (as shareholders). Their transition strategies resulted in a shift either to centralisation (e.g. providing a standardised image through regulations affecting the entire network) or decentralisation (transforming regional branches into independent firms), specialisation in terms of goods, and improvement and extension the range of retail services. However, each of the interviewed corporations (1994) sought stricter control over the distribution process by the combining retail and wholesale activities. Reorganisation involved investments of significant scale that could be realised at the expense of the reduction of the store network by selling or letting the premises. In this way, the adjustment of privatised corporations contributed to the organisational and spatial fragmentation of retailing. Until the mid-1990s, large scale development schemes were limited to discount stores and Metro malls; so the closing down of stores run by state corporations and opening of small shops en masse resulted in the spatial fragmentation of retail space. At this time international retailers were not very prominent, but is was significant that companies in which international retailers had a share or the ones that joined buying groups (particularly in the food sector) in the early 1990s, preserved the majority of their stores.

The primary scene for spatial fragmentation in the early 1990s was the city centre, particularly in Budapest and major county towns. Case studies of Győr and Szeged suggested that the rapid expansion of retail space and the fluctuation of shops stimulated by the enterprise rush slowly relaxed from 1994. Although, the main shopping street attracted major domestic and international retailers – with an increasing number of restaurants and banks – SMEs remained prominent in the main shopping areas; not least through the letting of space in former department stores and the transformation of stores belonging to SOEs into 'retail houses' comprising dozens of small shops. However, as a consequence of the competition and rising rents there was a shift towards specialisation among small shops in terms of the range of goods on offer, the quality of goods and services, the prices and the 'contributors' (i.e. the regular customers). The fragmentation process was also supported by the emergence of new areas convenient for daily shopping along arterial roads or at busy nodes in densely populated urban districts. Although the spread of retail shops and services resulted in a significant improvement in the access of convenience goods, the high number of closures and the permanently-changing structure of retail nodes required a permanent adaptation of the consumers until the late 1990s.

Consumer Attitudes towards Shopping and Retailers

While enterprise was stimulated by the liberalisation process and by the deficiencies of retail services, discouragement arose through a fall in demand – 25% during 1990–6 – arising from the decline in real wages. The structure of demand also changed with rising shares for electricity, food and water at the expense of durable goods, transport and communication and recreation. The real value of sales decreased year by year until 2000 (reckoned at comparative prices on a 1990 basis) due to the relatively high elasticity of demand for food and other convenience goods. Although an increase of real incomes stimulated domestic demand from 1998, the food business remained one of the least dynamic sectors of retailing (along with clothing) until 2001. The decline of domestic demand was a highly differentiated process in spatial and structural terms. Disparities in demand were rooted in the rising inequalities of household incomes: the standard deviation of personal income taxes (at county level) increased constantly during 1991–2001 and the difference in tax yield between the minimum (Szabolcs-Szatmár-Bereg county) and the maximum (Budapest and Győr-Moson-Sopron county) rose slightly as well (particularly the contrast between rural areas of the northeast and northwest). The fall in demand was particularly rapid in the northeast where recovery was generally slow compared to the rest of the country. Backward regions were characterised also by a higher elasticity of food consumption, with the decline in the demand for meat and dairy products partly compensated for by self-produced food. However, as a consequence of delayed investment, demand rose in all sectors from 1998, particularly for household durables, housing construction-related goods and services, and food and beverages. The counterpole of the northeast was the capital city and the northwest where the structure of consumption shifted rapidly towards the West European model from 1997, with rising demand for cars, telephones, recreation and cultural and health services. However, regional disparities in the consumption of convenience goods, clothing and household durables declined from 1997 – a process supported by the development of financial services, the spatially-differentiated rise of incomes and increasing consumer indebtedness (Simon and Szirmai 2002 pp.30–51).

In the early 1990s, attitudes towards consumption were conditioned by rapid adaptation to the changing conditions, especially rising prices that were particularly characteristic of Hungary within the ECE region (Jones Lang Wootton 1998). Consumers suffered from deficiencies of the regulation system and institutional control: poor information in general, low quality goods (especially in clothing and household durables) and inadequate hygiene in food shops, while consumer complaints were ineffective due on inadequate documentation and retailer inexperience in conflict management (while consumers took high risks through informal transactions in grey and black segments of retailing for which 'Comecon markets' and later open markets were important scenes. Before 1996 it was often impossible to trace shops due to the short lifespan of small enterprises. The informal segment flourished mainly in the major county towns of the east and south until the late 1990s as local people sought bargains and trade was boosted by the many visitors (shoppers as well as peddlers) who came from neighbouring countries (Ancsin 1997; Horváth-Kovách 1999).

Partly through deficiencies in retail services and the high number of closures, retailing was associated with unethical attitudes (e.g. quick profits) and even crime.[4] Retailers were aware of their negative image through increasing consumer distrust (indicated by the interviews conducted in Györ and Szeged in 1994) and so both consumers and many entrepreneurs saw the entry of international retailers as a route to new business ethics. For although the quality of retail services improved in the late 1990s, due to the tighter control over retail transactions and increasing efficiency of consumer protection, deficiencies (noted above) remained in the majority of shops. In 2001, 56% of the stores supervised by the Consumer Protection Inspectorate were fined for breaking regulations. However, consumers are usually not aware of their rights (e.g. over conditions of guarantee and access to information) and often did not exercise due care over the origin and quality of goods purchased. Although the number of complaints registered by the Inspectorate rose, most consumers still considered defective goods as their own responsibility – an attitude often exploited by retailers through limiting information and infrastructure to avoid and/or manage complaints (National Inspectorate for Consumer Protection 2002). Hence the arrival of international retailers and their introduction of consumerism came at a time when Hungarian shoppers were in the habit of accepting poor quality goods and services while voicing relatively few complaints. The time was ripe for the development of services and the introduction of new forms of retailing.

The Arrival of the International Retailers

In the later 1990s, the expansion of retailing was powered by the entry of new European and overseas competitors as a consequence of the emergence of the single European market and the liberal retail policies employed by the EU-member states. Under such circumstances, new corporate strategies were introduced in which retail location choices were highly influenced by national policies concerning competition, retail activities, international investments and environmental regulations (Dawson and Burt 1998; Marsden and Wrigley 1996; Wrigley and Lowe 1996). ECE markets were considered potential areas for expansion and the introduction of new corporate strategies. However, rapid transformation, the deficiencies of the legislative environment and the instability of national economies (marked by declining output in each sector, high inflation, rising unemployment and decreasing household incomes) hindered large scale developments. But Hungary experienced the early entry of international corporations. Most large international retailers – particularly those involved in the food and DIY sectors – established subsidiaries during 1989–92 with entry encouraged by Hungary's image as a relatively liberal CPE.[5] However, the most significant stimuli were the subsequent steps towards deregulation of the national market, such as the abandonment of price controls and restrictions on profit, liberalisation of external trade and freedom to combine different economic activities such as retailing and wholesaling (Jones Lang Lassalle 1999).

Reorganisation and privatisation of SOEs that controlled the process of distribution also attracted international agents from 1989. But in the early period of pre-privatisation (1990–2), domestic private enterprises and individuals were favoured while cooperatives and retail and wholesale companies in which the state was a shareholder were excluded from the process. Furthermore, the store networks of former SOEs (over 5,000 outlets of relatively small size) were sold one by one at public auctions. In this way, the pre-privatisation process contributed to the emergence of a fragmented market structure dominated by independent retailers until the mid-1990s. However, the privatisation law of 1990 facilitated the participation of international agents. Major retailers (mentioned above) participated in the privatisation of SOEs that were reorganised where they had capital goods (e.g. a store network and vehicle stock) and market share of a significant scale. Although, there were subsequent attempts made by the management of semi-state corporations to hinder the process and to pressurise the conservative government of 1990–4 to favour 'national capital' in privatisation transactions (Gál 1998; Karsai 2000), international agents did manage to acquire portfolios in domestic corporations. The demography of retail enterprises headquartered in Hungary suggest that the liquidity problems of domestic entrepreneurs – rooted in the relatively low level of capital accumulation, high inflation and interest rates and deficiencies of domestic financial services – encouraged the entry and the expansion (including buy-out transactions and large scale developments) by international retailers.

The prominence of international retailers can be better understood by considering the economic climate facing domestic and foreign business during the 1990s. The legislation governing individual enterprises and on corporations that was enacted in 1989 certainly encouraged business start-ups by domestic agents, although participation of international investors was also facilitated. The licensing procedure was quite liberal[6] and loans granted by the state were also available for new enterprises. But high interest rates and the deficiencies in the banking services involved high risks and discouraged capital-intensive activities. Hence the service sector – and particularly retailing – was the primary target for the surge of enterprise in the early 1990s. The process was calmed by stricter controls over business in general and retailing in particular from the mid-1990s when institutions and control mechanisms came into effect (such as IT-based registration for taxation, enterprise licensing and social insurance). The frequency of control increased along with the penalties for breaching the regulations. Also, the scope for tax avoidance by individual entrepreneurs was reduced while social insurance premiums were also raised under the 1995 'Bokros package'. Retailers were subject to tax auditing and rigorous hygiene control, while modifications to the Retailing Act became more demanding with respect to the the education skills of retail enterprises. Hence the number of individual entrepreneurs declined from 1996.

Enterprise was encouraged through economic restructuring in a liberal policy climate,[7] but the support system for domestic SMEs was both modest and inconsistent in the early 1990s: loans were too small (and the interest rate too high at over 10%) to finance costly investments, such as the (re)construction of a store network, the introduction of IT in retailing, or the setting up of a logistic basis. In any case, retailing was not favoured by SME-support schemes and manufacturing was given centre stage in economic policy-making until 2000. The acute shortage of capital in the retail sector could not be relieved by bank loans due to the high interest rates and to the reluctance of banks to take high risks and accumulate 'bad loans' during the years before the reorganisation (stabilisation) and the privatisation of the banking system. A new policy for supporting domestic SMEs introduced in 2000 provided government funds for bank loans at discounted rates in respect of certain development projects, including retail activities, but the amounts were again small (e.g. for IT–related developments) and the scheme made no significant impact on what had become a highly centralised retail market structure dominated by major MNEs, as in other transition states (Coles 1997; Drtina 1998). The perspectives for domestic retail enterprises were not improved by economic policies that lacked mechanisms for encouraging organisational integration. The only working post-socialist model of integration (the 'Áfész' network) was transformed following the 1992 law on cooperatives which gave cooperative membership shareholder status, while subsequent measures in 1994, 1996, 1997 and 2000 completed the shift from the original cooperative structure towards a corporate organisation. Since domestic models for integration did not emerge, European retail organisations (like Edeka and Intermarché) were studied and adapted to the domestic retail structure.

Meanwhile, national investment policy encouraged the influx of global capital to accelerate the structural transformation of the economy and this positive attitude contributed to the growing internationalisation of the Hungarian retail market. As early as in 1989, a law on foreign investment granted legal safety to foreign enterprises (legal protection from expropriation and a guarantee relating to free transfer of profits) and equal status with domestic organisations. Foreigners were allowed to purchase real estate for their enterprises while taxation consisted simply of an 18% corporation tax payable to the central government – and even this levy was softened by five-year tax breaks granted in 1989 to large investors in enterprise zones and backward or border regions without any sectoral consideration (Nagy 1995). Global agents might also receive fiscal concessions from local decision-makers, pressed by the permanent liquidity crisis[8] and also by local economic structural deficiencies to sell large plots of urban land (including many derelict areas) for development. Indeed, as the continuing economic crisis and the national government's drive towards the centralisation of resources made localities increasingly dependent on the decisions of large investors through the 1990s, local preference systems tended to favour major investors at the expense of local SMEs.[9]

It is also likely that foreign investors profited from deficiencies in the regulations concerning retail activities. The law on retailing dates back to 1978, although it was modified several times in the 1980s and 1990s. Yet although there are now more than 50 acts and regulations that must be respected, there are still several important matters concerning payment deadlines, pricing (selling below the suppliers' costs), opening hours, employees' rights – as well as many environmental aspects of retailing – that are not yet adequately regulated. The Hungarian retail regulation system was labelled as 'very liberal' even by the executives of MNEs (2002–2003) and it is evident that deficiencies are exploited by major retailers with regard to supply systems, prices and opening hours. Yet although deficiencies are recognised, a coherent legislative basis for retailing, this is not yet in place. Domestic retailers deplore both the shortcomings in national retail policy and the self-representation system for retailing that has disintegrated into several organisations[10] and is highly influenced by the international sector.

Corporate Strategies introduced by International Retailers

New corporate strategies arose out of the challenge posed by European and overseas competitors and their capital-intensive technological and technical developments seeking higher financial returns and the acceleration of the circulation of capital (Dawson and Burt 1998). Location choices were highly influenced by the national regulation systems and the business environment, as well as the social structure, the development of transport infrastructure and suppliers' networks and the structure of local land markets (Hernandez et al. 1998). However, the decision-making process (including location choices) was also conditioned by corporate strategies and culture (Hughes 1996). Interviews with the executives of major food retailers suggest that the entry of international agents was encouraged by the liberal regulation system, the privatisation process and also by the deficiencies of retail services discussed above; also by the structure, development and surplus capacity in the ECE food processing sector that provided a basis for a regional supply network (including the production basis for own-label goods). International retailers offered a safe but (globally) contested market for Hungarian producers competing for supplier status. In the opinion of domestic retailers, Hungarian food producers prefer the prices offered by international retailers, expecting an increased market share for their products in return. However, the selection of producers was rapid due to rising competition, retailers' requirements for quality control and flexibility, and retailers' price policies including the system of delayed payments that may take 60–120 days. In this way, the entry of international agents accelerated the selection of food producers, supported by the deficiencies of market regulations and by the weakness of the producers' lobby.

   In the mid-1990s, a new scheme for expansion was introduced into the corporate strategies of international retailers, stimulated in Hungary by the increasing financial stability of the national economy and rising household incomes. As the demand for convenience goods was characterised by a relatively high elasticity, income growth meant a significant rise in all segments of consumption particularly from 1998. Particularly in the food retail market, strategies shifted towards both large-scale greenfield developments and the expansion of store networks by buying and integrating established local retail chains. The scale of investments changed significantly, as the amount of invested capital doubled in 1995–6 and increases continued until 2001. The investment boom and the process of retail restructuring was also spurred by the entry of international real estate developers who channelled capital into new forms of trading, such as shopping centres (from 1998 especially) and retail and business parks (from 2000) (Nagy 2002). Of all investment 8.3% went into retailing during 1996–9, rising to 8.9% in 2000–1 (though the change in methodology used by the National Statistics Office should be noted: registered capital for 1996–9 and shareholders' equity capital for 2000–1). Enterprises with FDI (including all enterprises with a foreign capital exceeding 10%) played a decisive role by accounting for 83% of all investment in the sector during the first period and although the share fell to 65% in the second period the proportion of foreign capital in joint ventures actually increased significantly to 93.6%. Furthermore the amount of capital invested rose until 2001.

The spatial distribution of retail investments reflected the spatial preferences of particular retail corporations and a general shift towards a larger scale of provision. The primary target was the capital city (40%) but new developments were increasingly concentrated in the suburbs of Budapest (as indicated by data for Pest county) and county towns with a population exceeding 100,000[11] (Table 13.1). Medium size towns – especially county seats with 50,000–100,000 inhabitants – also attracted new stores, particularly from 1999. The market niches arising from structural deficiencies in the east and rising incomes generally were also exploited at this time (see Figure 13.1, based on National Taxation Office data). Although, the standard deviation of spatial distribution of investments was declining during 1997–2001, Budapest and its agglomeration along with the west and northwest gained a constantly-rising share.

The food and the DIY sectors were the first to experience the new corporate strategies that resulted in rapid concentration in terms of capital, organisation and space. From 1997, the most rapidly growing retail corporations were the subsidiaries of the Metro,[12] along with Auchan, Delhaize Group, Rewe, Spar, Tengelmann and Tesco. As a consequence, each of the food retail corporations included in the group of 'Top 20' Hungarian retailers and wholesalers (by value of net sales) were controlled by MNEs in 2001. In the late 1990s and the early 2000s, international agents introduced new forms of retailing were introducd and the number of hypermarkets, 'soft' discount stores and shopping centres rose dynamically, while the supermarket segment was highly contested and characterised by series of ownership changes and reorganisations. The entry of four international retailers into the hypermarket segment resulted in a rapid growth amounting to 27 stores during 1998–2001. The 'hypers' became the primary target for weekly shopping trips, accounting for one-fifth of retail turnover in 2001, and they had an increasing role in the market for durable goods.[13] In this latter segment, the price-preferences of Hungarian shoppers were exploited (e.g. by selling large quantities of low-price electronic goods produced in the Far East). The most dynamic expansion was accomplished by the Tesco-Glóbal Company that entered the

Table 13.1   Retail investments in Hungary, 1996–2001

County	1996–99 HUFmln	1996–99 HUFth/pc	2001–1 HUFmln	2000–1 HUFth/pc
Budapest	152851	84,4	152851	56,8
Bács-Kiskun	17207	32,3	17207	13,2
Baranya	11222	28,0	11222	11,0
Békés	8780	22,4	8780	20,4
Borsod-Abaúj-Zemplén	13369	18,3	13369	12,1
Csongrád	12181	29,2	12181	12,0
Fejér	14214	33,6	14214	19,0
Györ-Moson-Sopron	10192	24,0	10192	27,0
Hajdú-Bihar	14945	27,6	14945	14,7
Heves	5721	17,7	5721	10,9
Jász-Nagykun-Szolnok	6690	16,3	6690	23,6
Komárom-Esztergom	5916	19,0	5916	25,9
Nógrád	1325	6,1	1325	7,8
Pest	44492	43,1	44492	39,2
Somogy	7363	22,3	7363	11,7
Szabolcs-Szatmár-Bereg	14791	26,0	14791	11,6
Tolna	3667	15,0	3667	16,5
Vas	9589	36,0	9589	25,5
Veszprém	8933	24,0	8933	23,6
Zala	5844	19,9	5844	22,3
Total	369292	36,8	369292	26,5

Source: CSO 1996–2001.

emerging market by participating in the privatisation of a supermarket chain in northwest Hungary in 1990. Although the corporation held a significant stake in the supermarket business (particularly in Budapest and the northwest) the emphasis shifted to large-scale developments from the mid-1990s. This turn in corporate strategy was stimulated by rising competition and the commercial real estate crisis in the UK. The cornerstones of the new strategy were the restructuring and diversification of the store network, geographical expansion and the development of e-services (Wrigley and Lowe 1996). The first and the second elements of the strategy were much in evidence in Hungary since the first Tesco hyper was opened in Budapest and this innovation spread rapidly throughout the whole country, as well as the corporation's European network. The basis for expansion and for the organisational and technological innovation was the relatively large stock of liquid capital; increasingly important since the self-financing of new developments (e.g. land acquisition) facilitated a rapid expansion on a commercial real estate market characterised by rising land prices and rents until 2000. Tesco-Global Hungary Company held the best position for retailers in the list of Top 50 firms by shareholder equity in Hungary in October 2002 (Karsai 2003).

The rapidity of Tesco's spatial expansion outpaced the competition on the

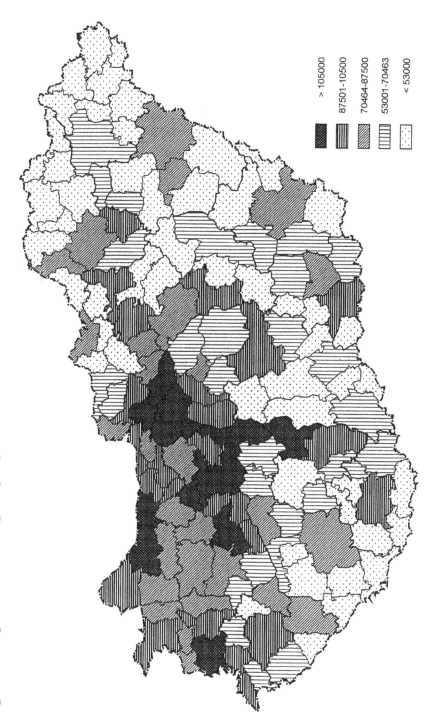

Figure 13.1 Spatial variations in per capita personal income taxation in 2001 (values in HUF)

Hungarian market. Auchan and the Delhaize Group (Cora) focused on Budapest and its suburbs and starting slow expansion across the provinces relatively late on (Figure 13.2, based on company websites). The Metro Group constructed stores only in the largest county towns (Debrecen, Györ, Kecskémét, Miskolc, Nyíregyháza, Pécs, Szeged, Székesfehérvár and Szombathely), while Spar's expansion was limited in time and space by its liquid capital. Although Tesco focused on Budapest and the major and medium-sized county towns in parallel, new developments were regionally differentiated by the lack of knowledge and experience in managing large-scale development schemes in the east. Yet despite early difficulties, eastern developments were highly successful since hypermarket shopping was considered as leisure in itself – particularly by low-income families – without the need for services such as restaurants, children's playgrounds or cinemas. This attitude was exploited by Tesco through the location of stores close to densely-populated urban districts with easy access by both cars and public transport.

The Delhaize Group used a different strategy involving all significant sectors in the retailing of convenience goods: hypermarkets (Cora), discount stores (Profi), supermarkets (Macth and Smatch) and Cash&Carry stores (Alfa). They entered the Cash&Carry and the highly contested supermarket segment by the takeover in 2000 of the Julius Meinl network when the Vienna-based company exited from ECE markets to preserve its family-enterprise character and reputation as a supplier of fine foods <http://www.meinl.com>. This transaction was the earliest major structural change in the food sector to involve international agents and it marked the maturity of Hungarian market. However, pursuing classical city centre food retailing in a price-oriented market proved a costly business, for the turnover of the supermarket chain has been declining since 2000. Subsequent steps towards concentration on different lines of retail business and reconstruction of stores have been made in the last two years under a new strategy towards discount and supermarket philosophies that targets all regions of the country (2003). Nevertheless, the spatial preferences of the corporation can be identified in the expansion process. The earliest greenfield investments (hypermarkets and discount stores) were concentrated in Budapest, followed by a shift to the eastern regions (but a westward shift in the case of Spar). Meanwhile, supermarket developments and new openings were focused largely on Budapest and the northwest.

Major corporations that remained under the control of the founders (family businesses) or were based on cooperation of retailers employed strategies that focused on particularly prosperous segments of the Hungarian market. The expansion of the Tengelmann Group was embedded in a Europe-wide strategy that evolved in the early 1990s. The market was tested by participation in the Skála-Coop privatisation (a major department store chain), but developments were increasingly focused on greenfield developments, in the food, textile and DIY sectors. The corporation employed a two-tiered food retail strategy resting on discount stores (Plus) and city centre supermarkets – all labelled 'Kaisers Supermarkets' but with each representing a particular spatial model adjusted to the spatial differentiation of household incomes and consumer preferences.

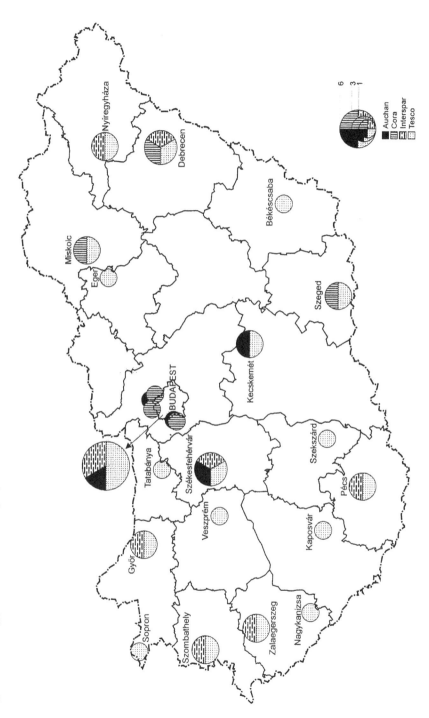

Figure 13.2   Hypermarkets in Hungary, 2002

Supermarkets were highly concentrated in Budapest and major county towns in the west (many established in Skála department stores), while the discount segment was characterised by a more dispersed structure (Figure 13.3, also based on company websites), including medium-sized towns in both eastern and western counties. Recently, Tengelmann has changed its corporate strategy by focusing on the discount segment and giving up the city centre supermarket network, a decision that will give a spur to the concentration of the food retail market. Meanwhile, the German Rewe Group – operating a highly structured store network in their domestic market in a cooperative framework – introduced a more cautious, step by step expansion in Hungary in 1994 involving a single segment of food retailing. Discount store developments targeted each region and all levels of the urban hierarchy Low-income households in urban regions and the population of rural areas characterised by a relatively conservative attitudes towards consumption were the primary target groups. Their 'Penny Market' stores attracted consumers ready to compromise over selection and quality of goods and services in return for lower prices.

Location choices of international food retailers were all driven by the highly liberalised emerging market of Hungary, and also by the poorly developed retail services in the early 1990s. The rapid expansion of MNEs in all segments of the food sector – as well as positive reactions of domestic retailers – demanded a permanent readjustment of corporate strategies to the changing and contested market conditions. As a consequence of the rising competition and the peculiarities of the domestic demand (lacking social models for consumption, price-orientation and propensity to compromise with low quality) the emphasis shifted to price wars and large developments. However, spatial preferences were increasingly important elements of corporate strategies. The geography of new developments reflected regional as well as rural/urban differences in purchasing power and attitudes towards consumption: hence the early expansion of discount store networks and the relatively late construction of hypermarkets and shopping centres in the east. Consequently, corporate strategies employed on the Hungarian retail market stimulated the rise of new dimensions in spatial (regional as well as rural/urban) disparities, such as the increasing concentration of shops in major urban centres, unequal access to new shopping facilities and advanced customer services; also the differentiation of consumption in terms of structure, frequency and space.

Concentration and Differentiation: Responses to the Corporate Strategies of Foreign Investors

Domestic Retailers

Domestic retailers were challenged by the expansion of international agents in all aspects of their corporate strategies, for new techniques of retailing and customer services greatly improved the shopping conditions and capital-intensive

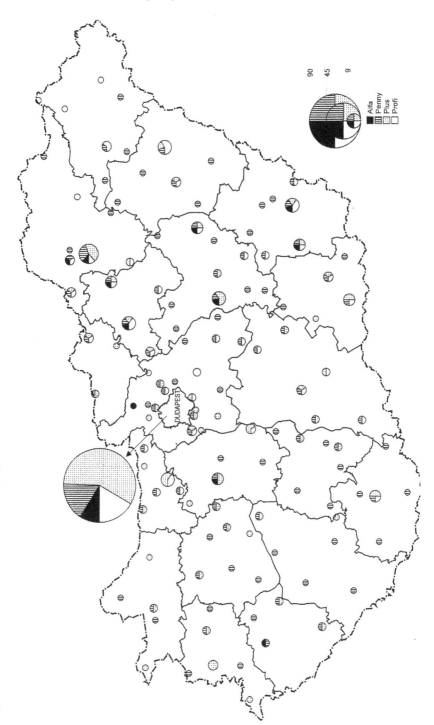

Figure 13.3   Soft discount stores in Hungary, 2002

technologies (e.g. in stock management) to raise efficiency in a way that small domestic retailers could not match. Constructing new models for vertical integration also provided advantages for international agents in the price contest. Furthermore, new forms of retailing introduced by MNEs attracted customers in increasing numbers through low prices and large operating scales as well as novelties, environments and services. Strategies were supported by effective lobbying to influence the policy-making process, so that domestic agents had to learn and respond. The changes stimulated a centralisation process in domestic retailing in terms of capital, organisation and space. The number of enterprises declined, particularly small family enterprises and food retail businesses. However, due to the rising domestic demand, the number of shops increased until 2003. The backward northeastern counties experienced slightly different trends, for the number of food shops was rose rapidly due to the relatively high elasticity of demand for convenience goods and the rather late entry of major developers.

The introduction of new customer services and shopping techniques by the internationals stimulated competition and supported the improvement of retail services provided also by domestic agents. Such changes were welcomed by consumers as well as by the majority of retailers in an emerging retail market characterised by instability and many negative side-effects of the transition process. However, capital-intensive developments accelerated the trend towards centralisation. The establishment of buying groups and the development of such organisations in terms of collective services and the number of members was interpreted by Hungarian food retailers as a positive response given by SMEs to the globalisation of retailing. Recently, four major groups (CBA, Coop, Honiker and Reál) have a gained a significant share on the domestic market. Although, the alliances employ particular strategies, the common supply system (including production and supply of own-label goods), development of a logistic basis and the introduction of IT-related technologies and the food quality control system are common to all these schemes of cooperation.[14] Since there are no particular groups of customers targeted by the buying groups' strategy, conflicts have arisen in the last few years that are rooted in the emerging (regional and rural/urban, as well as intra-urban) differences in the structure of consumption. This has stimulated highly differentiated demands and expectations in relation to the centralised distribution system. This cooperatives are running food stores throughout the whole country with goods selected to meet the needs of particular customer groups: conservative, slightly mobile households, living in rural areas or peripheral urban districts. Shops run by domestic retailers which cannot compete with the selection of goods provided by major international chains are putting more emphasis on a friendly, clean and human-scale environment. Although the CBA and Coop groups are included in the top ten food retailers in Hungary, they are constantly challenged by internationals and their alliances – in particular the largest buying group, formed by Metro and Spar. Therefore domestic retailers focus rather on preserving their market share by capitalising on their well-known brands and sites.

Consumers

Owing to the habits inherited from the economy of shortage, the swift, shock-like change in shopping conditions and also the deficiencies over consumer protection, MNEs have had a decisive role in articulating consumerism and structuring consumption in Hungary's emerging retail market. Patterns of consumption are being conditioned by spatial as well as by social factors. Since the changing pattern of shopping facilities has a hierarchical nature and the retail transition – involving the proliferation of small shops and the introduction of new forms of retailing through the development projects of international agents – has been concentrated in Budapest and the major country towns, 'rural' and 'urban' consumption attitudes have been identified through interviews with the chief executives of the top ten food retailers.The former is characterised by more conservative behaviour and readiness to compromise (e.g. accepting poorer selection of goods for saving time and money), while the urban consumer is more open to innovations and may fall victim to marketing manipulations. According to the Debrecen survey, these attitudes are conditioned also by household incomes, education skills and age. Both models have emerged in rural and urban settlements in the transition period and while regional differences were not considered significant, the strong price-orientation of shoppers living in the east was emphasised by retailers.

However, urban societies have also been restructured by the introduction of new forms of retailing. Large projects restructured the mental maps of local people and changed the direction as well as the frequency of shopping trips in cities. The Debrecen consumption survey suggested that the overwhelming majority of local people are aware of the changes in retail facilities and the larger scale of provision makes for effective advertisement. Attitudes towards new developments are highly conditioned by the social status: young people, well-off families and highly skilled urbanites appreciate new stores and malls while elderly and low-income people often have aversions to them (Nagy 2001). About two-thirds of consumers have changed their main food shopping trips over the past five years in favour of stores offering lower prices and/or a better selection of goods. But the transformation of shopping habits has been highly influenced by social status as well as by several personal factors and it is mainly the most affluent professionals and executives who have sought better shopping opportunities: taking advantage of their mobility and the detailed mental maps acquired in respect of local retailing. The elderly, the less educated and – in general – those with lower incomes have changed their shopping habits mainly because of lower prices.

Malls and new stores have increased their market share in almost all categories of goods: Metro became a serious threat to the hegemony of specialised shops (e.g. electronic goods) while shopping malls (especially Plaza Mall) expanding rapidly into the clothing market. Since their selection of goods was similar to that of traditional high street stores, Plaza was able to take over their functions while channelling demand in a new direction towards the northern edge of the city centre. However, while the Debrecen survey suggested a shift towards more 'settled' forms of retailing in the durables sector – predominantly hypermarkets and specialist shops – open markets remain the consumers' primary targets for clothing and convenience goods (including household chemicals) particularly for low-income families, while professionals and executives prefer new malls and specialised boutiques. The same applies to expensive durables: the affluent prefer individual specialised shops and malls, whereas the elderly and those with lower incomes still like open-air markets despite the risks over the uncertain origin and quality of the goods.

Groups of interviewers were also strongly differentiated over their choice of transport. The widest choice is available to the wealthy, mobile groups as well as the youngest shoppers who go for a flexible combination of walking and public transport. People in the lowest income categories and the elderly who resort to local shops are particularly affected by the transformation of city centre shopping facilities. But a key element in changing shopping habits is the time-budget of shopping since the majority preference for shopping several times a week covers different behavioural patterns. The well-off (belonging to the two upper income brackets) can afford the luxury of daily shopping completed by stocking-up at the weekend. New outlets and flexible (usually longer) opening hours are beneficial, chiefly to active earners and most importantly women who account for two-thirds of all shoppers. The role of multi-targeted shopping routes is also increasing but with dependence on social status (i.e. car-ownership) and also age, because the under-40s prefer multi-targeted routes to minimise the time spent on shopping. Overall the changes are partly related to household incomes, but age, physical conditions and employment (active/inactive) status play a major role in the timing and frequency of shopping. However, ability to exploit the increasing variety of shopping facilities, with flexibility over opening hours, has raised new inequalities in urban society.

Conclusion

This chapter has shown that FDI in retailing in Hungary has been very important in modernising a sector regarded most ambivalently under communism. Although comprising some HUF413.5bln which is only 8.6% of all investment by enterprises with FDI (i.e. enterprises with more than 10% of foreign capital in shareholder equity) during 1996–2001, it has had a disproportionate impact on the urban landscape through partial redevelopment in town centres and most impressively in the urban-rural fringe area of Budapest and major county towns. Arriving significantly at a time of retail fragmentation through the break-up of the inherited socialist networks, the foreign investors have spearheaded the inevitable follow-on process of consolidation and concentration. There is no doubt that handsome profits have been made and it is undeniable that these results have to some extent been boosted by inadequacies in the regulation system which has slowly caught up with European norms. But without the international agents the modernisation of

Hungarian retailing would have taken place much more slowly. However, the contested nature of the Hungarian retail market has produced a range of responses from 'eatailing' in the case of the aggressive expansionists to the exit strategy adopted by Julius Meinl.

From the late 1990s, major retail developers transformed shopping routes by their spatial decisions (e.g. over parking facilities) that have differentiated local society according to the characteristics of consumer groups. Social status (particularly with regard to income) and capacity to adapt (a function of age and qualifications) has a significant role in transforming consumer habits within cities. The differentiation of society is manifested in the use of new techniques and purchase of new goods (or refusal of them). Due to consumer habits inherited from the shortage economy, the swift, shock-like change of shopping conditions and the shortcomings of consumer protection, MNEs have shaped shopping habits as well as the consumer culture very forcefully in Hungary and other transition economies. But the location choices of the major developers suggest that the differentiation of the society according to rural and urban consumption patterns is likely to be remain a characteristic feature of the highly contested and globalised market. However, although the improvement of retail services, with increasing consumer safety and protection, is welcomed by retailers as well as shoppers, the closure of small shops and the increasing centralisation of retail space in major cities and in particular urban sites is a concern. Independent retailers run shops in the eastern and northeastern counties hit by unemployment and economic recession but survival strategies that rest on small family businesses are under threat. Furthermore, the supply of convenience goods for immobile social groups – particularly those living small villages in the northeast and southwest – is also endangered. Centralisation is not being addressed by national social programmes, given the scare resources for regional and rural development, although attempts by alliances of local governments to remedy the problem (e.g. by mobile shops) are important for a policy framework to reverse the process.

Notes

[1] In the 1960s the network of cooperative shops was expanded throughout the settlement hierarchy. It was followed by the extension of the store network of state corporations in the 1970s. The latter period of development was dominated by larger scales in the food and durables sectors and focused on towns and cities. Meanwhile, the distribution system was also reorganised by the establishment of specialisd wholesale corporations and their regional networks.

[2] The minimum level of registered capital was defined by this category (basically for corporations and limited companies). Therefore this group includes the largest enterprises in the national economy.

[3] The primacy of the size of the city and the population of its hinterland in locational choices was emphasised by the interviewed retailers, though retail investment also illustrated it. By the end of the 1990s, the most centralised store networks emerged in Csongrád, Hajdú-Bihar and Baranya counties, each dominated by a major city.

4 Before 1995 the weight of the informal (hidden) economy was estimated at one quarter of the national economy (Lacko 1995). Retailing was seen as one of the facades for money laundering to cover criminal offences (e.g. evading customs duties) and was much discussed in the media.

5 This image arose from the early introduction of the 'private sector' (1982) resulting in more than 10,000 new shops providing retail and other consumer services run by domestic entrepreneurs throughout the country in the 1980s. The new enterprises (agents of the second economy) had a decisive role in the growth of the national economy in satisfying consumer needs and accumulating financial and social capital for the transition period (Douglas 1995).

6 Anyone who had no criminal record and had qualifications (education skills) for the relevant activity obtained a license for individual enterprise. However, several restrictions were introduced in 1995, 1996 and 2000, such as the revocation of the license over public dues (taxation, social insurance or customs liabilities) and imposition of full financial responsibility for limited or joint stock companies. The Corporation Act (1990) also permitted individuals to join any limited or joint stock company or become shareholders in domestic corporations, basically without limitations. Conditions and institutions of legal control were not defined in detail and administrative (quantified) curbs were also scarce. Membership restrictions for economic enterprises were introduced by the new Corporation Act in 1997 to avoid tax evasion through cross-invoicing (whereby invoices are exchanged among associated companies in respect of fictitious services and notional prices in order to distort balance sheets).

7 Support for domestic enterprise and the rise of a new national bourgeoisie was a significant element in the ideology of conservative parties and governments from the early 1990s; evident in national economic policies during 1992–4 and 1999–2002.

8 The financial crisis was rooted in deficiencies of the Local Government Act of 1990 where the framework of redistribution (the proportion of resources centralised by the national government and the part left to local bodies) was not defined clearly enough. In the 1990s, the locality share of personal income tax revenue (the most significant element of local government income) fell from 50% to just 5%. Meanwhile, the range of public services (e.g. issuing licenses for enterprises, driving licenses and passports) provided by local governments was increased, with many activities delegated from central to local bodies. But resources redistributed by government to the localities failed to cover the increasing cost of services. The deficit could be eliminated by increasing local taxes and selling community assets. Although the range of national funds for local development increased, daily financial problems were not relaxed in most towns and in many cases the co-financing of development projects threw the communities concerned into debt crises.

9 Local tax revenues had an increasing role in stabilising community budgets. The level of local enterprise tax was raised to the maximum (2%) by the end of the 1990s and local SMEs had little scope for avoidance (e.g. by large investment projects).

10 Through the 1999 modification of the Act on Chambers of Commerce, compulsory membership ceased and in this way the Chambers lost not only the majority of their members but also their role in policing retail and wholesale actvities, representing retailers and reconciling conflicting interests.

11 Retail investment data is aggregated at county level. However, differences in the index (retail investments per 100,000 of the population) between counties reflect the urban concentration of retail capital.

12 Metro is considered a quasi-retailer by its competitors, since a major part of the population has access to a Metro card for shopping in Cash and Carry stores. In 2000, 'non-professionals' had a 40% stake in net annual sales revenues although the share has declined from 2001 onwards.

13 The share of hypermarkets in sales of household electronics (a highly dynamic sector from 1998 onwards) rose from 10% to 34% during 1996–2001; significantly exceeding the European average (Simon and Szirmai 2002).

14 The largest group is the CBA Retail Ltd (with 51 members and about 900 shops) whose stores are run through a franchise system while the flow of goods is provided by a central as well as regional logistics centres. The network is also expanding in Croatia, Romania and Slovakia.

References

Ancsin, G. (1997), 'Külföldi befektetok és bevándorlók bevandorlok Szegeden', *Tér és Társadalom*, 1997/3, 143–56.

Belusky, P. (1995), '*Magyarország kiskereskedelmi központjai*', *Földrajzi Értesítõ*, 2, 237–61.

Belyó, P. (1995), 'kereskedelmi vállalkozások helyzete és fejlõdési lehetõségei', *Statisztikai Szemle* 6, 485–500.

Coles, T. (1997), 'Trading Places: The Evolution of the Retail Sector in new German Länder since Unification', *Applied Geography*, 17, 315–33.

CSO (Central Statistical Office) (1996–2001), *Regional Statistical Yearbooks*, CSO, Budapest.

Dawson, J. and Burt, S. (1998), 'European Retailing: Dynamics Restructuring and Development Issues', in Pinder, A. (ed), *The New Europe: Economy Society and Environment*, Wiley, Chichester, 157–76.

Douglas, M.J. (1995), 'Privatisation Growth and Sustainability of the Retail Sector in Budapest', *Moravian Geographical Reports*, 3(1–2), 44–52.

Drtina, T. (1998), *Retailing in Europe: Crossing the Borders*, INCOMA, Prague.

Ducatel, K. and Blomley, N.K (1990), 'Rethinking Retail Capital', *International Journal of Urban and Regional Research*, 14, 207–27.

Dun and Bradstreet (2002–3), Magyarország nagy- és középvállalatai: 2002–2003 CD-adattár, Dun and Bradstreet, Budapest.

Gál, Z. (1998), 'A Centrum skálái', *Népszabadság*, August 1.

Harvey, D. (1989), *The Condition of Postmodernity*, Blackwell, Oxford.

Hernandez, T., Bennison, D. and Cornelius, J. (1998), 'The Organisational Context of Retail Location Planning', *GeoJournal* 45, 137–60.

Horváth, G.K. and Kovách, I. (1999), 'A fekete gazdaság (olajkereskedelem és KGST-piac) Kelet-Magyarországon', *Szociológiai Szemle*, 9(3), 28–53.

Hughes, A. (1996), 'Forging new Cultures for Food Retailer-Manufacturer Relations?' in N. Wrigley and M. Lowe (eds), *Retailing Consumption and Capital*, Longman, Harlow, 116–36.

Jones Lang Lassalle (1999), *City Profile: Prague Warsaw and Budapest*, Jones Lang Lassalle, Prague.

Jones Lang Wootton (1998), *Shopping for a New Market: Retail Opportunities*, Jones Lang Wotton, Prague.

Karsai, G. (2000), 'Az áruházi láncok privatizációja', *Külgazdaság*, 44(5), 62–71.

Karsai, G. (2003), 'A legnagyobb magyarok', *Heti Vilaggazdasag*, 25(32), 49–66.

Kornai, J. (1980), *Economics of Shortage*, North-Holland, Amsterdam.

Kovács, Z. (1987), 'Kereskedelmi centrumok és vonzáskörzetek Heves megyében', *Földrajzi Értesíto*, 36, 253–272.

Lacko, M. (1995), 'Rejtett gazdaság nemzetközi összehasonlításban', *Közgazdasági Szemle*, 42(5), 486–510.

Lupton, R.A. (2002), 'Retailing in Post-socialist Central Europe', *European Retail Digest*, 33, 45–9.

Marsden, T. and Wrigley, N. (1996), 'Retailing the food system and the regulatory state', in N. Wrigley and M. Lowe (eds), *Retailing Consumption and Capital*, Longman, Harlow, 33–47.

Mohácsi, K. (1998), 'A magyarországi élelmiszer-forgalmazás: a szövetkezeti kereskedelem lehetőségei', *Közgazdasági Szemle*, 45(5), 494–506.

Nagy, E. (2001), 'Winners and Losers of the Transition of City Centre Retailing in ECE', *European Urban and Regional Studies*, 8, 340–9.

Nagy, E. (2002), 'Fragmentation and Centralisation: Transition of the Retail Sector in ECE', *European Retail Digest*, 33, 41–5.

Nagy, G. (1995), 'A külföldi toke szerepe és térbeli terjedése Magyarországon', *Tér és Társadalom* (1–2), 55–82.

National Inspectorate of Consumer Protection (2002), *Annual Report 2002*, FVF, Budapest, p.28.

Nemes Nagy, J. and Ruttkay, É. (1987), *A második gazdaság földrajza*, n.p., Budapest.

Sárközy, T. (1989) 'A társasági törvényről', Baló, I. and Lipovecz, I. (eds), *Tények Könyve*, Computerworld Informatika Kft, Budapest, 711–20.

Simon, E. and Szirmai, P. (2002), 'Kereskedelem', in Nyomárkai, K. (ed), *Figyel"o Top 200*, Sanoma, Budapest, 89–98.

Sklair, L. (2001), *The Transnational Capitalist Class*, Blackwell, Oxford, 34–112.

Wrigley, N. and Lowe, M. (1996), 'Towards a New Retail Geography', in N. Wrigley and M. Lowe (eds), *Retailing Consumption and Capital*, Longman, Harlow, 3–30.

Chapter 14

Developing the Euroregions along the Polish-Slovak Border

Włodzimierz Kurek

Introduction

An important political change occurred in the ECECs in 1989 when transfrontier cooperation became a possibility (Bojar 1996). Movement has increased rapidly and in the case of the Polish-Slovak border there is much more economic activity with trade fairs in such border towns as Cieszyn, Jastrzębie-Zdroj, Racibórz, Ustrón and Wodzisław which attract buyers from Czech Republic as well as Slovakia: 100,000 or more foreigners may arrive in Cieszyn on fair days before Christmas and Easter. Euroregions (ERs) constitute the most developed form of such cooperation and 14 have emerged on Poland's frontiers since 1991; supported by various international organisations, including the AEBR, CoE and EU (Figure 14.1). The official basis for cooperation rests with government commissions, which operate along Poland's frontiers with all neighbouring states, while international treaties define detailed principles of the cooperation between the local (i.e. regional) authorities based on partnership, legal equality and good neighbourly relations (GUS 1999). Clearly, two basic principles must be combined: (a) the notion of self-government, with scope for local community initiative and (b) national interest which must condition the range of local actions (Kozanecka 1999). There are already many examples of mutually-advantageous cooperation to consider and both local authorities and communities in border areas are keen to make progress, although the process is still constrained by financial and legal barriers (Molendowski and Szczepaniak 2000). The chapter will concentrate on the Carpathian, Tatra and Beskid ERs, lying along the Polish-Slovak frontier, with emphasis on the conditions and development opportunities in the relevant Polish territories (Table 14.1).

General Characteristics of the Euroregions

The Carpathian Euroregion (CER) was created in 1993 according to the agreement signed in Debrecen between local authority representatives of Poland and those of Hungary, Slovakia and Ukraine (joined by Romania in 1997). The total area is very

Figure 14.1   Euroregions along Poland's frontiers

large and there is a predominance of mountain terrain with considerable value for tourism, given the relatively unpolluted natural environment. Agriculture and industry are also prominent. The Polish sector comprises Carpathian territory in the southeast of the country: 180 communities in the two voivodships of Małopolskie and Podkarpackie (18,686sq.kms, inhabited by 2.38 million people). There is an extensive forest cover and basic resources include natural gas and oil. The main industrial branches deal with glass, wood and food while agriculture is of fundamental importance and international road and railway lines link Poland with Ukraine. The Slovak part embraces Košice and Prešov regions (10,459sq.kms and 1.11 million population): another mountainous region with tourism, manufacturing (chemicals, foodprocessing and light industries) and a good infrastructure of road

Table 14.1   Selected data for Euroregions along the Polish-Slovak border, 2000

Criterion	Poland	Euroregions		
		CER	TER	BER
Urban populations (percent)	61.8	40.9	35.0	49.5
Population density (persons/sq.km)	124	118	126	242
Natural increase (ptp)	0.3	2.6	4.5	1.4
Unemployment rate (percent)	15.1	16.1	10.6	12.5
Apartment floor space (sq.m/pc.)	19.2	18.1	17.9	19.3
Tourist beds (ptp)	17.3	10.3	92.3	31.3
Forested area (percent of total area)	28.4	36.0	39.1	45.1
Protected area (ditto)	32.5	48.0	90.0	41.0

Source: USK 2000; USR 2000.

and rail links with Poland and Ukraine. Hungary's five northeastern provinces (Borsod-Abaúj-Zemplén, Hajdú-Bihar, Heves, Jász-Nagykun-Szolnok and Szabolcs-Szatmár-Bereg) have an area of 28,639 sq.kms and a population of 2.61mln. In this lowland area (part of the Great Hungarian Plain) – with some mountains of medium height in the north – there is metallurgy, horticulture and viticulture (the latter at Tokaj). The Ukrainian sector, of 56,600sq.kms and a 6.43 million population comprises the provinces of Cernovcy, Ivano Frankivsk, Lviv and Transcarpathia where the East Carpathian mountains dominate. There are oil, gas and mineral reserves and engineering, food processing and light industry is represented on transport axes which are oriented both north-south and east-west, but tourism is underdeveloped. The Romanian part consists of the five counties of Bihor, Botoşani, Maramureş, Sălaj and Satu Mare with 27,104sq.kms and a 2.27mln population. There are bauxite and iron mines as well as industries concerned with engineering and food/timber processing. The CER wants to cooperate by exchanging of information and experiences; sustain regional development and spatial planning; develop trade, tourism and other branches; protect and improve the natural environment; enhance education, cultural exchange and protect the common cultural heritage; and extend mutual relations with exchanges of professional, social and youth groups. Many economic, cultural and sporting events have already been organised along with courses, seminars and fairs which have helped to highlight economic potential. The frontiers are now much easier to cross: in 1999 four new tourist walking-cycling routes were opened across the Polish-Slovak frontier at Balnica-Osadne, Czeremcha-Czertižne, Ożenna-Niżna Polianka and Roztoki Górne-Ruskie Sedlo.

The Tatra Euroregion (TER) was created in 1994 among Polish and Slovak communities to further transfrontier cooperation over environmental protection and tourism (for which the region is already famous in view of the mountains and folklore), including the spa potential based on a range of mineral waters. The Polish sector comprises 18 communities in the Małopolskie voivodship covering with 1,952sq.kms and 245,000 inhabitants. In addition to the Tatra (which include the mountain of Rysy, 2,499m) the area covers the Podhale Basin, the Pieniny Mountains, with the Dunajec gorge and Czorsztyn lake, and adjacent parts of the Beskid Mountains. The Slovak part (with 63 localities in nine districts of 6,572sq.kms and 223,000 people) extends to the historical regions of Spisz and Orava, and parts of the Nižne Tatry and Wielka Fatra Mountains. The outstanding natural and cultural resources include many castles and architectural attractions in many towns and villages – already well-developed in terms of tourist infrastructure. The aims of the TER lie with regional development including spatial planning and improved frontier crossings and technical infrastructure; the development of trade, tourism and other branches; protection and enhancement of the natural environment (including protection against natural disasters and cooperation over mountain services); and cooperation over sport, health protection and cultural heritage through professional, social and youth groups. TER's achievements include the PHARE 2000 Polish-Slovak Transfrontier Cooperation Programme and its common fund for small projects concerned with cultural exchange, local democracy, manpower resources, transfrontier studies and development concepts including economic growth and tourism. Also, TER's Polish-Slovak Information System can now offer a database to support its work in cultural-historical, economic and tourism matters.

The agreement setting up the Beskid Euroregion (BER) was signed in 2000 to cover 3,928sq.kms and a population of 992,000: 605sq.kms and 161,000 people in Czech Republic; 13,000sq.kms and 566,000 people in Poland; with 2,023 sq.kms and 264,000 people in Slovakia. The Czech part comprises Frýdek-Místek district and the border part of Karviná while the Slovak villages are drawn from parts of Bytča, Čadca, Kysucké Nové Mesto, Námestovo and Žilina districts. The Polish part includes the Żywiecki Beskid and parts of the ʻSląski Beskid and the Mały Beskid, which rise to over 1,000m (districts of Bielsko-Biała, Sucha Beskidska and Żywiec). There are particularly close links in the Babia Góra where the association of Polish villages – ʻStowarzyszenie Gmin Babiogórskich' – works closely with its Slovak counterpart. Cooperation is related especially to the labour market, through promotion of economic growth particularly transfrontier traffic and tourism; protection and enhancement of the natural environment; spatial planning including a common infrastructure in the border region; and cultural exchanges (Kozanecka 1996). Much has already been achieved, especially in the fields of culture (folklore) and sports – including gliding, skiing and yachting which the ER has actively promoted – as well as the development of tourism, supported by the improvement of the road system and the opening of frontier posts to serve the tourist trails. As with the other regions success depends heavily on mobilising financial resources (particularly those available from PHARE CBC for environment protection, tourism, handicrafts and infrastructure) and improved marketing for cultural and sporting events and tourist fairs (Dziadek 2002).

Carpathian Euroregion (CER)

Southeastern Poland shows outstanding natural values through an unpolluted natural environment and an abundant flora and fauna (with rare and endangered species) which make this region extremely valuable at the European scale. There are two national parks (Bieszczadzki and Magurski), nearly 70 nature reserves, 11 regional parks and other protected areas which together cover over half the region. The cultural heritage derives from a group of coinhabiting nationalities and their respective religious affiliations involving Judaism and the Roman Catholic, Uniate and Orthodox churches. Although some ethnic groups are now very small (Bojkos, Dolinians, Lemkos, Pogórzans and others), folklore traditions continue and the old customs and rites are still transmitted through the generations thanks to active folk groups concerned with song and dance; not to mention folk artists interested in traditional weaving, basket and toy making, embroidery and lace making, painting and sculpture. There are many architecturally-valuable places of worship for all four confessions, though the remnants of Jewish culture (synagogues and cemeteries) are often in the poor state of repair. Some Catholic churches are richly decorated – as in the case of Haczów village where the church has international significance – but most distinctive are the brick or timber Orthodox and Uniate churches in the Bieszczady and Beskid Niski Mountains and the Przemyskie Foothills.

Population dynamics are stronger than across Poland as a whole; with a natural increase index of 2.6 compared with 0.3 nationally. Only 40% of the population lives in urban areas (62% nationally) and it is estimated that over half the total population lives on pensions, while the rest depend on agriculture or industry. The farmers are poorly educated because half of them did not progress beyond primary school. There is much out-migration and the unemployment rate is high: 17–22% in 2000 in many districts (where young and poorly-eeducated people are particularly vulnerable) compared with 15.1% nationally. Another rural problem is 'hidden unemployment' arising from low productivity among farmworkers who are often family members. There are also numerous non-viable SOEs – still not privatised – which retain a considerable number of underemployed workers. Thus the indices profiling living standards in the CER are below the national level. Incomes are also low: in 2000 the monthly net salary averaged PZŁ1,500–1,700 compared with about 2,000 nationally (GUS 2001).

The agrarian structure is unsatisfactory, being a legacy of past socio-economic conditions. In addition to a manpower surplus, farms are small and fragmented with insufficient market orientation. The average farm area is below four hectares (seven in Poland) and usually comprises more than ten plots. The present transition has so far witnessed little change and there is no land turnover to facilitate greater land concentration. Farms tend to supply their households with food while cash is obtained from other sources. Farms have a general profile – cereals and potatoes, with cows, pigs and sheep (and many horses) – with insufficient specialisation. Meanwhile yields are below the national average due to poor soils and the low use of fertilisers. There are zones of the specialised farming (e.g. horticulture) but instead of intensifying in line with holding size and available labour the tendency is to economise on fertiliser and other inputs. Meanwhile the food industry is concerned with meat, cereals, fruit and vegetables, milk and sugar processing. More could be done to increase labour intensification through ecofarming and greater capacity in marketing and processing. As agricultural marketing becomes organised on West European lines it is important that farmers should organise themselves through cooperatives. In this way they could take greater advantage of the opportunity to sell food at border crossing points (UMWP 2000).

Forests cover about 36% of the Polish sector of the CER compared with 28.2% nationally. But the forest share rises to 50% in Sanok and 70% in Bieszczady where there has been much afforestation in sympathy with the decline in agriculture since the Ukrainian population was displaced after the Second World War for security reasons (see below). Wild animals include rare species, like elk and fallow deer (in addition to protected species like beaver, lynx and Polish bison) which provide opportunities for wealthy hunters from abroad. Industry consists largely of food processing, followed by chemicals (Dębica, Jasło, Sanok and Nowa Sarzyna) and engineering (Rzeszów and Sanok). Growth has occurred in wood processing, especially furniture, due to growing domestic and foreign demand which the numerous small factories can satisfy. The main problem for industry is poor adaptation to prevailing market demand. Obsolete technology and poor quality standards, with low productivity and profitability, reflect the habit of working for a market in ECE that has now largely collapsed. Some privatised enterprises with FDI are doing better than the national average. For example the Good Year tyre factory at Dębica with an investment of $112mln followed by United Biscuits in Rzeszów ($25mln); Ropczyce Sugar Plant ($16mln.) and Owens Illinois ($17mln) in Jaroslaw; though unfortunately the region lacks attractive greenfield locations. So FDI has not yet proved a significant influence in the production and export structure. Strong economic units are needed with up-to-date technology and a capacity to innovate which would stimulate existing local industries, but investment attractiveness is limited by the small regional market and the inadequate infrastructure. Local authorities need to be proactive in using funds from abroad to enhance the image of the region in Europe.

The road network – managed by both the region and district authorities – has a high density but a generally low quality which restricts the speed of travel. North-south and east-west motorways are needed, along with urban ring roads and new bridges over main rivers like the San. Presently the national roads handle the international traffic on the routes from Wrocław, Kraków and Rzeszów to western Ukraine (Lviv) and from Warsaw and Rzeszów to eastern Slovakia (Košice) – served by frontier crossings at Medyka and Barwinek respectively. Additional frontier crossings have been opened on the road system since 1989: two for Ukraine and one for Slovakia. The railway network is adequate with a main line – of a good technical standard – linking Western Europe with Ukraine, with transfer from standard to broad gauge at the Medyka border crossing east of Przemyśl. However the technical standard of the local railways is highly unsatisfactory and so their competitiveness is poor. Rzeszów airport is geared to international traffic and has space for expansion athough at the moment it is underused. A well-established tourist industry takes advantage of a largely unpolluted natural environment with attractive cultural landscapes. There are good opportunities for hiking and skiing – also riding and cycling which are becoming more popular – while the rivers and reservoirs cater for sailing, canoeing and fishing. There are spas at Horyniec, Iwonicz and Rymanów and sightseeing draws visitors to the historic city cores with their strongholds, places or worship for Judaism as well as the Catholic, Orthodox and Uniate churches – including some 15th-16th century wooden churches – and fine residential architecture; not to mention the 'skansens' (open-air museums) dealing with folk architecture. However the tourist infrastructure is still underdeveloped, with an accommodation base consisting mostly of campsites, hostels and rooms to let. There is great need for further investment to improve standards, particularly for sanitation which constitutes a pollution hazard in places such as the Solinski Reservoir. The underdeveloped technical infrastructure limits the services available and compromises the image of the region's resorts. And there is some rather irrational management by SOEs that are still very prominent in the business. However, agrotouristical farms are increasing especially in the most popular areas: the Bieszczady and the Beskid Niski Mountains and the Iwonicz and Rymanów spas. There are many waymarked paths and a good provision of ski lifts although most are rather short.

The proximity of the frontiers provides potential for international tourism geared to recreation and transit for people from urban areas in Slovakia and Ukraine, but the foreign component – which includes some Germans and Russians – is presently minimal and it is the domestic market that must be relied upon for the immediate future (USR 2000). The Uście Gorlickie area has good possibilities for tourism to supplement agriculture and forestry. The area has cultural attractions (Orthodox chuches and folk events) as one of the main centres of the Ruthenian 'Lemko' population, a mountain group akin to the Ukrainians and subject to dispersal under the communists for wartime collaboration with the Germans and alleged connections with the Ukrainian Insurgent Army (UPA). The frontier crossing was opened in 1994, initially for pedestrians and car travellers, but for larger vehicles the following year (Birek 1995). It is hoped to upgrade the road as an international route and improve telecommunication links. Already, a number of agreements have been reached between the Polish and Slovak communities. In addition to the regular weekly markets, annual fairs have been inaugurated in Uście Gorlickie as well as Bardejov on the Slovak side. Tourism has enhanced prospects: the Watra Festival of Ruthenian Culture at Zdynia is much more heavily attended by Ruthenians and others on both sides of the frontier. In addition, Uście Gorlickie boasts a 'hucul' mountain horse stable and the Klimkowka Lake (dammed in 1994). Meanwhile the Slovak side offers the historic town of Bardejov and skiing grounds at Giraltovce. There is a coordinated development of tourism on the Polish side between Krynica, Muszyna, Plwniczna and Uście Gorlickie; while cooperation is developing among the three spas in the border area (Bardejovske Kupele, Krynica and Wysowa) which have similar natural environments and natural characteristics but different tourist infrastructures. There has been some revival of traditional handicrafts and there are plans for further tourist facilities, including accommodation at the frontier itself (Konieczna). However, developments will be slow because local budgets are limited and access is poor. Moreover, the lack of young, well-qualified people limits awareness of the potential for further diversification of the agricultural economy.

Tatra Euroregion (TER)

Only about a quarter of the territory is in Poland but it includes the mountains in the southern end of the Małopolskie (Kraków) voivodship which are outstanding for biodiversity with over 1,500 plant species (compared with 2,300 nationwide), 150 for birds and some 70 for mammals (90 in Poland as a whole), many of which are unique or rare. Mineral waters containing chlorine are used in five spas and exist in other localities where the potential could be exploited in the future. Moreover in the Podhale Basin there are geothermal waters (estimated at some 109cu.m and lying at a depth of 1,600–2,600m) which could meet the heating needs of the whole region (Manecki 1998): a geothermal plant operates at Biały Dunajec. 35% of the population is urban (living mainly in Nowy Targ and Zakopane) and the natural increase is 4.5% compared with only 0.3% nationally, while there is also a positive migration flow of 0.5% and the age structure is relatively young. An unemployment rate of 10.6% in 2000 (below the national average of 15.1%) completes a relatively favourable profile for the human resources of the region.

However the agrarian structure of the region is highly unsatisfactory through smallholdings (averaging six hectares) which are also fragmented. But half the land is agricultural and 60% of this is arable – used particularly for potatoes, oats and fodder plants – despite the poor mountain soils. Cattle and sheep breeding predominates in a situation when production is generally for the owners' needs. Only some 30% of farmers have produce for sale. With the reintroduction of a market economy it is necessary for agriculture to become more efficient and competitive; and also better adapted to the natural conditions. Although it is threatened in some areas by development pressures on arable land, agriculture is central to any conservation strategy in the region for both social-cultural and economic reasons. Given the emphasis on environmental protection and the low level of fertiliser use, the future seems to lie with ecofarming which could also influence the profile of the food processing industry and thereby penetrate more regional, national and international markets (Turnock 2002b). However, outside help will be needed for reorganisation and modernisation. And there is a further constraint because the dispersal of buildings detracts from the landscape value and complicates the rationalisation of farmland for economic growth. It also makes infrastructure more costly, including the provision of primary schools.

Tourism is a major source of income in the view of the two national parks (Pieniny and Tatra) and the spas of Muszyna, Rabka and Szczawnica. Moreover,

Czorsztyn lake – on the Dunajec below the Pieniny – could become the focus of an attractive cross-border tourist region with coordination to avoid wasteful duplication of facilities. A new bridge has opened over the Dunajec at the custom house between Červený Kláštor (Slovakia) and Šromovce (Poland). There many interesting architectural monuments in wood, like the church at Dębno, the buildings at Chochołów and the open-air museum of local architecture at Zubrzyca Górna. The rich folklore also includes language, costume, literature, music and painting on glass; richly-decorated furniture and tools; and old grazing customs. Proximity to the Slovak frontier, with only moderate distances to Austria and Czech Republic, enhances the potential of this tourist base where 1,693 units of accommodation offer a total capacity of 46,800 places – much of which is provided in Zakopane and the adjacent villages. There are however some ecological problems in the national parks, which could be solved by improving facilities in other attractive areas so that visitor pressure could be reduced (Ptaszycka-Jackowska and Baranowska-Janota 2003). At the same time there is a general constraint arising from the inadequate capacity of the roads, including the provision of motels and parking places. Developing the transport infrastructure in terms of quality and density is a basic task for the local authorities, while improving the frontier crossings and the access roads is fundamental for transfrontier cooperation. There is also a need to adapt the resorts to the needs of a market economy in order to increase the demand for spa treatments. There is a need for partial privatisation of the sanatoria to allow modernisation and a higher standard of service. However FDI is not yet prominent in the tourist industry and funds are supplied largely by commune budgets and Polish companies (while exploitation of the geothermal springs in the Podhale for domestic heating is financed partly by PHARE and the World Bank). Industry is still predominantly in the hands of SOEs – acting as quasi monopolies – which should be privatised, but there are also some 19,000 SMEs which operate in industry, transport and tourism while also dominating commerce and construction. They are very important for employment and economic growth generally, but further development requires access to capital, technology and consultancy services; facilitated to some extent by TER activities over fairs, exhibitions and other promotions which need to be increased at all levels. Once again, improvements in transport are crucially important to accelerate the cross-border flow of goods and services (USK 2000).

Beskid Euroregion (BER)

This is another mountain region with a forest cover of 41% (as in the TER). It shares with the other two regions a concern for flood protection, as a consequence of which (allied with the need to augment the water supply to the Upper Silesian Industrial Region) the Żywiecki and Międzybrodzki reservoirs were built on the Soła River. Much of the region is protected, most notably in the case of Babia Góra National Park, one of Poland's most valuable natural areas (included to the World Nature

Reserves system). Of the population 49% live in the cities and demographic structure is relatively favourable with 1.5% natural increase and only 12% unemployment. The largest city – Bielsko-Biała – has many industrial plants operating in the textile and automotive industries (specifically FIAT in the latter case) while Żywiec boasts the famous brewery (named after the town) and the timber industry is promiment across the region. FDI is considerable in the industries, petrol stations and supermarkets of Bielsko-Biała: in addition to FIAT there is investment from Aral, British Petroleum, Du Pont (also at Kozy and Koziegłowy), EBRD, Philips, Shell and Tesco; along with Delphi Automotive Systems in Jeleśnia, Gerolsteiner in Koziegłowy, Kraft Jacobs Suchard in Cieszyn and PepsiCo in Żywiec – mostly arising from proximity to both the Upper Silesian Industrial Region, with its huge market, and the frontiers of Czech Republic and Slovakia.

Agriculture is again based on small farms (80% are smaller than five hectares) which are also fragmented and geared to subsistence: only half have produce for the market. Low incomes and high rural unemployment (due to the lack of jobs outside agriculture) make for low demand for consumption whch discourages investment; though the infrastructure is again underdeveloped – partly due to settlement dispersion – and the natural conditions are also difficult (Górz and Kurek 1999). Tourist capacity is estimated at 30,000: private rooms again predominate but 17% falls to hotels. The industry is of national importance in this area – being well-situated for weekend tourism linked with the cities of Bielsko-Biała, Kraków and the Upper Silesian conurbation. It thrives on attractive, unpolluted mountains, forests and lakes as well as important cultural-spiritual features (Kurek 1966). There is much scope for hiking on Babia Góra, Pilsko and the Żywiecki Mountains but the whole region is covered by a dense network of waymarked tourist paths and there are many tourist shelters. Skiing finds the best natural conditions (for medium skiers) on the slopes of Pilsko, while there is increasing provision for riding, cycling, sailing and canoeing – along with sightseeing in the historic town centres. Traditional accommodation categories are being extended through agroturism. Agreements are operating between villages on the two sides of the frontier: Pcim with Breza, Ujsoly with Novo (involving a new border crossing point) and Zabierzow with Hruštín (Drgona 2002; Hasprova and Mihailikova 2002).

Concluding Discussion: Development Priorities

Taking into consideration natural conditions, the existing economic potential, the manpower resources and transfrontier links (Table 14.2), socio-economic growth should consist of diverse developments in rural areas through modern agriculture and food processing; and the creation of new jobs in industry, commerce and services – especially tourism based on the natural and cultural resources (where there are good possibilities in agrotourism to use available accommodation and 'ecological' food supplies). This calls for more national and foreign investment, the

Table 14.2   Strengths and weaknesses of Euroregions on the Polish-Slovak border

STRENGTHS	CER	TER	BER
Unpolluted environment	X	X	X
High natural-cultural value for tourism	X	X	X
Abundant natural resources including spa waters	X	X	X
Developed tourism with further growth potential		X	X
Favourable demographic structure (age and migration)	X	X	X
Skilled manpower surplus	X	X	X
Privatisation of the economy advanced		X	X
Investment attractiveness	X	X	X
Low unemployment			X
Economically-active and enterprising population	X	X	X
Low share of declining industries		X	X
Presence of high-tech industries			X
Strong and uniform regional/cultural identity		X	X
WEAKNESSES			
High level of dependence on agriculture	X		X
Agriculture fragmented with low productivity	X	X	X
Inadequate municipal/transport infrastructure	X	X	X
Inadequate environmental protection infrastructure	X	X	X
Low level of urbanisation	X	X	X
Low level of locally-generated municipal income	X	X	X
Excessive local variations in cultural development	X		X
Small number of enterprises with FDI	X	X	
Low quality of life and per capita GNP	X		
Inadequate regional promotion	X		

Source: Fieldwork.

development of SMEs, modernisation of transport and other services (water supply and sewage installations), with safeguarding of the environmental and cultural values, the development of SMEs and enhanced international cooperation (Koter and Heffner 1998; Turnock 2002a). The modernisation of farms is of fundamental importance: enlarging the holdings with the best potential while reducing the total number and increasing productivity through fewer workers. Improvements in food processing, marketing and distribution are also needed – and cooperation has a part to play in this. Better adjustment to the natural conditions is necessary, especially since it is being argued that the Polish-Slovak transboundary zone merits protection in its entireity (Więckowski 1999). Forests have a critical protective function – with regard to water supply and tourism – which calls for the further afforestation of wasteland and farmland with poor soil as well as a restoration of woodland diversity in accordance with natural conditions.

Economic development calls for further industrial restructuring and privatisation to improve competitiveness for export penetration into the EU as well as neighbouring countries, while SMEs are needed to complement the larger units (assuming that financial, legal and informational barriers can be overcome). This is turn requires a search for markets backed by better information, promotional efforts through missions and fairs (facilitating the direct contacts between Polish and foreign businessmen) and the opening of the local exhibitions and trade centres. And it is again emphasised that, notwithstanding some recent improvement, initiatives by domestic and foreign investors – in both industry and tourism – require a more adequate technical infrastructure with road traffic especially in mind as to as to permit safer and faster travel. Significant here is the Poland-Slovakia Project for a 'Cross Border Communication System' to supply a comprehensive information flow to stimulate enterprise and intensify cooperation in terms of administration, scientific research, culture and training. Local transport improvement will need to integrate with an expanding motorway network which may enhance the potential of the eastern regions in general (Grimm 1999). Bearing in mind the need to protect the tourist resorts, the water supplies and the natural and cultural resources of the mountains, preferred sectors for development would be food processing, car accessories, clothing and electrical engineering. But as concerns local enterprise in peripheral regions, the psychology of the local population may be such as to require a programme of stimulation and education by the local government and NGOs to mould attitudes positive to investment (Burzyński 2000); while local authorities should adapt local regulations to the needs of the tourist industry, help to create tourist associations and promote tourism through a distinctive product in each region which could be linked with selective improvements for particular activities (Mika and Pawlusiński, 2002).

Foreign investments are needed to create more jobs but also to introduce modern management methods and improve access to new technologies and new markets (Domański 2001) as in the case of ICN Polfa at Rzeszów where the quality and range of pharmaceutical products has increased. They can also lead the way in environmental protection as indicated by the activity of Fiat Auto Poland at Bielsko-Biala where the company obtained ISO14001 certification because of the elimination of air and ground water pollution and the solution of waste management problems. And improvements in building design can be seen through McDonalds restaurants and the expanding network of petrol stations. Links between foreign and local enterprises have been very beneficial. Thus the Alima-Gerber plant in Rzeszów, producing nutritious food for children – in an old factory modernised through a $60 million investment by Gerber Products – has not only reduced pollution through a modern boiler room, sewage treatment and anti-acoustic screens but has entered into agreements with fruit and vegetable producers which serve as a stimulus to agricultural improvement. While the number of employees is 800, the number of farmers and agricultural workers affected by the plant is even higher (and the firm also plays a role in the community through support of schools, orphanages and health care foundations). In order to boost investment the border regions can offer skilled labour, supportive local authorities and a relatively high economic development level. Nevertheless there are major variations in attractiveness. The cities – like Bielsko-Biała and Rzeszów – have attracted a disproporionate share, while in regional terms the western areas, and especially Bielsko-Biała, are likely to appear more desirable than the eastern zones comprising Podkarpackie voivodship (Swianewicz and Dziemianowicz 1999); while on account of the rugged mountain terrain and the priority for nature protection, the Tatra is only suitable for the development of tourism.

References

Birek, U. (1995), 'The opening of a road crossing as a factor in the economic development of border communities using the example of U'scie Gorlickie community in the Beskid Niski Mountains', in V. Baran (ed), *Boundaries and their Impact on the Territorial Structure of Region and State*, University M. Bel, Faculty of Natural Sciences, Banská Bystrica, 101–4.

Burzyński, T. (2000), 'Uwarunkowania rozwoju turystyki przygranicznej: stan i propozycje dalszego rozwoju', in Urząd Marszałkowski Województwa Podkarpackiego (ed), *Kierunki aktywizacji gospodarczej przez turystykę na obszarach przygranicznych*, UMWP, Rzeszów, 63–70.

Domański, B. (2001), Kapitał zagraniczny w przemy'sle Polski: prawidłowo'sci rozmieszczenia, uwarunkowania i skutki, Jagiellonian University, Kraków.

Drgona, V. (2002), 'Euroregion Cooperation: An Example of the Nameštovo Border District', in A. Dubcova and H. Kramarekova (eds), *State Border Reflection by Border Region Population of V4 States*, Constantine the Philosopher University, Department of Geography, Nitra, 133–43.

Dziadek, S. (2002), 'Szanse i zagrożenia wspó łpracy transgranicznej na obszarze Euroregionu Beskidy', in Wyższa Szko ła Bankowo'sci i Finansów (ed), *Rozwoj Euroregionu Beskidy: Do'swiadczenia Oczekiwania Perspektywy*, WSBF, Bielsko-Biała, 54–63.

Górz, B. and Kurek, W. (1999), 'Variations in Technical Infrastructure and Private Economic Activity in the Rural Areas of Southern Poland', *GeoJournal*, 46, 231–242.

Grimm, F-D. (1999), 'Strukturen Beziehungen ünd Perspektiven des östmitteleuropäischen Verdichtungsbandes Sächsen-Schiesen-Sudöstpolen-Westukraine', *Europa Regional*, 7 (3), 23–36.

GUS (Główny Urząd Statystyczny) (1999), *Euroregiony w nowym podziale terytorialnym Polski*, GUS, Warsaw and Wrocław.

GUS (Główny Urząd Statystyczny) (2001), *Powiaty w Polsce*, GUS, Warsaw.

Hasprova, M. and Mihalikova, J. (2002), 'Kysuce: a part of the Beskydy Euroregion', in A. Dubcova and H. Kramarekova (eds), *State Border Reflection by Border Region Population of V4 States*, Constantine the Philosopher University, Department of Geography, Nitra, 144–51.

Koter, M. and Heffner, K (1998), 'Borderlands or Transborder Regions: Geographical Social and Political Problems', University of Łódź /Silesian Institute in Opole, Łódź/Opole.

Kozanecka, M. (1996), 'Pogranicze Polsko-Slowackie i Aktualne Problemy Współpracy', in J. Kitowski (ed), *Problemy Regionalnej Współpracy Transgranicznej*, WEFUMCS, Rozprawy i Monografie Wydziału Ekonomicznego 10, Rzeszów, 307–16.

Kozanecka, M. (1999), 'Otoczenie Euroregionalne Polski jako przejaw Procesów Integracyjnych w Europie', in J. Kitowski (ed), *Problematyka Geopolityczna Europy ʾSrodkowej i Wschodniej*, WEFUMCS, Rozprawy i Monografie Wydziału Ekonomicznego 18, Rzeszów, 219–29.

Kurek, W. (1996), 'Agriculture versus Tourism in Rural Areas of the Polish Carpathians', *GeoJournal*, 38, 191–6.

Manecki, A. (1998), 'Strategia Zównoważonego Rozwoju (Ekorozwoju) Karpat w Granicach Województwa Nowosądeckiego', PAN, Kraków.

Mika, M. and Pawlusiński, R. (2002), 'Możliwoʾsci Działania Władz Samorządowych na rzecz Rozwoju Turystyki na Obszarach Przygranicznych', in Wyższa Szkoła Bankowoʾsci i Finansow (ed), *Rozwoj Euroregionu Beskidy: Doʾswiadczenia Oczekiwania Perspektywy*, WSBF, Bielsko-Biała, 118–28.

Molendowski, W and Szczepaniak, M. (eds.) (2000), 'Euroregiony: mosty do Europy bez granic', *Elipsa*, Warsaw.

Ptaszycka-Jackowska, D. and Baranowska-Janota, M. (2003), 'Tourism within the Polish and Slovak Transfrontier', in W. Kurek (ed), *Issues of Tourism and Health Resort Management*, U.J. Prace Geograficzne 111, Kraków, 43–60.

Swianiewicz, P. and Dziemianowicz, W. (1999), 'Atrakcyjnoʾsʾc Inwestycyjna Miast', Instytut Badaʾn nad Gospodarką Rynkową, Warsaw.

Turnock, D. (2002a), 'Cross-Border Cooperation: A Major Element in Regional Policy in East Central Europe', *Scottish Geographical Journal*, 118, 19–40.

Turnock, D. (2002b), 'Ecoregion-Based Conservation in the Carpathians and the Land-Use Implications', *Land Use Policy*, 19, 47–63.

UMWP (Urząd Marszałkowski Województwa Podkarpackiego) (2000), 'Strategia Rozwoju Województwa Podkarpackiego na lata 2000–2006', UMWP, Rzeszów.

USK (Urząd Statystyczny w Krakówie) (2000), 'Rocznik Statystyczny Województwa Małopolskiego 2000', USK, Kraków.

USR (Urząd Statystyczny w Rzeszówie) (2000), 'Rocznik Statystyczny Województwa Podkarpackiego 2000', USR, Rzeszów.

Więckowski, M. (1999), 'Natural Conditions for the Development of the Polish-Slovak Transboundary Ties', *Acta Facultatis Rerum Naturalium Universitatis Comenianae Geographica Supplementum*, 2/1, 257–63.

Chapter 15

Foreign Direct Investment and Social Risk in Romania: Progress in Less-Favoured Areas[1]

Remus Creţan, Liliana Guran-Nica, Dan Platon and David Turnock

Introduction: Social Risk in Romania – Unemployment and Stress

The geographical imbalance in FDI in the states of the region is a striking feature and România is no exception (Guran-Nica 2002). Because of the social risks that arise it is appropriate to consider both the opportunities and the barriers presented by some of the poorer regions. There were substantial inequalities within Romania under communism but they were largely contrasts between urban and rural areas in terms of living standards and job opportunities, along with problems in Moldavia in general on account of a relatively rapid population increase which outstripped the available employment and necessitated migration to other regions, especially Transylvania. Under transition the decline in employment has impacted heavily on industrial regions, especially those dominated by mining and heavy industry where the highest salaries were paid under communism. The proportion of industrial employees within the total active population fell from 36.9% in 1990 to 28.8% in 1994 and 27.1% in 1997 and 23.2% in 2000 through the industrial restructuring and risk minimisation policies of the transition years. Some of those dismissed found employment in the tertiary and primary sectors, but the unemployment rate rose from 3.0% in 1991 to 10.9 in 1994 and then fell back to 9.1 in 1997 before rising again to 10.5% in 2000 after the government embarked on the restructuring of the mining sector during 1998–9: exacerbating an already-serious poverty problem (Puwak 1992) – for although the rates are modest by ECE standards a major consequence of restructuring in Romania has been the growth of subsistence farming. Stress has given rise to increased criminality and crises in labour relations involving strikes and sometimes-violent demonstrations (Ianoş et al. 1996). At the same time, the predominantly rural-urban migration flow was reversed and there is a substantial movement from the rural areas abroad (Grigoraş 2001).

Until 1994, the highest unemployment rates occurred in Vaslui and Bistriţa-Năsăud counties (over 20%), followed by Neamţ, Tulcea and Botoşani (over 15%) (Figure 15.1). The list included counties which had benefited disproportionately

Figure 15.1   Romania's development regions, industrial parks and free zones

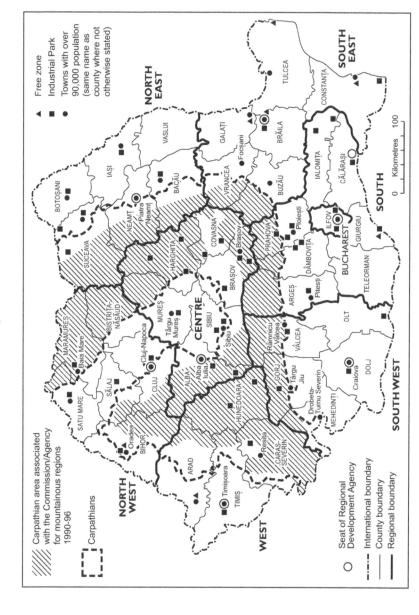

from communist industrialisation in the later years (post 1966) and were now losing traditional markets and encountering raw material supply problems. Meanwhile, the mining sector (still heavily subsidised by the state in the early transition years) helped to keep rates low in Hunedoara (8.3%) and Gorj (4.2%). But the decision to restructure the mining industry pushed Hunedoara to the top of the table (with over 15% unemployment during 1997–2000), followed by five counties with over 13%: Botoşani, Neamţ and Vaslui in Moldavia, Brăila in Muntenia and Vâlcea in Oltenia. So the distribution of labour conflict changed. In 1993, the highest levels were recorded in Covasna, Dolj, Harghita, Maramureş and Tulcea counties with many small wood-processing enterprises that were privatised relatively early and suffered losses in employment that provoked resistance. Strikes also occurred in the large engineering and textile industries. But in 1997, the highest levels of conflict occurred in Bihor, Constanta, Dâmboviţa, Dolj, Hunedoara and Timiş counties: a major transformation in the case of Dâmboviţa and Timiş which had recorded no conflicts at all in 1993. After 1997 conflicts continued at a high level with continuing reorganisation linked with pressure on government to meet targets set by the international financial institutions. However there was only a partial correlation between unemployment and labour conflicts. In 1993, the highest unemployment rates occurred in Vaslui and Bistriţa-Năsăud (over 20%), followed by Neamţ, Tulcea and Botoşani (over 15%), yet labour conflicts reached their highest levels (in relation to the population) in Covasna, Dolj, Harghita, Maramureş and Tulcea counties. Only Tulcea appeared in both lists. So it seems that some conflicts arose through anticipation of redundancies and the desire to defend jobs; in turn reflecting variations in militancy and the effectiveness of the trade unions (Guran-Nica and Turnock 2000).

Risk became very evident in the social sphere through family instability and criminality. Nuptiality was encouraged under communism by full employment and housing assistance – with stability bolstered by judicial 'counselling' and taxes imposed on divorces. But divorces increased sharply from 17.1% of all marriages in 1990 to 25.7% in 1994 before falling back only slightly to 23.6% in 1997. High divorce rates of over 30% occurred in the southern half of the country where nine counties were experiencing severe economic problems. Meanwhile, criminality rose from 160 per 100,000 of the population in 1990 to 421 in 1994 and 496 in 1997, reflecting both economic difficulties and an element of voluntarism following the demise of the former authoritarian regime (Fenn and Keil 1994), though there are now signs of a decrease. The spatial distribution remained fairly steady in highlighting a group of counties in the south with the highest values: Constanţa, Gorj, Hunedoara, Ialomiţa and Mehedinţi. Again, there was only a partial fit with the unemployment pattern for while dire poverty – arising from job losses and low welfare levels – could be invoked for Gorj, Hunedoara and Ialomiţa – other factors applied in Constanţa and Mehedinţi, including illegal border traffic. Yet there was by no means a perfect correlation between unemployment and the indicators selected to measure social stress. Although Moldavia experienced heavy unemployment in the early 1990s and clearly had a severe poverty problem, it has remained relatively stable. Part of the explanation may be a large rural population benefitting from land restitution which has restored peasant proprietorship, while union activity was reduced as unemployed people returned home with severance pay – often from factories in distant parts of the country: hence the difficulty of focusing on a specific grievance. On the other hand, Braşov was the scene of unprecedented violence in November 1999 when workers from the 'Roman' truck factory attacked the prefecture, although this city (and county) was experiencing relatively low unemployment and has always been among the most developed regions, with a strong and complex economy (dominated by engineering) attracting migrants from all parts of the country. The demonstrations were mounted in support of wage increases (to bolster falling living standards) and withdrawal of redundancy threats by the management of a near-bankrupt enterprise. It is evident that Braşov has a capacity for militancy because a challenge was mounted in the city against the Ceauşescu regime in 1987.

However there is close correlation between social instability and unemployment in Hunedoara county in southeastern Transylvania, where the secondary sector accounted for over 40% of the active population. The Jiu valley is Romania's major hard coal producer and the metallurgical industry is also very strong. But, although protected under communism – when the growth of employment attracted many young people from Moldavia, as well as the surrounding areas of Oltenia and Transylvania – the mining industry will only be viable in the market economy context after drastic reorganisation and down-sizing (Săuleanu and Pârlea 1997). Yet the miners have deeply resented the loss of their privileged status and have demonstrated a propensity for violent action. Living standards began to fall in the early 1990s (though earnings remained above-average and there was relatively little unemployment), while the amenities of the Jiu Valley and the limited qualifications of the workforce discouraged diversification. Opposition to reform was demonstrated on the streets of Bucharest on three occasions in 1990: in January and February when counter-demonstrations were staged in support of the conservative Iliescu presidency; and again in June when the opposition's long-running 'occupation' of Piaţa Universităţii was brought to a violent end. In September of the following year the miners again arrived in special trains to protect their wages and defend their industry against the reforms envisaged by the Roman government: the outcome was the dismissal of the prime minister. Although it is possible that the miners were manipulated by conservative political forces – including the demagogic union leadership of Miron Cosma ('the black prince') high social risk rendered the miners amenable to intervention. In 1998–9 a fifth 'mineriadă' was launched against the Vasile government's policy of mine closures to press demands for higher wages and enhanced redundancy payments of $10,000. On this occasion however the establishment was wholly opposed and – after public transport was denied – the march of 15,000 miners was eventually halted by force in Vâlcea county well short of the capital. A further (sixth) expedition occurred in February 1999 after Cosma was convicted (in his absence) by the Supreme Court and sentenced to 18 years imprisonment for actions against the state dating back to 1991. Instead of staying at home where he might have been protected, he chose to make a defiant gesture by leading a further march on the capital, but only to be arrested along with his entourage when his 'black army' of 3–4,000 was again halted soon after it had passed beyond the limits of Hunedoara county. Despite limited concessions, the rationalisation in the Jiu Valley has continued and many miners have returned – with their redundancy payments – to their places of origin. At the same time, the government demonstrated its authority in the eyes of the global institutions and foreign investors. But social risk will remain high while sufficient alternative employment is lacking. However, the homogeneity of large mining communities has been undermined by restructuring programme – as many young people find their way abroad, with Italy and Spain as particularly popular destinations – while the centre-right government of 1996–2000 introduced regional development policies in line with preparation for EU membership.

Regional Development Approaches

The problems of transition have provoked deep misgivings over Romania's European aspirations and provided minority support for an isolationist stance rooted in the belief that Romania should nurture its own values and protect its physical and human resources from harmful external pressures (Marga 1993; Tismăneanu 1997). But since 1996, Romanian governments have shown a readiness to undertake major reforms while complying with the guidelines set by international financial bodies (IMF and the World Bank) in order to retain a stable business environment and avoid the hyper-inflation crisis suffered by Bulgaria in 1997. And despite a strong nationalist showing in the 2000 election, the European agenda has retained centre stage through greater emphasis on social needs and enhancement of human capital. Attempts to limit social risk have included retraining and encouragement of SMEs through incubators and subsidised credits which provide coping strategies for some redundant workers. Regional policy was launched in 1996 – an election year – with programmes for certain deprived rural areas (Apuseni Mountains, Danube Delta and the counties of Botoșani, Giurgiu and Vaslui) covering infrastructure (including salary premiums to retain professional staff), agricultural development and job creation through SMEs and fiscal concessions. But in line the EU system of cohesion funding, the new government created eight development 'NUTS II' regions in 1998 (Guvernul României 1997). Subject to certain specific constraints – such as Alba county's desire not to be in the same region as its powerful neighbour, Cluj – Romanian social scientists identified functionally-coherent areas polarising around major provincial cities and near-equality in population (in excess of two million) as the basis for durable economic and social development including CBC (Figure 15.1). Meanwhile local government remains grounded in the inherited county ('județ') system, which has not been seriously challenged apart from the possible restoration of (a) some of the old counties suppressed after 1950 and (b) an intermediate tier between county and municipality (Săgeată 2003). The regions have their own development councils and agencies with headquarters in each regional centre (selected from the available county towns) and they have responsibility for the formulation and implementation of development programmes. Councils – consisting of representatives of the constituent counties with respect to county councils, municipalities, other towns and communes – determine policy, while the corresponding agencies (acting as non-governmental, non-profitmaking bodies) enhance attractiveness and encourage investment by combining promotional work with the administration and monitoring of PHARE programmes. The regional organisations are backed by a national council and a national agency to approve EU structural funding and allocate resources from 'Fondul Naţional de Dezvoltarea Regională' (the national regional development fund: NRDF). Cognisance must also be taken of the National Development Plan which (for 2004–6) seeks to increase competitiveness and develop private sector; modernise infrastructure; develop human resources and social services; modernise agriculture and the rural areas generally; support research and technology; and protect the environment.

While there are historic contrasts between regions – with Ilfov (covering the Bucharest area) and the North East marking the extremes – the planners have been particularly careful to emphasise the variations that exist within most regions between stronger and weaker counties. Hence the new programme is commended in part as a means of limiting the danger of polarised sub-regions through action to combine stronger and weaker counties (Ramboll Group 1996). For example, while the West region as a whole is doing relatively well and is poised to develop its international relations, there are also serious problems of unemployment and the internal contrasts are very evident when Caraş-Severin and Hunedoara are compared with Arad and Timiş. Hence, "growth poles located at the border between centre and peripheral sub-regions could play an important role in solving regional problems" (Ramboll Group 1997 p.5). This means addressing unemployment in depressed industrial regions but also assisting agriculture where this is the key to the poverty problem. Understandably, some of the weaker counties may be unhappy to see their regional centre established in a strong city: so the South East preferred Brăila to Galaţi, while the Central region opted for Alba Iulia (despite its position at the western extremity) and in the South rejection of both Piteşti and Ploieşti was followed by a contest between Alexandria and Călăraşi in which the latter prevailed. While the regions encourage enterprise from all quarters, with a strong emphasis on the promotion of SMEs domestically, FDI is a key consideration. Investments of over $1.0mln during 1990–2000 total $2.86bln, equivalent to $127pc nationwide but with regional variations between $605 in Ilfov and $17 in the North East (Table 15.1). Ilfov (comprising mainly Bucharest and its suburban fringe) has taken 59.9% of the total, with particular emphasis on the tertiary sector (79.8%) compared with light industry (58.8) and heavy industry (30.4). But a number of provincial cities are also doing well, particularly Cluj-Napoca, Craiova, Ploieşti and Timişoara, and two small towns (Mangalia and Sebeş) have per capita rates well in excess of Bucharest which scores $789. Of course, only in the case of greenfield investments is there real locational choice, but in such cases investors certainly appreciate qualified

Table 15.1 Distribution of large* foreign investments in Romania, 1990–2000 ($mln)

Region	Sector A	B	C	Total	$pc	D	City@	Total	$pc
Central	55.42	128.59	55.99	239.99	90.8	20	Sebeș	55.28	1876.4
Ilfov	233.27	536.53	941.02	1710.82	604.8	13	Bucharest	1586.61	789.2
North East	8.29	55.94	2.64	66.88	17.5	13	Iași	15.89	46.0
North West	47.47	58.21	126.99	187.52	81.8	16	Cluj-Napoca	118.94	361.2
South	68.14	111.69	7.68	187.52	54.1	6	Ploiești	154.18	619.1
South East	58.91	21.40	23.34	103.65	35.3	11	Mangalia	55.17	1252.8
South West	184.78	11.86	2.39	199.04	82.9	4	Craiova	170.75	546.6
West	110.47	103.87	19.13	233.42	114.4	12	Timișoara	250.99	458.2
Total	766.72	912.09	1179.19	2858.00	127.4	95			

* over $1.0mln
@ town with the highest total investment in the region
A heavy industry (mining, metallurgy, engineering, chemicals, building materials)
B light industry (food processing, wood processing, textiles and clothing)
C services (banking, commerce, insurance, transport and other sectors)
D number of locations (by municipality)
Sources: Romanian Development Agency and National CCIA.

labour and location in transport corridors. Even so there may well be a need for linkages with firms already established, as in the case of a German-owned Schweighofer furniture enterprise which located in Sebeş because the existing Frati (Italy) board factory absorbs the waste material.

Apart from conventional economic profiles, investment decisions may also be affected by incentives enhancing the attractiveness of certain locations. First, a programme for industrial parks is under way (2002–5) to improve regional economic infrastructure: a budget of ROL583mld covers technical assistance and non-reimbursible help for local authorities and/or domestic and foreign private interests providing at least 70% of costs for a fully serviced industrial park or at least half the cost of a software park (subject to a maximum grants of ROL20mld and ROL10mld respectively). The network will thus depend on local initiatives and in particular progressive local authorities generating capital to establish a greenfield park or redevelop derelict and polluted sites (where owners are agreeable). In the Central Region the regional development agency (RDA) has combined with local authorities to create the greenfield Mureş Park (49ha) at Vidrasău (Ungheni) near Târgu Mureş airport; while the local authorities of Sibiu and Sura Mica along with the county council and the CCIA – supported by USAID, a society for technical cooperation ('Gesellschaft fur Technische Zusammenarbeit') and Sibiu's Romano-German Foundation – are developing the 95ha Sibiu-Şura Mică Park where factory lots of 2.8–10.9ha are available. Another local authority venture concerns a site close to the future motorway between Alba Iulia and Sebeş. These ventures should be very successful; being located – as the developers point out – in the centre of the country in an area 'bursting with a renewed spirit of free enterprise and entrepreneurship [with] business-friendly local governments and agencies' as well as good transport and telecoms, a skilled workforce, competitively-priced land, training facilities and consultancy services. Meanwhile, in addition to Şelimbar near Sibiu, private ventures are proceeding on greenfield sites at Odorheiul Secuiesc and Sfântu Gheorghe, well away from the Eurocorridors, while the parks at Cugir, Gheorgheni and Zărneşti are based on existing factories than have excess space. So it is evident that some small towns are also getting involved and the park programme may be helpful in drawing capital to high unemployment areas.

Second, following precedents established on the Danube in the 19th century and again under communism in the town of Sulina in 1978, an agency for free zones ('Agenţia Zonelor Libere') was set up in 1991 to administer sites where economic activities would be free of customs duties. Starting in 1992, the first zones appeared on the Danube and Black Sea coast (Brăila, Constanţa-Sud, Galaţi, Giurgiu and Sulina) but another now exists at Curtici-Arad, with further possibilities at Iasi, Oradea, Otopeni and Timişoara – the latter associated with the restoration of the Bega Canal expected in 2007 (Caraini and Cazacu 1995). The aim is to attract business connected with exports since all the zones are in border areas on major international transport routes but Guran-Nica (1997) also refers to a strategy of boosting employment along the Danube which tends to lie in the shadow of the metropolitan growth zone around Bucharest. So far, foreign and mixed investments have been quite small except at Galați where there are three involving more than $1.0mln and eight others greater than $100,000: in food/light industry, engineering, wood, chemicals, commerce and tourism. Furthermore, the government negotiated a RICOP programme for industrial restructuring and professional reconversion in certain counties, followed in 2001 by 11 'Zone de Restructurare Industrială' (industrial restructuring areas: IRAs) covering groups of towns and industrialised rural communes in several counties. All the eight regions apart from Ilfov have interests in these areas: two each in the North East, South East and South West and one each in the other four, with one shared between the Centre and North West (Figure 15.2). Substantial financial assistance for SMEs became available in these zones in 2003, with the RDAs permitted to promote particular sectors. However it must be emphasised that this scheme does not exhaust the help given to SMEs which also benefit from a range of programmes, each of which usually applies to particular parts of the country – and not least the rural areas which have been benefitting since 2003 from the EU SAPARD.

Third, the term disadvantaged areas ('regiuni defavorizate') or less-favoured areas (LFAs) has been adopted for areas affected by the government programme for restructuring in the mining industry which suffered particularly badly through the opening up of the Romanian economy to global forces. After powerful support under communism – when the country's natural resources were used in preference to imports, with virtually no regard to the real costs incurred – the massive losses involved under market economy conditions had to be addressed by radical measures taken by the centre-right government in 1997, combining redundancy packages for miners with efforts to provide alternative employment. The World Bank provided a $44mln loan to finance the closure programme which was a significant factor in reducing employment in mining from 175,000 in 1977 to 67,000 in 2003. According to arrangements made during 1998–9, developers in LFAs could benefit through fiscal concessions for 3–10 years in respect of new business where the unemployment rate was 25% higher than the national percentage; a single industrial branch accounted for over half the salaried population working in industry; and/or where massive lay-offs (due to the liquidation, restructuring, or privatisation) affected more than a quarter of the permanently-domiciled active population; and where an adequate infrastructure was lacking. New enterprises (whether trading companies with majority private capital, family associations or individuals) obtaining a 'certificate de investitor' were exempted from profits tax and all taxes (import duties and VAT) on equipment, buildings, transport and land – provided the relevant activity (farming, services, trade or environmental protection) was located within the zone, along with the headquarters. Also, a development fund was provided for the LFAs to stimulate exports, underwrite external credits and finance investment programmes to enhance social capital. Although the law made no specific reference to mining, all 25 zones created during 1998–9 were linked with restructuring in this industry and had ten years duration (Borcoș and Vîrdol 2002). And although 13 non-mining areas subsequently gained recognition these were all given just three years duration except the four designated in 2000 (which have a life of ten years, though five in the case of Hunedoara) (Figure 15.2 and Table 15.2). The scheme was looked on with some concern by the EU which desires a 'level playing field' for economic development. Hence new rules for LFAs imposed in 2003 restricted the fiscal concessions to customs duties on raw materials (except for the meat industry) and removed the concessions on profits tax from new ventures. Furthermore, since 2000 the qualifications for LFA status have changed to total unemployment three times the national average (as well as isolation and poor infrastructure). Unemployment in the LFAs in mining areas nowhere exceeds double the national average while nine are actually below average (Timofticiuc 2003). Hence some areas approved for designation in 2001 were not confirmed. It would appear that LFAs have in a sense been overtaken by the broader IRAs – within which most of the LFAs are now situated. However, the handbook for applicants seeking finance for SMEs in the IRAs is quite categorical about the LFAs having extremely difficult socioeconomic conditions (MDP 2003 p.87) and the SWOT analysis for each IRA highlights the relevant LFAs as offering particularly attractive opportunities.

It should emphasised that the mine restructuring programme by no means exhausts the scope for recognition of LFAs. For the concept has great relevance to the problems of rural areas in ways removed from levels of unemployment and infrastructure (Nica 1993; 1999). Ianoş (2000 p.180) referred to some 'complex disadvantaged areas which ought to benefit by a special development strategy' – with six deserving cases in the Central Region alone – and went on to refer to 'arii profund dezavantajată', distinguishing between areas satisfying one criterion of difficulty and those with several (e.g. isolation, unemployment and poor infrastructure) (Ianos 2001). F. Bordânc (see Bordânc and Nancu 1999) worked for several years at the Romanian Academy's Institute for Agricultural Economics on a viable classification that will recognise different categories of difficulty, while Nădejde (1999) has taken a range of demographic, economic, geographical and social criteria to highlight 15 problem regions across the country. Although no blueprint has emerged that is likely to be politically acceptable as the basis for a special aid programme, these surveys are useful to local authorities and ministries in prioritising their investments. The problems regarding local roads and water-sewage services were addressed by the centre-right government's programme supported by the World Bank (Drogeanu 2000) and the effort continues through SAPARD. But the relief of poverty in areas where the vast majority of people are working in agriculture (often on a largely-subsistence basis) is a longer-term challenge, although there is now a nationwide agricultural advisory service ('Agenţia Naţională pentru Consultăţii Agricole': ANCA) to stimulate marketing and food processing. But the pre-accession funding is very modest and it will hardly create the diversified, sustainable rural economy envisaged in the early transition years (Paşcariu 1993) or overhaul run-down local services to realise the vision of modern community facilities in key villages (Voiculescu 1999). The full range of possible criteria has been laid out in a regional development strategy for Timiş county and illustrates the wide potentials of the LFA approach (Coifan 1999

Figure 15.2   Aspects of regional development in Romania

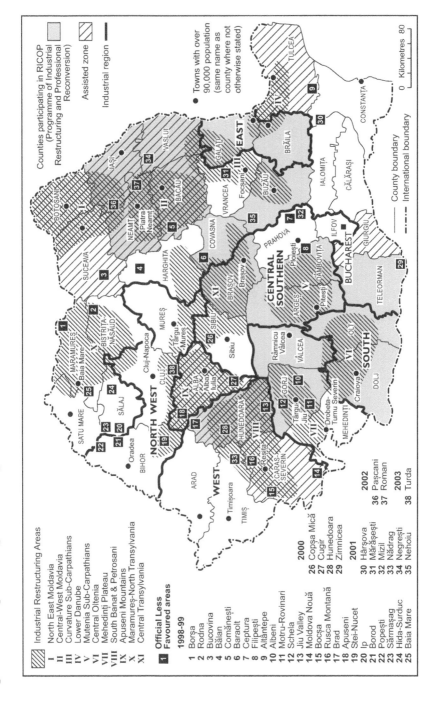

Table 15.2   Distribution of LFAs and IRAs by region

Region	No	Sq.kms	LFAs #	Pop'n	#	Yr.1	Yr.2	IRAs Towns	Pop'n	#
Central	5	1567.5	4.6	104.5	3.9	1998	2000	24	80.2	29.7
North East	5	3983.0	8.1	296.7	7.9	1999	2002	25	1143.0	30.5
North West	10	4276.9	25.2	446.5	15.3	1999	2003	21	602.8	20.7
South	4	329.5	1.0	65.4	1.8	1999	2001	13	501.0	14.1
South East	4	355.9	1.0	38.0	1.3	1999	2001	13	1051.6	35.5
South West	3	1282.4	4.4	106.6	4.3	1999	1999	20	845.0	34.4
West	7	5098.9	15.9	426.9	20.23	1998	2001	34	802.8	38.0
Total	38	16894.1	7.1	1484.6	6.5	1998	2002	150	5748.5	25.2

: Percentage of the regional/national total.

Yr.1/Yr.2: years when each region's first and last LFAs were established

Population figures – in thousands – are for 1991 (LFAs) and 2000 (IRAs)

Source: Ministry of Development and Forecasting, Monitorul Oficial and Statistical Yearbooks.

pp.123–36). Apart from declining industrial areas, others were characterised by economic backwardness (some specifically for agriculture); poor integration with county and national road systems; inadequate electrification, transport or water supply; heavily depopulation; remoteness in frontier zones; exposure to environmental risk; or priority for conservation.

The official LFAs have done quite well. While all sectors are represented, 1,624 of the 2,475 certificates issued before July 2001 concerned the construction industry. During the four years 1999–2002, 54,153 new jobs were created (of which 62% were taken by the unemployed): annual totals rose from 7,670 in 1999 to 10,904 in 2000 and 21,142 in 2001, but fell back to 14,437 in 2002 despite further increase in designated areas. However for the years 1999–2001 each of the 39,716 jobs required investment capital of €9,014 and fiscal concessions amounting to €3,147 which critics find difficult to justify when most of the country is disadvantaged in some way. A particular anomaly has arisen in the case of meat processing because subsidised imports undercut producers outside LFAs who use Romanian meat and the national meat market is thereby compromised (Timofticiuc 2003). Hence the changes in the rules in 2003 were particularly evident in this industry. Clearly FDI is an important consideration, requiring a 'strategic approach to inward investment as well as an integration with a broader longer-term regional development strategy that focuses on existing industrial strengths' (Amin and Tomaney 1995 p.218). However, when it comes to levels of domestic and foreign investment in individual LFAs, figures provided by Popescu et al. (2003) for the original 25 mining areas and also Hunedoara (which includes a small mining industry) point to a highly varied picture. €357.98mln were invested during 1999–2001 which is equivalent to €303.9pc, but €243.54mln went to just four LFAs (Baia Mare, Bucovina, Comăneşti and Stei) where the per capita equivalent was €671.6; a further €76.45mln went to Baraolt, Bocşa, Brad, Filipeşti, Hida and Valea Jiului (€230.9pc) while the remaining €37.99mln went to the 16 other LFAs at an average rate of just €78.5pc. The foreign component (including the domestic component of enterprises with mixed capital sources) was €40.21mln or 11.2% and it is interesting to see that this share was greatly exceeded in several LFAs which attracted little investment overall: Apuseni, Bălan, Borşa and Sărmăşag – all in Transylvania where contacts with Hungary and/or Germany are strong. Otherwise the same polarisation is evident with the top four LFAs attracting €27.81mln (€76.7pc), the following group of six €7.70mln (€23.3pc) – and the rest €4.70mln (€9.7pc). However it should be noted the whole picture for FDI is distorted by the €16.52mln invested at Stei (€1,028.8pc) – e.g. in a modern biscuit factory – which accounts for 41.1% of all FDI in the LFAs. At the same time half the LFAs in the top ten were markedly less attractive for foreign investors than for domestic investors, because the FDI share was only 6.3% in Bocşa, 4.0% in Brad, 3.1% in Jiu Valley, 2.0% in Bucovina and 0.3% in Comăneşti. The technology used by some of the foreign ventures appears to be quite low for Ianoş (2000 p.180) referred to interest from a number of firms from Israel, Netherlands and Switzerland concerned with wood processing, ready-mades and dairying in LFAs in the North East. On the other hand in the Central region during 1999–2002 there was a strong foreign interest in Cugir LFA (accounting for 36.4% of all projects) and across the region as whole the foreign contribution was strongest (relative to domestic projects) in chemicals and engineering.

Two examples may be cited to illustrate different aspects of FDI. First, Guran-Nica (2002 pp.145–60) examines a $1.0mln Italian investment in Moldofil, the Câmpulung Moldovenesc spinning mill (in Bucovina LFA) which was established in 1972 and privatised by EMBO in 1993. An Italian holding company – which included the Romanian-Italian joint company Romalfa (with the Săvineşti synthetic fibre company 'Melana') in its portfolio – was interested in extending its operations into Romanian spinning capacity in a way that would allow Săvineşti to act as a 'mother factory' (although under communism Moldofil imported natural cotton and obtained chemicals from Brăila, Dej and Iaşi). However, although proximity to Săvineşti was a factor in the decision to acquire the Bucovina factory, quality considerations have resulted in only one percent of the acrylonitrile coming from this source. Instead, virtually all of it comes from Italy, while raw cotton arrives from Syria, Tadjikistan and Uzbekistan. The Italian investment – complemented by Moldofil's contribution through land and buildings (also valued at $1.0mln) – has provided new technology and opened up global markets. Before 1989 the yarn was used in domestic textile industries in Oradea, Panciu and Satu Mare, with some export to Italy and Turkey. But now, not only are the strong customer links within Romania consolidated ('1 Iunie' Timişoara as well as 'Tricoton' Panciu, 'Croitex' Oradea and 'Ardeleana' Satu Mare – all of whom are now exporting to Western Europe and North America on the strength of improved quality) but a third of the output goes to Hungary and Italy. The workforce has declined to 42.5% of the 1989 level (from just over 800 to some 350) – part of a general decline in employment in the area – but salaries have increased in line with productivity and labour relations have been placed on a more progressive basis.

The second example concerns a potentially very large foreign investments by Roşia Montană Gold Corporation (RMGC) <www.rosiamontanagoldcorp.com>, based in the Apuseni LFA on a 4,282ha concession awarded to the Canadian mining company Gabriel Resources in 1997 (Turland 2002). As a source of gold and silver since Roman times, mining was being undertaken in the 1990s by a state company which had abandoned its traditional underground operations in 1985 and employed 775 workers in an open pit operation in the Cetate sector of Roşia Montană, started in 1970 with obsolete technology. It has now been agreed that the state operation will be replaced by RMGC's modern plant (owned 80% by Gabriel Resources, 19.3% by the SOE 'Minvest' and 0.7% by small investors) which will operate not only at Cetate-Găuri but also in the Cirnic, Jig and Orlea sectors during 2006–2024, following the pre-construction (1996–2003) and construction (2003–5) phases. Proven/probable reserves of 225.7mln.t are expected to yield 331t of gold and 1,600t of bye-product silver each year, using a cyanidation process. The employment and capital investment is very substantial: (a) direct employment of 200–550 (and a multiplier of 400–600) during the pre-construction period – with

$50mln spent on exploration, pre-feasibility and final feasibility work plus ongoing investment of $125–150mln; (b) employment of 2,000 during the construction phase, plus a multiplier of 6,000 jobs and capital expenditure $350–400mln; and (c) employment 500 during the operational phase plus the multiplier of 3,000 jobs through contracts with local business in respect of building, supplies, services, repairs and maintenance. This is very encouraging for young people who are being forced to work abroad for lack of local opportunities. The fact that Gabriel Resources will take the bulk the profit while the state will get comparatively little (especially in view of LFA incentives) should not obscure the merits of the scheme in terms of economic and social security in the area in the immediate future. RMGC believe they are implementing 'the first of several new mining projects that could be developed in the Apuseni Mountains [ensuring] that no single community is overburdened by too much development too fast' (Ibid p.16). The commune should become 'a catalyst for long-term sustainable development throughout the region' (Ibid) through the regional development strategy arising out of the economic and social impact assessment (ESIA).

However, there is a threat to over 2,000 people in some 880 dwellings (especially in the Corna Valley where land is needed for the dumping of tailings and overburden) although the resettlement action plan involves new homes and land parcels at three locations in the immediate vicinity – with provision of services and small business development – or monetary compensation allowing people to move where they choose (including apartments in urban areas like Alba Iulia). The production process is also a cause for concern in view of the pollution in the Tisza valley through cyanide released at Baia Mare in 2000. But if the processing is well-managed there could be environmental benefits considering the pollution of streams which has been going on for many years on account of acid rock drainage. Nevertheless the project remains highly controversial and communities in the area have literally been torn apart. It is a tragedy that the desperate need for employment in a LFA is driving people to accept what will be Europe's largest opencast gold mining project (1,600ha) that threatens to reduce a picturesque landscape into a toxic, sterile desert through blasting, pulverising and the use of cyanide leaching technology. There is naturally a cleavage between those who expect to get jobs and the subsistence farmers – particularly in the Corna Valley, threatened by a 185m earth dam (needed for a tailings pond) – who are determined to defend their land. The two sides are represented by rival NGOs: 'Pro Rosia Montana' – established by the company – and 'Alburnus Maior' <http://www. rosiamontana.org> which has maintained a resolute opposition since its formation in 2000 (Buza et al. 2001). But there is a wider dichotomy among the public at large, including politicians, which balances employment (without viable alternatives) against the inevitable damage to the archaeology (especially the Roman galleries) and the destruction of churches and cemeteries. The company has tried to maintain momentum by the acquisition of property and premature relocation, including work on a new settlement in advance of a formal go-ahead, but has encountered delays in delivering its long-awaited ESIA and appears to be facing financial difficulties through the failure to secure loans (notably from the World Bank's private lending arm, the International Finance Corporation) and also through mounting project costs. At the time of writing the company remains unable to convince a majority that environmental and social issues have been adequately addressed while the prime minister has issued cautionary words against automatic agreement to become 'an economic colony' and the European Parliament is carefully monitoring the country's environmental capacity in the run-up to Romania's EU membership.

West Region and its Sub-Regional Variations

Here the RDA seeks to promote a region (32,034sq.kms in area with a population 2.05mln: 64.7 persons/sq.km) with the best business opportunities arising through location; not to mention a diversified economy with a qualified workforce, good communications (with immediate access to Eurocorridors) and high agricultural and tourist potential. The modernisation of the region through diverse ethnic groups introduced through the Habsburg colonisation of Banat greatly extended the cultural profile based on earlier contacts between Romanians and Hungarians. Although not the region with the largest non-Romanian population, Romanian Banat must now be area with the largest number of different non-Romanian ethnic groups. And it is apparent that the political climate has been predominantly one of tolerance and mutual respect to the point where all groups are valued as parts of the region's economic and cultural identity. Nationalism has not significantly detracted from the positive contribution of ethnicity for the human resources of the region. However, the emphasis on agricultural colonisation on the plain was complemented by priority on extractive industries and metallurgy in the mountains. Reşiţa (Caraş-Severin) has a history of ironworking dating back to 1771 which makes the region one of the longest-established industrial zones in SEE: a situation reflected in urban expansion, early electrification, the growth of a relatively dense railway network and the emergence of food processing industries based mainly on the rich agriculture of the Banat Plain. The capacity of the Reşiţa metallurgical and engineering complex to act as a mother factory, stimulating a series of production transfers to new locations, also points to a high level of experience and skill. The historic strength of the extractive industries is further extended by Hunedoara's mineral wealth in coal in the Jiu Valley, complementing the long-established working of non-ferrous ores in the Brad area of the Apuseni Mountains to the north. But the model of maximum self-sufficiency adopted under communism has made for painful restructuring because the polluted mining areas have been largely overlooked by foreign investors. Hence the acute sub-regional problems in West Region – including an old infrastructure, non-viable SOEs and low productivity labour – which are central to the development strategy. The development organisations established in Arad (ADAR), Caraş-Severin (ADECS) and Timiş (ADETIM) with help from Germany's Nordrhein-Westfalen province helped to provide expertise before the wider West Region came into existence. But the attractions of Arad and Timişoara – with a diverse industrial profile and lowland countryside 'revitalised by reinstatement of private property [whereby] farmers increase their revenues, develop local services and intensify the village/town relationship' (Ramboll Group 1997 p.31) – have worked to the detriment of Caraş-Severin and Hunedoara where the need for new jobs is particularly critical. The region's per capita GDP is above the national average but the Timiş figure is almost 50% greater than that of Caraş-Severin (although Timiş still scores only 40% of the EU average, it is above the average for candidate countries).

Timişoara, the 'technopolis' and obvious centre of Romania's West Region, is clearly best placed to sustain self-reliant internationally-competitive development with good education/training facilities and research institutes (including a university seeking the status of 'centre of excellence'); and environmentally friendly industry (at an advanced stage of restructuring) with a stock market, business centre and supply networks. It has a network of SMEs to serve large enterprises, good banking and local government support, fiscal incentives and low cost sites and premises. In the early years of transition there were investments in the city from Alcatel, Coca-Cola and Proctor & Gamble, with a second wave at the turn of the millennium which has included the American enterprise Solectron and such German companies as Continental, Draxlmeier and Siemens. Foreign business organisations in the city include a Romanian-German Economic Forum, a Romanian-Dutch Business Centre and 'Biroul de Promovare Economică Timişoara-Karlsruhe'. The Italian presence has been growing strongly through four Italian banks which contribute to one of the largest Italian business communities in Romania (sustaining an Italian language high school, a 'Biroul Italia' within the Timiş County CCIA and the prospect of Italian consulate). A technology park is favoured by investors from the Veneto region, but more generally a new study of Timişoara – with all-party support – promotes the city as a growth pole in the context of an expanding Europe and a growing Central European Free Trade Association (Ciuhandu 2001). The city could act as a pilot for the adoption of the EU 'acquis' and the process of Romanian integration into the EU generally. An exhibition centre and commercial park are anticipated as part of a satellite town which will help to gear up Timişoara for the creation of 15,000 new jobs during 2000–7 and hopefully a reduction in unemployment from 8.6 to 3.5%; plus a boost in average income by 20%. All this reflects the dynamism of the 'Timişoara model' based on a cosmopolitan mentality of peaceful cohabitation in a multi-national environment; attractive facilities (including land for development) in a privileged location – enjoying a positive 'frontier effect' – which is being professionally exploited by progressive councils in the both the city and county. Further transport plans will help the city, including motorway construction and the reopening of the Timişoara-Szeged railway (not to mention the benefit of a restored Subotica-Kaposvar link in providing an east-west rail route from the Black Sea at Constanţa and the Adriatic at Rijeka. Moreover, the Danube is a major international river and a Eurocorridor in its own right, while the Tisza is also navigable with a modernised port at Szeged in Hungary (also a logistical centre with railway 'piggy back' services to Wels in Austria). Romania could plug into this system through the reopening of the Bega Canal from Timişoara to Zrenjanin (disused since 1958), perhaps linked with a free zone in the east of the city already mentioned. Finally, integration between transport modes will be enhanced through a new 'Interporto' facility: a 70ha road-rail facility with 10,000–30,000sq.m of floorspace.

With extensive foreign contacts, the RDA is seeking to promote innovation and its transfer through links between business and science. Arising out of meetings of experts on the business infrastructure and annual fora contributing to the region's innovation strategy, a current preoccupation is the development of industrial clusters through business incubators, technology transfer and industry parks in Arad and Timişoara – assisted by €4.8mln through PHARE as a supplement to the 2001 programme mentioned below, not to mention the Arad-Curtici free zone. But while this is the motor for a developing region, linked with the university faculties in the region, the intra-regional tensions also direct attention to improved environmental standards (especially in waste management) and better use of the human resources through realising the unexploited tourist potential and in other aspects of rural development. The West Region therefore seeks a restructured agriculture and rural diversification under the guidance of a regional information centre for agricultural problems and a rural development agency. And county plans are supporting such arrangements in respect of small towns and communes with central place functions as centres for non-agricultural functions in predominantly rural area. Integrated local programmes, implemented by SMEs in agriculture, food processing, marketing and tourism (reflecting in part the diversity of ethnic minorities), may be encouraged. The small town of Deta is seen as a suitable location for a specialised livestock market and for fruit processing, including production of a range of brandies, while Făget could be a tourist centre for the Poiana Ruscă and a suitable location for a small business incubator (generally linked with the larger towns). Each town should also have a organisation for business people: 'Club al Oamenilor de Afaceri'. Stronger rural centres could also emerge in the Timişoara area at Bethausen, Ciacova, Fârdea, Gătaia and Recaş – where local halls could be refurbished as multifunctional centres for educational and social-cultural functions – although deficiencies in the infrastructure (medical services, natural gas, roads, telecommunications, water and sewerage) are also highlighted. But the main focus for concern is the IRA of South Banat and the Petroşani Coal Basin which covers four urban groupings (Deva-Simeria-Orăştie; Hunedoara-Călan; Petroşani with Aninoasa, Lupeni, Petrila, Uricani and Vulcan; and Reşiţa with Anina, Bocşa, Oraviţa) and a scatter of towns and rural areas: Brad, Caransebeş, Haţeg, Margina, Nădrag, Oţelu Roşu, Tomeşti, and Topleţ (the last-named being grouped with towns in Oltenia). The region also has seven LFAs. Most of the programmes administered by the RDA relate in part to these areas: PHARE 1998 for industrial restructuring, through SMEs, tourism and development of human resources; PHARE 2000 for the development of human resources through restructuring, creation of SMEs and the improvement of local and regional infrastructure; and PHARE 2001 for infrastructure, social services and education with particular concern for economic and social cohesion in the restructuring regions. There are also PHARE programmes for CBC (currently for 2004–6), development programmes for Caraş-Severin and Hunedoara counties financed by the NRDF and special programmes for LFAs dealing with business development and the stimulation of investment and rural activity. In the latter instance, potential entrepreneurs in rural areas may be aware of local opportunities but have difficulty in getting all the information their require: agricultural prices; land availability; and markets. In this connection CCIAs are drawing up lists to documents the sources of capital, equipment and possible partners.

Restructuring Problems in West Region

The IRA and LFA strategies arise from massive reduction in employment in the large SOEs. In the towns of Caraş-Severin alone – taking the Reşiţa engineering and metallurgical enterprises (CSR and UCMR, including Arsenal and Reşiţa-Renk), the Caransebeş engineering works (Caromet), the fabrication company 'Construcţii Metalice Bocşa' (CMB), the Oţelul Roşu steelworks (Socomet, now Gavazzi Steel), the Topleţ engineering firm 'Societatea Economică Magheru' (Semag) and the Anina mining company 'Miniere Banat', the employment in 2000 was 16,495 – only 35.9% of the 46,440–strong workforce of 1989. Unemployment rates have soared although they are difficult to calculate realistically because benefit is restricted: 'ajutor de şomaj' provides 75% of the minimum wage for six months to a year depending on social security payments made; 'ajutor de integrare profesională' offers vocational integration allowances of half the minimum wage for six months for graduates (aged over 18) from education institutions who cannot find jobs within 60 days. Because the chances of getting a new job through official channels are slight many unemployed people do not register with the Labour Office ('Oficiul Forţelor de Muncă') and the official rates are usually much too low because they only cover those receiving benefit and others who remain on the records: 'înscrişi în fise (care cauta de lucru)'. Hence there can only be intelligent guesses as to the 'real' unemployment rate which could extend to the ultimate extreme of considering all elements in the active population (aged 18–62) not known to be in waged employment. In Jimbolia the official unemployment rate is only 5.7% but consideration of people known to be without work raises the level to 39.6% (according to the local authority). And in the seven main industrial centres of Caraş-Severin (Anina, Bocşa, Caransebeş, Oraviţa, Oţelu Roşu, Reşiţa and Topleţ) the official rate of 8.1% compares with a theoretical maximum of 54.9%. In the individual towns there is at least a difference of 40% and in the case of Oravita the official rate of 12.4% compares with theoretical maximum of 74.0%.

People may return to their rural roots, especially when they have land to provide subsistence, but for many this is not an option and the end of unemployment benefit or redundancy money (generally by 2000 in the latter case for miners) brings hardship which may only be alleviated by money from other family members or means-tested 'minimum income guarantee' (MIG) benefits – usually €15 monthly – payable by local authorities in return for casual labour – replacing the old social aid ('ajutoral social') programme in 2002. Hence a frequently desperate situation arises, as at Călan where lack of work is 'punctul generator de explozii sociale'. The smaller towns often suffer from unmodernised railways and poor local road and fixed phone systems, partly because decline associated with a 'zona monoindustrială' (Chiribucă et al. 2000) means a reduced local budget inhibiting provision of good business environment attractive to investors. In Călan the defective water system has been blamed for outbreaks of enterocolitis while in Bocşa the football stadium lies derelict since the local engineering enterprise (CMB) is no longer able to support the local team and the adjacent campsite ('tabără') has been disappeared through the theft of wooden huts taken for firewood! Local authorities therefore seek better housing and school buildings and investment in sewage, ecological waste management and more green spaces.

Consolidation of the inherited industry is challenging without private capital since the state sees the remaining SOEs as a drag on the economy and has only limited resources to modernise in the hope of attracting buyers. The experience at Reşiţa has been most unfortunate since the privatisation of the metallurgical works – after investment in a mini-mill involving an electric steel furnace and continuous casting – in favour of the American company Noble Ventures broke down over the division of 'responsibilities' between the government and the company in the context of challenging world market conditions for steel. The plant is now back in state ownership and production has resumed pending a further privatisation deal with another company (Hillinger and Turnock 2000). Meanwhile the old 'Combinat Siderurgic Victoria' at Călan has been restructured and broken down to comprise the 'Sidermet' (the core business, modernised to produce cast iron tubes) and 11 other companies now employing 1,500 in all. Other heavy industries include rolling stock repairs in Simeria. Building materials and construction enterprises have also slimmed down but light engineering firms – e.g. producers of car accessories, like 'Volane' in Brad and 'Sogero România' electrical assemblies in Orăştie – have good opportunities if they can meet the quality and logistical requirements of the assembly plants. Logging and wood processing including furniture, textiles and knitwear, and leather and footwear all have growth potential since most of the raw materials are locally available from Carpathian farms and forests. There are also opportunities in livestock farming and food processing (meat, milk and alcoholic drinks, with some processing of medicinal plants). Simeria would like to introduce a guarantee scheme to take farm produce and process to European standards, along with low interest loans to enable purchase of farm equipment, a consultancy service to achieve high cultivation and veterinary standards and overhaul of drainage and irrigation systems.

There are also small settlements in the rural areas where key industrial plants have closed completely: 'Solventul' at Margina (1999) due to outdated technology and the failure to find new customers when business with Serbia's Pančevo petrochemical plant was lost through the UN embargo; 'Ciocanul' rolling and galvanising mill at Nădrag (1999) and the 'Stitom' glassworks at Tomeşti (2000): the latter privatised but unable to survive without capital for modernisation when sales proved insufficient to meet the wage and energy bills. These places suffer from lack of managerial expertise; lack of markets for traditional local industry production; a workforce narrowly specialised; no local funds for modernisation; poor services e.g. there are no digital telephones away from the main communication axes, no sewage systems and little outside help apart from World Bank support for the Nădrag heating system. By contrast a fortunate situation has arisen at the frontier town of Jimbolia where high unemployment arising from the liquidation of several state companies ('Soceram' for building materials; 'Canabis' for flax processing; and the 'Comnutrin' fodder plant) has been balanced by FDI not only in the privatisation of 'Pantera' footwear (where a Romanian-German venture supports 500 jobs) but in several new ventures: Canadian capital in the 'Valova' bakery (50 jobs); Croatian capital in 'Bolia' footwear (210 jobs); German capital in 'Vogt' electronic components (800 jobs); and Italian capital in 'Fagi' footwear (109). This is in addition to Romanian capital in the privatisation of 'Venus' plastics employing 75. In another border town, Sânnicolau Mare, the creation of over 7,000 jobs in the production of electrical goods by Delphi Packard (Austria) and Zoppas Industries (Italy) has eliminated unemployment altogether. Meanwhile, the development of the private sector through SMEs operating in local industry is very slow and a more active strategy is needed to complement the legacy of heavy industry where modernisation will mean further redundancy through higher productivity. On the other hand, tourism tends to attract unrealistically high expectations. Small resorts of Geoagiu near Orăştie and Vaţa near Brad (the latter recently refurbished by Romtelecom which now has a share of Greek capital) can benefit from the biodiversity and scenic resources of the Carpathians as well as historical-cultural values evident at Sarmizegetusa – the Dacian citadel in the Orăştie Mountains. But the town of Simeria also wants to promote tourism based on bathing in the Strei river and its 18th century 70ha arboretum (containing 150 species and recently refurbished according to a conservation plan). Brad aims at a 100–bed hotel but accepts that tourism is constrained by inadequate promotion and information and also needs improvements in transport systems and the conservation of monuments and nature as well as more accommodation. Orăştie therefore seeks a business centre ('Centru de Afaceri') as well as the development of a local airfield nearby at Aurel Vlaicu.

The LFAs of West Region

The seven LFAs comprise: Brad and Valea Jiului (1998); Bocşa, Moldova Nouă-Anina and Rusca Montană (1999); Hunedoara-Călan (2000); and Nădrag (2001) (Figure 15.4). Seven other areas have tried unsuccessfully to gain LFA status: the individual settlements of Jimbolia, Margina and Tomeşti and the entire mountain region in Arad county where four more extensive LFAs were proposed: Sebiş-

Figure 15.3   Less-favoured areas in Romania

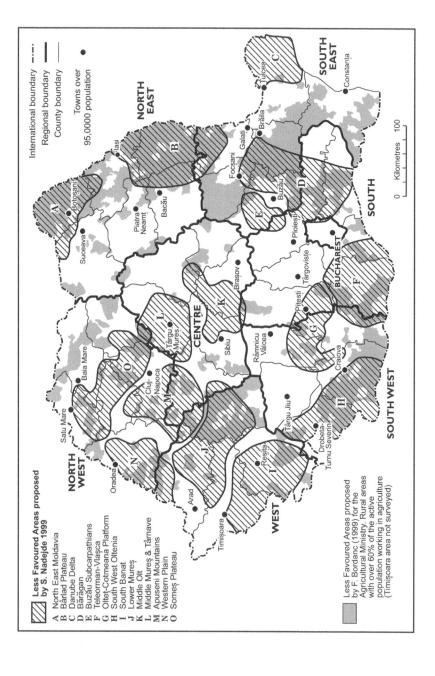

Figure 15.4 Less-favoured areas in Romania's West Region

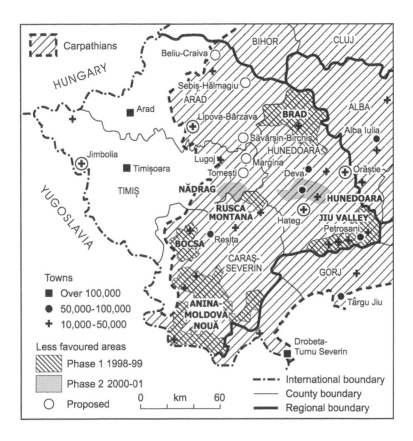

Hălmagiu (one town and 15 communes: 41,127 population); Lipova-Bârzava (one town and six communes: 27,096 population); Beliu-Craiva (seven communes: 17,175); and Săvârşin-Birchiş (five communes: 10,553); all with potential in the forests (including accessory products), tourism, food processing, building materials and light industry. Publicity material for Caraş-Severin's LFAs has been produced through Soros Foundation support for a Romanian organisation promoting democracy ('Fundaţia pentru o Societate Deschisă') which cooperates with the County Hall ('Prefectură') and the CCI in Reşiţa. Opportunities are seen in industry, particularly in food processing, timber and furniture and textiles, while agriculture features with regard to livestock rearing, milk production with cereals, fruit growing and viticulture in appropriate areas. The organisation responsible for promoting the mineral exploitation ('Inspectoratul Zonal pentru Resurse Minerale Caransebeş') is trying to develop interest in this domain with respect to coal, gold-silver ores, refractory sand and mineral water. Interest has been expressed in the mineral waters of the Anina and Bocşa areas (at Prednicova and Calina near Dognecea respectively), but there has been little progress at Rusca Montană despite considerable potential at the Ruşchiţa marble quarry: dating back to 1886, with attractive stone similar to that of Carrara in Italy. Overall, RDA staff have tried to build good relations with investors and backed by strong political pressure the LFAs have made a modest impact. However, the results during 1999–2001 were very disappointing with total investment of €53.64mln or €126.7pc – well below the average of €303.9 for all LFAs nationwide (Popescu et al. 2003). Furthermore the FDI component was only €2.07mln (3.9%) or €4.9pc: often restricted to small investments by entrepreneurs connected with the region who find suitable premises for low-tech operations. Thus a lingerie factory operates with Italian equipment in a former vegetable store in Brad to supply the Italian market, while elsewhere in the Brad LFA an empty building at Lunciou de Jos is used to produce furniture from laminated panels imported from Italy and another empty building in Băiţa turns out cosmetics from imported materials. Meanwhile Rusca Montană LFA is the only one nationwide to have attracted no FDI whatsoever.

An LFA based on Ironstone Mining and Heavy Industry: Bocşa

Bocşa LFA is situated at 170m in a depression in the Dognecea Mountains (drained by the Bârzava) which lies 18kms west of Reşiţa on the road and rail route to Timişoara. The area includes the town (population 19,171 in 2001) and the three communes of Dognecea, Lupac and Ocna de Fier, lying to the south with Croat, German, Hungarian and Roma elements in addition to the Romanian majority. Lying within the Carpathians, close to the contact with the Banat Plain, there is some activity associated with the meadows and oakwoods of the Dognecea Mountains, but of much greater significance for the development of the region was its contribution to the Reşiţa metallurgical complex by virtue of the mineral endowment arising from small areas of basalt (within what is predominantly granite and limestone country) containing iron ore. The minerals were appreciated when the Habsburgs wrested control of Banat from the Ottomans in 1699 and started a mercantilist enterprise in their new border territory. A phase of expansion began with the arrival of mining families from Bohemia in 1718 to establish the new settlement of Bocşa Montană as a community of woodcutters and charcoal burners beside the older community of Bocsa Română. This provided the basis for the smelting and forging complex built by Baron von Rebentisch (known as the 'Altwerk') which was superseded after the floods of 1722 by the 'Neuwerk' – a name that gave rise to the present Naiverc quarter of the town (Figure 15.5). Activity continued after a brief setback of renewed Habsburg-Ottoman hostility (1737–9) and although new capacity was installed further up the valley at Reşiţa in 1771, Bocşa continued to act as an independent centre of iron production. The community became quite diverse ethnically for in addition to the local Romanians and the German and Hungarian colonists there were Romanians from Sasca who came to take up forest work ('Săscani') and Roma associated with the Sf.Ilie monastery of Vasiova who settled in the Godinova and Măgura districts.

Figure 15.5   The town of Bocşa in Romania's West Region (the lower map extends the one above on the right side)

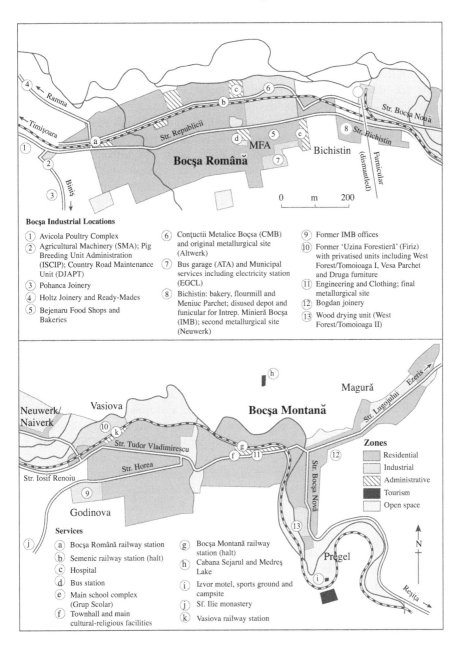

Bocşa Industrial Locations

① Avicola Poultry Complex
② Agricultural Machinery (SMA); Pig Breeding Unit Administration (ISCIP); Country Road Maintenance Unit (DJAPT)
③ Pohanca Joinery
④ Holtz Joinery and Ready-Mades
⑤ Bejenaru Food Shops and Bakeries

⑥ Conţuctii Metalice Boçsa (CMB) and original metallurgical site (Altwerk)
⑦ Bus garage (ATA) and Municipal services including electricity station (EGCL)
⑧ Bichistin: bakery, flourmill and Meniuc Parchet; disused depot and funicular for Intrep. Minieră Bocşa (IMB); second metallurgical site (Neuwerk)

⑨ Former IMB offices
⑩ Former 'Uzina Forestieră' (Firiz) with privatised units including West Forest/Tomoioaga I, Vesa Parchet and Druga furniture
⑪ Engineering and Clothing; final metallurgical site
⑫ Bogdan joinery
⑬ Wood drying unit (West Forest/Tomoioaga II)

Zones

▮ Residential
▯ Industrial
▨ Administrative
▮ Tourism
▯ Open space

Services

ⓐ Bocşa Română railway station
ⓑ Semenic railway station (halt)
ⓒ Hospital
ⓓ Bus station
ⓔ Main school complex (Grup Scolar)
ⓕ Townhall and main cultural-religious facilities

ⓖ Bocşa Montană railway station (halt)
ⓗ Cabana Sejarul and Medreş Lake
ⓘ Izvor motel, sports ground and campsite
ⓙ Sf. Ilie monastery
ⓚ Vasiova railway station

In 1873 the narrow gauge industrial railway from Reşiţa to Bocşa and Ocna de
Fier was completed and the following year the standard gauge public railway from
Timişoara arrived. This was the basis for the expansion of the industrial complex as
a whole through steel making and engineering, with concentration of iron and steel
making at Reşiţa and closure (probably in the 1880s) of blast furnaces built at Bocşa
and Dognecea during 1858–61. But mining continued at Dognecea and Ocna de
Fier, while the production of agricultural machinery began at Bocsa Română in
1904 and other installations included a sulphuric acid factory and the Colţan
limekilns. By this time the whole complex has been privatised through purchase by
the Austrian company 'Staatseisenbahngesellschaft' (STEG) in 1854, reconstituted
as 'Uzinele şi Domeniile din Reşiţa' (UDR) when the Banat Carpathians passed to
Romania after the First World War (Graf 1997). There was further expansion under
communism when the agricultural machinery works grew into the large fabrication
complex (CMB) in the district known as MFA ('Ministerul Forţelor Armate')
arising out of a military firing range which was eventually absorbed by the
expanding complex and its adjacent housing. The result is an extremely linear
settlement involving a fusion of Bocşa Română and Bocşa Montană with Vasiova at
the point of contact where the main railway station and most of the services
(including the town hall) are situated. The 'Altwerk' was located in Bocşa Română
while the 'Neuwerk' was placed in Vasiova before the late 19th century furnace was
placed still further upstream on the edge of Bocşa Montană where a head of water
– to drive the bellows – was obtained by building a leat of about one kilometer in
length along the Bârzava from a weir at Pregel. Although closed some time after the
Ocna de Fier-Reşiţa industrial railway was built – in favour of a consolidated iron
and steel industry in Reşiţa already noted – the tower used to hoist raw materials to
the top of the furnace (and subsequently used as a small hydropower station that
exploited the dam and waterfall – 'stăvilar' – until the 1960s) remains in situ. And
it merits conservation, especially since it stands in a pleasant wooded environment
known to the locals as the railway park since Bocşa Montană station lies adjacent.
In Bocşa Română heavy industry has been perpetuated through CMB while the
Bichistin area (close to Neuwerk) developed under communism as a milling
complex and a depot for 'Intrep. Minieră' handling iron ore brought by funicular
from Ocna de Fier nearby (replacing the industrial railway). A large sawmilling
complex ('Uzina Forestieră' but more usually known as 'Firiz') developed at
Vasiova station and further capacity exists in Bocşa Montană, notably the Egyptian-
owned 'Stejar Forest' company which combines sawmilling at Vasiova (adjacent to
the rest of the state installation which now lies derelict) with steam drying of sawn
timber at a separate site supported by a furnace burning waste material.

Deindustrialisation has greatly reduced employment at the old SOEs. With
origins as an agricultural machinery factory (enlarged under communism), CMB
really took off when most of Resita's bridge-building section was relocated in 1950,
followed by a tower crane section in the 1970s (when CMB became Romania's
biggest producer of both building cranes and river bridges, as well as a producer of
swing bridges and various welded structures for the mining industry and for thermal

and nuclear power stations). Employment eventually reached 5,000 (30% German and Hungarian), but a quarter of the workforce travelled up to 20kms six days per week from villages – such as Biniş, Berzovia, Doclin, Fizeş, Gherteniş, Ramna and Vermeş – with bus services available for all three shifts; while 1,000 young Moldavians and Oltenians lived in the 'Căminele Tineretului': comprising two poorly-built four-storey apartment blocks adjacent to the factory. The enterprise worked on contracts for the Cernavodă nuclear power plant and export orders from China and the USSR as well as Austria, Canada and Japan. Problems mounted after 1989 despite quality certification in welding and collaboration with a range of foreign companies such as Ansaldo, Chyoda-Mara, Krupp, MAN Babcock and Mitsubishi, although the situation was not helped by a large debt of $3.35mln arising from goods supplied to an iron ore plant in Krivoi Rog (Ukraine) in 1989–90. Political leadership continued – involving the centre-left Party of Social Democracy and its predecessors during 1990–1998 and again since 2001, contrasting with influence from the centre-right National Liberal Party during 1998–2001 (though without significant differences in management style). Some sections were closed completely and sold as scrap, but there has been some retooling – e.g. a Messer machine was installed in 1998 for electronically-controlled cutting of sheet iron – and links have been renewed with Cernavodă while industrial halls have been built for foreign companies – European Drinks and Mitsubishi – and there is some fabrication for hotels. There was a sharp reduction in the workforce from 1,408 to 600 during 2001 (leaving 250 in bridge building, 110 in material preparation, 57 in scaffolding and 43 in painting – plus 120 in management), but on the whole the decline in employment has been gradual with the pressure of small wage increases (below the rate of inflation) to encourage employees to leave voluntarily or take early retirement. This has avoided major unrest, notwithstanding some tension over high managerial wages and discrimination over redundancy and recall on the grounds of family rather than performance. While older workers are generally happy with deals that secure pensions and the young are prepared to take up commerce, middle-aged workers are anxious because they are too young to retire but lack the energy to work abroad or in cross-border trade. The German-Hungarian element is now less than a tenth, but there is still a strong commuting element (84 people travel from Berzovia, Biniş, Ocna de Fier and Ramna) involving families with two to three members still working at the factory on the basis of their performance and/or relations with management. CMB was privatised in September 2002 with sale to a former manager who hopes to rejuvenate the enterprise – but with even fewer workers. Wages have improved and the development of an industrial park to make use of empty space is seen as a way forward.

The iron ore company closed completely in 1998 with 150 redundancies. The poor quality iron ore at Ocna de Fier ceased to be profitable after 1989 and activity declined soon after the revolution when the Reşiţa furnaces were closed and the funicular from the mine in Ocna de Fier to the to the processing plant in Bocşa Română was abandoned: from 1994 only poor quality residual ore was handled. Meanwhile, the 'Avicola' poultry complex was built in the 1950s to provide eggs and chicken for local consumers and the others in the northwestern part of Caraş-Severin; employing people from Biniş, Dulău and Valeapai who left the collective farms for a commuting way of life before building their own houses in the northern part of Bocşa Montană or buying property there cheaply in the 1980s. From 800 in 1990 the workforce declined to 200 in 2002, after several waves of redundancy, although there are still some commuters from Berzovia, Biniş and Ramna – the nearest villages to the factory on the western side. The enterprise was privatised in 2000 in favour of the manager of the former communist enterprise ('Comtim') and – with a new slaughter house – the enterprise has gained a reputation for quality, especially in Arad, Lugoj and Timişoara (although an air pollution problem remains unresolved). The new chicken rearing enterprise employs 40 of the former SOE workers, while others have found work at a new meat processing enterprise which has opened with LFA fiscal advantages, although it now faces a difficult future since most concessions were withdrawn in 2003.

Restructuring of wood processing has generated some FDI through the Egyptian owned 'Stejar Forest' producing sawn timber, with steam drying. The original 'Firiz' mill (the name still used locally for the main surviving unit) was built in the 1960s opposite the Vasiova railway station to exploit the oak-birchwoods of the Bocşa Hills and at its maximum 1,200 were employed in the town (mainly Vasiova) and surrounding villages: Berzovia, Moniom, Ocna de Fier, Ramna and Vermeş; but with sawdust pollution a significant problem. Since 1989 the complex has been restructured in sections. Sawmilling was privatised as 'West Forest' but continued to be subordinated to Caransebeş through a parchet enterprise interested in the export of steamed beech timbers from Bocşa to Africa (as the result of a contract with the trading company 'Romanel' owned by the family of former prime minister Petre Roman). The business continued to operate on two sites – in Vasiova and Bocşa Română – often known locally as Tomoioaga I and II, after the manager who was a former 'Firiz' employee. In order to survive through the 1990s it was necessary to sell much of the machinery and progressively lay-off workers (with some retirements) until 1996. However in 2002 the business was acquired by Egyptian-owned 'Sand H Trust Construction' to supply steamed beech timbers to the Middle East and small boxes for fruit-packing in Greece. Later in the year almost 100 employees were laid off, reducing the workforce to 70. The new company has scrapped more machinery without any modernisation, but wages are linked with productivity (and overtime is available) and the old habits of drinking in work-time have been eliminated. Meanwhile some 30 people are employed in two other enterprises: the furniture section was acquired as 'Druga Mobila Bocşa' in 1994 and is now based partly in Bocşa Română while the parchet section – operating as 'Vesa Parchet Bocşa' within the otherwise-derelict West Forest complex in Vasiova – is owned by a Roma citizen of Bocşa who spent time in Germany and can now employ other members of the local Roma community.

Whereas there were some 6,000 industrial and construction workers in the town's SOEs in the 1980s, barely 750 remain today. Only about 30 persons (from 12 families in Bocşa Montană and Vasiova) commute to the Reşiţa engineering and metallurgical complexes compared with 120 under communism. New jobs have been created – partly as a result of the LFA – but the results have been modest. Some 235 jobs have been provided in wood processing, joinery and parchet; 65 in restaurants, bars and tourist services; 43 in tailoring; 32 in oil distribution and filling stations; 22 in five separate bakeries; 13 at Caritas (see below) and 10 in engineering: a total of 420 (not all of which are official, because some workers are taken on by Romanian and even foreign SMEs without proper documentation – 'carte de muncă' – and are paid very low wages in order to boost profits). Small foreign investments use surplus space at CMB: an Italian clothing firm and a French firm making car parts. There are also some Romanian investors who have returned home after making money abroad – hence the references in the town to 'belgianul', 'italianul' or 'neamţul' (German). One 'francezul' was actually elected mayor in 2000 but had to resign two years later when a new law denied office to persons holding dual citizenship – though he was already a disappointment after his failure to achieve the anticipated surge in business under an entrepreneurial 'primar'! Other Romanian investments include two within CMB: the Mazzolin (Timişoara) printing works and the workshop of an individual entrepreneur (Florentin Cârpanu) who makes aluminium and wood panels. Elsewhere, printing ink cartridges are produced (based on Italian technology) and a jewellery workshop operates in a converted apartment. Several small joinery firms have developed: Pohanca (on the Biniş road), Holz or HUB, owned by the German Kurt Bluml (near Ramna) and Bogdan (on the Ezeriş road); also small furniture factories by the Drugă and Meniuc families; family associations concerned with domestic gas heating systems; and an enterprise working in iron products. All are profitable SMEs with fewer than 50 employees, including many redundant 'Firiz' workers who have doubled their wages by switching from eight to 10–12 hour days. It should also be noted that since LFA status allows foreign purchases to be imported customs-free, second-hand cars can be imported and sold very profitably. Small businesses (up to ten employees) connected with motor vehicles is another domain largely in the hands of Germans or Romanians with work experience in Germany or Italy.

As a result of restructuring poverty has become a significant problem affecting around 40% of the population since unemployment and MIG benefits are low (and the former is available only for a limited period), while monthly pensions will only secure food purchases for two weeks (unless a simple bread and milk diet is adopted) quite apart from housing and heating costs. Young and middle-aged workers facing redundancy from the SOEs have often taken up cross-border trade while a few lost money (and sometimes their houses) in the Caritas pyramid investment scandal of 1991–3. During the UN embargo against Yugoslavia – when petrol was smuggled across the Danube at night in the Moldova Nouă area to help maintain the Pančevo petrochemical complex (which operated throughout the period of sanctions) – some Bocşa people fitted larger petrol tanks to their cars and drove into Serbia three or four times a week during 1993–4. By bribing the customs officials at Jimbolia or Stamora Moraviţa, a selling price in Serbia six to eight times the Romanian pump price of DM0.5/l could generate enough income to build a new house. The end of sanctions led to new forms of commerce such as the purchase of clothing and leather goods in Hungary for sale in Reşiţa, or the sale of Romanian nuts and household tools in Serbia and Montenegro. Work abroad was obtained with travel documents by stowing away in lorries or the roof of a railway carriage in order to reach Germany and Italy but especially Spain where $1,000 per month could be earned on the black market ('la negru') – in competition with Albanians and Bosnians – by picking fruit or harvesting maize. The first people to get to Spain would then help others to follow. Each year some 50 people from Bocşa (almost all women) are now hired unofficially to work in Germany on asparagus or strawberries for just two Euros per hour compared with the four that would be expected by Poles, six by the Portuguese and seven by Germans. They are organised by reputable German firms and travel in special coaches (with proper documentation) and return with electrical goods. Meanwhile the people who used to commute from the villages tend to work only in subsistence farming. Indeed some Bocşa people have moved into the rural areas; buying houses cheaply in depopulated villages like Bărbosu (which has no permanent population but only families from Bocşa who make visits in spring and autumn to do the essential farm work) or Dulău where extensive cherry and plum orchards could generate some business.

Religious organisations – particularly Baptists and the Pentecostals – help the poor by running shops (selling second-hand clothes and time-expired medicines) and maintaining kindergarten and primary school facilities: they also assisted people to emigrate during the early transition years. Their petty commerce is tolerated through personal contacts with local officials established under communism (a benign influence previously demonstrated through permission to build the Baptist Church in Vasiova pre-1989, as well as the second church opened in Bocşa Română in 2001). Much aid was provided during 1989–1992 through Caritas (not to be confused with the pyramid investment scandal of the same name) with which the German Forum ('Forumul Germanilor din România') was connected, along with the Baptist and Catholic churches (which therefore attracted new members at the time!). But while food, clothing and equipment (including bicycles) was given away to poor Catholics and the Măgura Roma, much was sold through commerce organised by the Catholic priest. However, there is continuing support through Caritas involving food, clothing and toys from Netherlands for handicapped children (formerly accommodated at Sf. Ilie monastery and since 1994 at a local gymnasium school), while the organisation – often through family connections – also supplies clothing and medicines from Germany (especially Karlsruhe), Netherlands and USA to maintain the services provided by the Baptist or Evangelist Churches. Meanwhile Roma people, who still live near the monastery in the Bocşa Izvor/Godinova suburb, sell firewood (usually stolen) and the business continues despite the availability of gas heating since 2000 because the high fuel price has forced the poorer people to retain their wood-burning stoves. Local Roma women also sell pumpkin or sunflower seeds and cheap toys while men make brandy stills, golden ear-rings and work as musicians. The Roma also get money abroad by selling flowers but more normally by begging (initially in France but later in Finland and the UK) for fictitious community projects. Those repatriated by the French authorities for theft or other criminal acts were denied passports and placed on police and customs authority lists as undesirables, but this did not prevent the proceeds of begging from financing new villas for some Roma families on the main road or even the town centre. Social integration remains a problem for the Godinova Roma who retain their language, but less so in the case of the Măgureni Roma – living in Bocşa Montana in the vicinity of the Ezeriş road – who came with the 'Bufeni' or 'Săscani' Romanians and speak only Romanian: like the Roma community in Vasiova who moved from Dognecea in the communist period, they have received help from Caritas and the Catholic Church – as well as social assistance from the townhall following redundancies from CMB. The remaining Germans are helped by relatives abroad – also through the German Forum which provides money for 'Vershank' in February, the Wine Festival and 'Oktoberfest'.

How can the town be made more attractive? Housing is good in the sense that about 70% of the stock consists of brick buildings – while 20% use stone and other substantial materials compared with only 10% for wood. Most houses have two or three rooms and are roofed with tiles. 55% have gas heating while all have had electricity since the 1970s, but it is only the better-off one third of households that have running water, while waste water treatment is unsatisfactory to the extent that animal sewage may flow along gutters and constitute a pollution hazard. People working abroad are able to extend their houses, perhaps through additional storeys, while those with pensions or local salaries cannot afford improvements and struggle to secure modest bank savings. Flats are quite cheap (about $2,000 for two rooms) but none are under construction since only a few young families still lack their own apartments and have to remain with parents or grandparents. It is a pity that people living adjacent to CMB are poor and unemployed and cannot afford to buy the flats which therefore go to people from other regions e.g. Moldavia. Meanwhile, services are substantial and changes since 1989 show a balance of positive and negative trends. The telephone network has been enlarged and waste water treatment has improved, while gas is available for domestic use. There has been some restructuring of secondary education, while the local hospital survives along with a clinic and pharmacies. Retailing has improved, through a number of small shops and there is increased provision of banks and insurance houses (often linked with offices in Reşiţa). Several small hotels, restaurants and guest houses survive along with facilities for music and dance and adult education. On the ethnic side, the German Forum still organises activities, but the emigration of much of the German community had accelerated the demise of the culture house in Bocşa Română which has also seen the privatisation of the '1 Mai' recreation ground and the closure of the children's playground. The crisis at CMB caused the closure of the Workers' House ('Casa Muncitorească') in 1990 as well as the football stadium ('Bocşa Izvor') in the Pregel area (named after the former German owner of the park). Deterioration can also be seen in the destruction of the campsite through theft of the wooden cabins for firewood in an area with good air and attractive woodland that was recognised as a climatic station in 1931 with a strand and facilities for boating on the Bârzava. Pollution remains an issue with 17 sources of air pollution, 32 for water pollution and 30 for soil pollution.

The town certainly has possibilities to enhance its attractiveness through local recreational amenities in view of the good quality of the woodlands which cover the surrounding hills and provide a base for agrotourism. The reinvigoration of tourism may be seen in the restructuring and modernising of Stejaru Inn after purchase in 1998, while the Pregel area of Bocşa shows some signs of revival through a small foreign investment in the Hotel Izvor, privatised in 1992 in favour of a local man returning from Germany with resources to establish a small private establishment, rebuilt after a major fire caused by lightning. Moreover on the northern side of valley opposite Vasiova, the small Izvor lake – in wooded surroundings beside an empty swimming pool and abandoned pleasure ground – shows signs of revival through a local boarding house and a new business to operate rowing boats on the lake. The Sf.Ilie monastery (1905) – restored to the Orthodox Church in 1990 after being requisitioned by the state in 1959 for use as a children's home – attracts Orthodox Romanians and Catholics from Caraşova for the 'hram' festivities in July. The monastery introduced textile workshops in the 1930s and also houses an important collection of religious books. The town's 14 different religious cults generate visitor interest through their festivals and churches: Bocşa has one of the oldest Roman Catholic churches in România while Orthodox churches in the area were built in 1755, 1808 and 1815: the most notable is that of Sf.Nicolae, built 1795, painted in 1810 and restored in 1938. Archaeological sites include the prehistoric jewellery found in the Aron valley, Coţofeni vestiges at Colţan (where the quarry closed 1990) and the fortified settlement of Dealul or Gruniul Cetăţii developed by the Turks in 1658 – from what Hungarians claim to have been 'their' fortress three centuries earlier! – and hence named Buza Turcului (but gradually reduced to ruins after the Turks left). Vasiova's famous peasant poet – Tata Oancea – who launched the first local literary journal 'Vasiova' is remembered through a memorial house in Strada Horea, the public library named after him as well as a poetry festival which began in 1976. Since 1977 Bocşa has also organised a national 'Aurelia Fătu Răduţu' festival following the death of a renowned folk singer, with the possibility of establishing a museum in the house in Vasiova where she was born. Since 1984 the Culture House ('Casa de Cultură Orăşănească Bocşa') has used it own funds to organised international camps – usually in Ocna de Fier – for painting and sculpture.

Reference should also be made to the problems of three declining rural communes included in the LFA, especially Dognecea and Ocna de Fier where 90% of professionally active men and 30% of active women used to work in mining. The young have departed for Timişoara (though the remnants of the Roma community remain) and there is a female majority (some women work in shops and bars in Reşiţa) but no solutions have been found for the older people of working age. Lupac has lost its coal mining (apart from one small pit) and there is a slow migration by the Caraşoveni minority back to agricultural villages where there is a better basis for subsistence and where most of the men work nine months of the year – i.e. excepting the winter months – mainly in the Zagreb area of Croatia rebuilding houses damaged the homeland war. As Caraşoveni or Croats (depending on their declaration in 1992) they have Croatian passports and double citizenship: a substantial advantage when Croats gained visa-free access to the Schengen zone ahead of Romanians. They live comfortably at home but they receive lower wages than Croatians and while they evade taxation through their black market work they have no claims to social security either. The villages have electricity and fixed telephones – and a few bars and shops selling food and second-hand clothing – but Dognecea and Ocna de Fier need proper water and sewage systems and all need piped gas. There could be a basis for tourism in the culture of the Caraşoveni, with Lupac claiming 16th century origins as a Caraşoveni settlement sponsored by Hungarian kings to defend the western front of the Semenic Mountains. Coming from Albania, Bulgaria, Croatia, Macedonia and Montenegro, these people (who consider themselves Croats because of their Catholic faith) continue to demonstrate their rich cultural heritage. The biodiversity provides potential through the forest reservation at Dognecea and the opportunities for hunting (hares, pheasants, wild ducks and pigs) in the Areniş and Dognecea Mountains and Bocşa Hills.

There are tremendous opportunities over industrial archaeology which has been discussed across Caraş-Severin as a whole in several publications (Graf 2000a; Hillinger et al. 2001), including one which commends the promotion of 'Calea Fierului din Banat' (Olaru 2000). While there are some opportunities in Bocşa, most arise in Dognecea and Ocna de Fier where several lakes (Danila, Lacul Mare, Nuferi and Vârtoape) and quarries (Arhanghel, Iuliana, Paulus, Stros and Terezia) are among the areas protected. In Dognecea the legacy covers both 18th century non-ferrous working (1722) and 19th iron smelting. Meanwhile, Moraviţa – usually known as Ocna de Fier, or Piatra de Fier on the basis of the German and Bohemian names of Aisenstein and Steinechka respectively (referring to the rocky landscape) – has a history of copper and gold smelting in the old part of the village near Paulus and Terezia quarries; along with iron ore, limestone and marble working which continued until recently (Graf 2000b). A tourist project advocated for the Dognecea-Ocna de Fier area (inspired by experience at the Wieliecka salt mine near Kraków) would make use of six lakes (including two at Dognecea refurbished in the 1980s) – created in connection with the washing and sorting of ore – and an underground transport passage between Dognecea and the former narrow gauge industrial railhead at Ocna de Fier. In Dognecea the local furnace site is available for reclamation because although most of the old buildings are covered by slag, it would be possible to present the ruins of one of the furnaces, closed in the late 19th century in order to concentrate production in Reşiţa. But for the moment an excellent impression of the geological interest of the area is conveyed through a private museum ('Muzeul de Mineralogie Estetică Constantin Gruescu') which lies adjacent to Paulus Mine and comprises a remarkable collection of beautiful rocks collected from the local mines.

There is no doubt that LFA status for the Bocsa area is absolutely justified. The official unemployment rate is well above the national average while the real rate (allowing for people who have not registered) is thought to have been 67% since the official figure of 3,211 people officially unemployed compares with 5,126 jobs known to have been lost at CMB and Firiz alone, without taking account of Avicola and Miniera (the discrepancy may arise in part from the loss of some records in the Labour Office during 1990–4). Clearly more entrepreneurs are needed; for it is unfortunate that CBC through the DCMTER seems to bring most benefit to large towns with 'municipality' status where NGOs tend to thrive most conspicuously. Efforts are being made by Bocşa's cultural 'cămin' to improve IT skills as well as traditional crafts like tailoring and shoe-making, while the townhall has been involved in some professional conversion courses. There is opportunity in the mineral water (e.g. at Calina near Dognecea), while ornamental rocks could support some local economic diversification – not to mention 250mln.cu.m of reserves of marble at Dognecea and Ocna de Fier. However, pressure is relieved by a slow decline in population, and because most of the unemployed work abroad earning more than would be possible in Bocşa, the poverty rate has moved inversely to the notional unemployment level and it is now estimated at 25% (down from around 40%). It is also helpful that Bocşa is close to the frontier and many families now have relatives abroad.

Less-Favoured Coal-Mining Areas: Jiu Valley

The Jiu Valley in the southern part of Hunedoara County is a major problem area, despite being Romania's most important pitcoal producer. The 50km basin is fragmented by faults into 18 mining fields with thick coal seams – many of them vertical – of 6,000–8,000cal/kg (Gruescu 1972). The mines developed from Lonea in the east in 1840 to the central area (Vulcan in 1857 and Petroşani in 1858) and further east in 1890 (Aninoasa) and 1892 (Lupeni) where the coal ceases to be of coking quality. Although 'Westsiebenburgische Montan Verein' invested in the industry from 1858 (and Petrila coal was supplied for coking at Călan furnace in 1863), the arrival of the railway from Simeria in 1867 provided the major breakthrough. Production rose from 2.23mln.t in 1913 to 2.76 in 1943 (after a set-back during the depression when only Lupeni colliery remained open) to reach 4.19mln.t in 1960; 7.82 in 1970; and 9.40 in 1981, boosted by several new shafts e.g. Bărbăteni, Câmpu lui Neag and Uricani. However, due to growth of lignite fields the share of total Romanian coal production fell from 69.1% in 1938 to 51.3% in 1960 and 28.8% in 1975). Demand increased in the metallurgical industry through the Hunedoara coking plant (1950–5) and the Călan semi-coking works (1956–7), while some coking coal was also sent to Reşiţa for mixing with the local (higher quality) production. Power generators are also supplied: Vulcan (21.5MW) and Paroşeni (300MW) locally and others further afield where further cooling water is available (e.g. Mintia near Deva: 840MW). But demand has fallen since 1989 and imported coal is cheaper. So the government has been making heavy cut-backs, combined with relatively generous redundancy terms under the mining industry programme of 'disponibilizarea' and attempts to introduce alternative employment. Employment in the mines declined from 53,446 in 1989 to 42,000 in 1997 and 18,200 at the end of 1999 when benefits were fast running out. The national pitcoal company ('Compania Naţională a Huilei') has closed several mines since 1997: Câmpu lui Neag, Dâlja and Petrila Sud. But there was also modernisation at Valea cu Brazi in 1991 which meant that 70 men then produced the same amount of coal as the 3,000 at Aninoasa and the three mines subsequently closed! A new preparation plant opened at Coroieşti (Vulcan) in 2003 (rationalising the processing of coal previously handled at Livezeni and Uricani as well as Coroieşti) and this unit now complements the facility at Petrila for the eastern part of the coalfield. While modernisation is a mixed blessing in terms of jobs, it cannot seriously be resisted whenever finance is available: not least because the dangerous condition of the mines was blamed for tragic accidents at Vulcan in 2001 and 2002. There has also been a decline in the related machinery businesses e.g. at Umirom, Upsruem and Upsprom from over 3,000 jobs to just 1,000; at Gerom International and Electroutil (now 125 jobs); and the construction companies Consmin and Conpet (reduced to 628 jobs). As already noted, the run-down of the industry was deeply resented by a workforce privileged under communism. Exhaustion of severance money has increased poverty levels and is a factor in the increase in crime. In 1997 Hunedoara county sought special legislation to encourage private enterprise through fiscal incentives – leading to LFA programme with Valea Jiului among the first three areas to be declared in 1998.

Diversification has been an extremely slow process. When the miners challenged the communist leader N. Ceauşescu in 1977 with demands for six hour shifts, early retirement, improved equipment, better living conditions (housing and amenities – like hot water – and food) they also sought better work opportunities for women (going beyond the 'Vîscoza' artificial fibre plant built before communism). And although there was some repression by the authorities for the workers' demonstration – while coal production was boosted by use of the army and the arrival of a poorly-motivated workforce recruited from other part of the country – the Vulcan clothing factory (1979) was followed by a furniture factory at Petrila (1982). Since the revolution, the deep crisis in the region has meant that 'a concerted effort in jobs and job retraining, health care, infrastructure, is absolutely critical' (Kideckel et al. 2000 p.154). Despite the LFA in 1998, the miners continued to vent their anger with a five week general strike, re-enactment of the famous 1929 Lupeni strike on its 70th anniversary (partly by miners whose severance pay was quickly evaporating through purchase of household durables and luxury goods) and attempted marches on Bucharest already referred to. Some success in diversification has arisen from the Lupeni cigarette factory of 1998: a DM7.0mln project by Bulgar Tabac/Romned International, using Romanian capital and Dutch/Italian technology. A modernised two-storey building employs 200 locals (plus 70 key employees from the Târgu Jiu cigarette factory) packing cigarettes imported from Bulgaria, though the firm intends to invest in a tobacco plantation in Gorj which lies immediately to the south. A storage battery enterprise ('Acumulatorul') is linked with mining industry while an industrial joinery complex uses imported Italian PVC. EU assistance in 1999 focused on SMEs with total funding of €850,000 to provide grants of up to 75% of the total cost for approved schemes in both the Jiu Valley and Gorj (a strip-mining area which has also experienced a major cut-back). Fourteen projects were selected in the Jiu Valley with funding of up to €20,000 for 'simple' projects and 70,000 for 'complex' projects to create some 200 jobs in all. There was also a follow-up programme of €10mln from PHARE for a Mining Area Reconstruction and Rehabilitation Fund. The Jiu Valley projects covered two Petroşani firms making double glazing units (which need to hold adequate stocks of wood, aluminium and PVC materials); a sawmill at Vulcan which has successfully built up a customer base; a turning and machining workshop in Petrila turning out quality products for export; and other businesses concerned with baking, frozen food, knitwear, printing, elecro-mechanical work, waste recycling and land registry services (all in Petroşani); horticulture in Lupeni; and baking and office services in Uricani. The scheme demonstrates capacity but the need to encourage an entrepreneurial spirit. Much more diversification is needed: the town of Vulcan sees opportunity in wood processing; food processing (meat and milk), light industry; glass and ornamental rocks.

Foreign investors have effectively stayed away from the area although in the economic sense it is 'ill' and needs FDI to offer a cure. However the area's reputation for violent action against the authorities (both local and central) is extremely off-putting; not to mention the relatively poor qualifications of a workforce that nevertheless expects salaries comparable with mining (although they are hardly justified by prevailing productivity standards). Also there is no large, relatively prosperous town in close proximity to provide a sense of security: Hunedoara is 60kms to the north and is also a high unemployment area and the more affluent city of Deva is 20kms further on, while to the south Târgu Jiu lies 50kms away through the Jiu defile. Measures have to be taken to improve the area's attractiveness. Recent research has looked into the coping strategies for redundant workers and progress made in retraining and small business creation; highlighting the need for confidence building in the creation of partnerships and the need for small incubators and subsidised credits. A professional training and reconversion centre is operating and a business consulting service is also being provided in Petroşani by a local foundation established in 1997 for the promotion of SMEs and supported by the UNDP and the local business community. The local authority is backing an industrial estate at Livezeni Colliery which will mean the redevelopment of the old preparation plant. The main road along the Jiu Valley is being upgraded in connection with the projected Calafat-Vidin bridge over the Danube and the related Eurocorridor. Intersecting east-west road links (7A Petrila-Râmnicu Vâlcea and 66A Uricani-Băile Herculane) – which require heavy investment in modernisation and realignment, especially within the Retezat Naţional Park – will enhance the tourist potential of the area. Local studies highlight the potential on the routes to Valea Tăii and Lake/Cabana Şureanu (actually in the territory of the adjacent town of Cugir) and the Jiu valley route to holiday homes around Cabana

Voievodu. But on an altogether bigger scale, Petroşani Council is supporting a comprehensive plan for 'Zona Turistică Valea Jiului' which includes a greatly enlarged skiing resort in the Parâng Mountains (PMP 2003) at the present Cabana Rusu/Staţia Meteo Complex where the separate units of Maleia and Cabana ANEFS are already linked with skiing grounds at 1,600–1,750m around Releu TV by a 2,232m 'telescaun' (1973), a 317m 'teleski' (1989) from the former and a 382m 'teleski' from the latter in 1994. Additional skiing areas known as 'Pârtia A' and 'Pârtia B' were opened up by a 1,280m 'teleski' in 1984. The expansion includes new skiing areas extending from (a) ANEFS to Coama Parângul Mic (1,790–2,060m) with an 853m 'teleski', Slima (1,620–1,995m) with a 1,250m 'teleski' and possibly Gruniu; while Pârtia A and Pârtia B will be developed by 'Europarâng' with a 440m 'teleski'; and from Maleia there are propoed extensions at Dragu Petrisor (1,450–1,520) with a 550m 'teleski', Poiana Mare (1,250–1,500) with an 875m 'teleski' (with Saivane a further option). The complex would expand almost four times from its present 12.7ha to 46.0.

Petroşani sees the need to promote a positive image for the region. This has been helped by World Bank funds for SMEs and 'ecologizare' which covers both farmland and urban infrastructure while the EU has also financed SMEs and socio-economic development (Alexandrescu 2001). A water scheme started in 1998 helped to improve local infrastructure and also provided temporary employment (e.g. at Petrila during 1999–2000) while dyking of the Jiu river was also undertaken after the 1998–9 floods. But despite some World Bank assistance for roads, housing repairs/community facilities and waste management, much more remains to be done. Petrila has no park or childrens' playground and dust blown from waste tips is a considerable nuisance, although there is a prospect of afforestation being undertaken in connection with the World Bank 'Prototype Carbon Fund' (PCF) – linked with the Kyoto Protocol – which will support over 6,700ha of acacia and poplar plantations on degraded land in a number of lowland counties (Brăila, Dolj, Mehedinţi, Olt, Tulcea and Vaslui) (Blujdea et al. 2003) based on an experimental model for the Danube Valley showing that poplar trees would reach an industrial diameter in ten years and also provide certain non-timber forest products. The PCF provides a mechanism to 'purchase the net carbon sequestered by the newly established plantations' (Abrudan et al. 2003 p.16) – otherwise the work would not be economically viable on land ruined by irrigation and mismanagement. The scheme has relevance to tips in Jiu Valley where eight hectares were planted by the local 'Ocol Silvic' from 1987 (pine and 'cătină' scrub) ultimately with World Bank money which made it feasible to provide a soil depth of 20–30cms on the tips. Once the methods have been perfected it is anticipated that work will be continued with World Bank support for the mines to continue the work themselves.

Less-Favoured Coal-Mining Areas: Anina

This study concerns part of the Moldova Nouă LFA, originally delimited to include the Anina and Oraviţa areas and all the communes of the Almăj Depression and the Danube Defile, extending eastwards to Iablaniţa and Mehadia. However, in order to concentrate the resources in the areas of greatest need, the area was reduced to two separate districts within this zone: (a) Anina and Oraviţa with Bozovici, Ciudanoviţa, Mehadia and Prigor; and (b) Moldova Nouă, with Berzasca, Cărbunari, Coronini, Sasca Montană and Sicheviţa communes. Oravita, which offers additional fiscal concessions and free professional help, has already diversified through 'Normarom', a Franco-Romanian enterprise producing garden furniture, and a German enterprise producing furniture based on the local beechwood is a possibility. Meanwhile, with commitment to the area by the EU and the German Land of Nordrhein-Westfalen, Moldova Nouă wants to develop its port on land previously part of the local copper mine and then establish a 'free port' regime. But the infrastructure is a problem since the road from Orşova through the Danube Defile remains unsurfaced, though it has the potential to provide a new route from Bucharest to Belgrade, with a frontier post at Socol. Moldova also lacks a rail link although the reopening of the route from Iam (near Oraviţa) to Baziaş and its extension along the Danube could solve this problem and, at the same time (given modernisation of the line to Berzovia in the north) make it easier for the Reşiţa metallurgical and engineering works to despatch rails and heavy equipment (Hillinger 2000). But the main concern here is with Anina which has a long history as a coal mining area (Feneşan et al. 1991). The town stands at 645–780m and arises out of the settlement by people from Styria at Steierdorf in 1773 in order to produce charcoal for the Oraviţa copper smelter. The transformation began after 1790 when coal was discovered by a local woodcutter and this proved highly significant in the next century in the context of steam navigation on the Danube.

The Habsburg mining authorities began prospecting on their own account and, from 1845, implemented a programme to retrieve the coalmines already working under concessionaires. They also contemplated an underground tunnel to Lişava from where a surface railway would run the Danube but this plan was changed to an all-surface route looping northwards to Gârlişte and completed in 1863. By this time the mining was in the hands of STEG who eventually operated a string of mines from the railhead in the north (Anina and Thinnfeld) to Colonie, Ponor and Uteriş around Steierdorf in the south, with a cluster in the centre around Valea Teresia (including Colovrat, Friedrich, Gustav, Hildegard and Kubeck). STEG also worked blackband ironstone and the three furnaces established at the Anina railhead by 1867 continued to work until 1927 when the local ore was exhausted. Moreover, bituminous schist formed the basis of an oil distillation business which operated until Romanian mineral oil became available in 1882. However Anina continued to operate as the leading coal producer of what became the UDR complex in 1918; supplying both the company's coke ovens (relocated in Reşiţa in 1935) and a local 12MW power station. Coal mining was given every support under communism – and new mines were opened at David and Miniş to supply refractory clay (not to mention uranium mining close by at Ciudanovita and Lisava) – although the real costs rose with increasing depth (currently 1,500m) despite rationalisation to the point where coal was wound only at the principal shaft in Anina with standard gauge rail access. Under the programme for rationalisation of mining all the other mines have been closed completely apart from two (Gustav and Colovrat) retained for ventilation of the eastern and western flanks of the coalfield respectively. This followed a decision taken immediately after the revolution to stop work on a vast opencast mining project to exploit bituminous schist as a power station fuel; supplying in the first instance a 990MW station at Crivina, south of Anina (where one 330MW unit was already operating in 1989). With over 2,000 miners receiving redundancy packages the employment at Anina has fallen to 850 and saleable annual production is 0.42mln.t compared with over a million in the communist period.

With mining accounting for just over two-thirds of all employment in the town, the restructuring programme increased unemployment to 35%, exceeding the national rate by 20%. There was little manufacturing apart from a long established screw factory and a sawmill (originally connected with the STEG/UDR industrial operations). Jobs for women have increased with a bakery and the 'Unisport' shoe factory in the railway station area: the latter employs 100 women – of whom 40 commute by bus from Oraviţa – and training has achieved high quality production. But the situation remains difficult for men who depend heavily on temporary work on road schemes and other infrastructure projects, though reference should also be made to ecological work at the Ciudanoviţa and Lişava uranium mines as well as local coal mines, while the vast Crivina schist quarry is being reclaimed by a Greek company. With the establishment of the LFA an Italian capitalist expressed interest in a new sawmill producing for export, while an Austrian firm considered a furniture factory and a Maltese entrepreneur thought about a clothing enterprise. However the main success has been 'Vartex Textil', a Romanian-Italian enterprise involving Bellandi of Florence which produces woollen goods from pieces i.e. a finishing and parcelling operation employing 100 (and possibly doubling) in a former state warehouse adjacent to the railway station. There is a strong export business especially to France. The Italian knitwear firm 'Ame Damasa' has set up in a former restaurant and the Romanian-Swedish company 'Rom-Impex' used premises near the bus station to produce pallets from waste wood and also parchet for which there are orders from Germany. These enterprises have done well although further training is needed to increase motivation and productivity.

The attractiveness of Anina is compromised by a degree of isolation which can be eased through road improvements currently being undertaken. The local environment has been badly scarred by mining; a legacy which is now being addressed. The town has been well endowed with local resorts in the surrounding hills where clean air and high landscape quality. But on the western side Sommerfrische has been degraded and partly demolished in connection with the bituminous schist quarry, while Brădet once favoured as a place of convalescence was taken over by the army during the Yugoslav crisis of the early 1950s and the new apartment accommodation was subsequently occupied by the immigrant population needed to boost the workforce in local mines. The facilities on the eastern side at Maial and Mărghitaş have not been compromised in the same way but both need refurbishment and this was noted at the latter in 2003 in a largely unspoilt area in the wooded Buhui valley with extensive meadows on the higher ground. A local building company ('Alutus') is working on a hotel which will supplement the existing chalets and campsite and should be suitable for all-year use in view of the winter sport potential. Diversification should also provide a role for agrotourism based on the scenery of the mountains and the Danube defile, with particular interest attaching to the national parks and protected areas including many natural monuments (Deliman 1998; Drugărin 2003). The Anina Mountains (reaching 1,160m at Vf. Leordiș) display karstic forms and an interesting vegetation which includes Mediterranean species. But there is also scope for hunting and cultural/religious interests (through churches, monasteries and festivals) especially in the context of a traditional ethnic diversity. The industrial history is greatly undervalued. The Anina railway could certainly be developed as a major tourist asset with interpretation to cover the phases of construction, the use of horse and locomotive traction, the historic stations at Oravița (1849) – the oldest in Romania – and Anina (1864) and the locomotive depot at Oravița (1898) along with vintage equipment retained at wayside stations. Industrial archaeology also offers substantial opportunity through the mining legacies in Anina, although several former mines have been completely dismantled in recent years. Oravița also offers the oldest theatre in the whole of SEE: designed by Viennese architect Johann Neumann and built between 1789 and 1817 in a town that was once the centre of Caraş County (Radu 2000).

Conclusion

This review – which has attempted to combine several threads of research[1] – has deliberately concentrated on the more challenging locations which have been examined against the background of more dynamic centres attracting a disproportionate share of FDI (in the context of the modest share which România as a whole has attracted). To what extent can the problem regions overcome their disadvantages? Romania's official LFAs have enjoyed substantial concessions for several years and it is appropriate that different aspects of disadvantage should be researched as thoroughly as possible: not all problems can be addressed by fiscal concessions but a comprehensive database will help to ensure that priorities are established on a rational and equitable basis. FDI in the LFAs has been very restrained with hardly any large investments of over $1.0mln. Interviews with foreign investors in Timişoara point to an acute sensitivity over circumstances that could compromise the chances of securing a profit on capital invested. Timişoara is appreciated for a well-qualified and committed workforce (accepting wages that are only marginally higher than in the remaining SOEs), easy of supply of raw materials (such as leather from local livestock farms to support high-tech footwear factories) and direct air flights to many European cities, especially in Italy (a major concern for a 5,000–strong Italian community in the city). By contrast, LFA workforces are perceived are being relatively strike-prone and unproductive on account of unacceptable work practices once endemic in the SOE sector – and particularly in mining through the high status and relatively good wages enjoyed under communism, with the Jiu Valley an extreme case. At the very least this makes for higher training costs and an element of risk that most entrepreneurs are unwilling to take except where family ties exist and premises are readily available at low cost (although this points to the potential benefit of more industrial parks in LFAs). Thus FDI cannot be expected to take an enlightened attitude and spread evenly across each national territory. There are far greater incentives to cluster in places where links between foreign businesses and businessmen can be fostered. And with so many virtual 'no go' areas for FDI the realistic approach may well be to concentrate on the development of indigenous enterprise through new SMEs in the IRAs, with the now-fading LFA programme providing a rescue for areas suddenly hit by extremely high unemployment. However, more concerted efforts are needed to break down negative images of mining areas. And it will always be down to communities themselves to frame their own solutions by expanding development-oriented NGO networks to intensify promotion of economic potential and enhance skill and competence levels.

Note

[1] Collaboration over this chapter was supported by a British Academy grant in aid of joint projects and formalised in the case of Remus Creţan's Bocşa study through West University of Timişoara Contract 0012634.

References

Abrudan, I.V., Blujdea, V., Brown, S. et al. (2003), 'Prototype Carbon Fund: Afforestation of Degraded Land in Romania', *Revista Pădurilor*, 118(1), 5–17.

Alexandrescu, V. (2001), 'Depresiunea Petroşani: Consideraţii Economico-Sociale în Perioada de Tranziţie', *Terra*, 31, 70–3.

Amin, A. and Tomaney, J. (1995), 'The Regional Development Potential of Inward Investment in the Less Favoured Regions of the European Community', in A. Amin and J. Tomaney (eds), *Behind the Myth of European Union: Prospects for Cohesion*, Routledge, London, 201–20.

Blujdea, V., Abrudan, I.V. and Pahontu, C. (2003), 'Scenariul de Acumulare a Stocului de Carbon în cadrul unui Proiect de împădurire a Terenurilor Degradate în România', *Revista Pădurilor*, 118(2), 1–4.

Borcoş, A, and Vîrdol, A. (2002), 'Aspecte Metodologice privind Ierarhizarea Zonelor Miniere Defavorizate din România', *Revista Geografică*, 8, 178–83.

Bordânc, F. and Nancu, D. (1999), 'Implicarea Cercetării Geografice în fundamentarea Politicilor Sectorale de Dezvoltare Rurală', *Revista Geografică*, 6, 34–40.

Buza, M., Dimen, M., Pop, G. and Turnock, D. (2001), 'Environmental Protection in the Apuseni Mountains', *GeoJournal*, 55, 631–53.

Caraiani, G. and Cazacu, C. (1995), *Zonele Libere*, Editura Economică, Bucharest.

Chiribuca, D., Comşa, M. and Rotariu, T. (2000), *The Impact of Economic Restructuring in Monoindustrial Areas*, SOCO Project Paper 87, Vienna.

Ciuhandu, G. (2000), *Concept strategic de Dezvoltare Economică şi Socială a Zonei Timişoara*, Editura Brumar, Timişoara.

Coifan, V. (ed) (1999), *Strategia de Dezvoltare Economică a Judeţului Timiş*, Editura Orizonturi Universitare, Timişoara.

Drogeanu, P-A. (2000), *Reforma şi Lucrărilor Publice în România ultimilor ani 1997–2000*, Editura Universalia/MLPAT, Bucharest.

Drugărin, C. (2003), 'Tourist Potential of the Anina Mountains', in R. Creţan (ed.), *Fifth Edition of the Regional Conference of Geography: Geographical Researches on the Carpathian-Danubian Space: Reconsideration of the Geographical Approaches in the Context of Globalisation*, Editura Mirton, Timişoara, 637–47.

Fenesan, C., Graf, R., Zaberca, V.M. and Popa, I. (1991), *Din Istoria Carbunelui: Anina 200*, Muzeul de Istorie al Judeţului Caraş-Severin, Reşiţa.

Fenn, J.M. and Keil, T.J. (1994), 'Ingrijorarea faţa de Creşterea Fenomenului Criminalităţii in România Post-Revolutionară 1990–1992', *Revista de Cercetări Sociale*, 1(4), 51–71.

Graf, R. (1997), *Domeniul Banatean al StEG 1855–1920*, Editura Banatică, Reşiţa.

Graf, R. (2000a), 'Judeţul Caraş-Severin: O Regiune Industrială Veche cu un Potenţial Industrial-Cultural şi Turistic care asteapta sa fie valorificat', in H. Bonninghausen et al., *Cale Fierului din Banat: Un Proiect de Dezvoltarea Regională pe Baza Turismului industrial*, Editura InterGraf/Friedrich Ebert Stiftung, Reşiţa, 149–61.

Graf, R. (2000b), 'Dognecea: Un Posibil Punct de Atracţie Turistică in Banatul Montan', in E. Sabiel (ed), *Turism Integrat: Banat şi Maramureş*, Editura InterGraf, Reşiţa, 105–12.

Grigoraş, V. (2001), 'Strategii de Mobilitate în Sâncrai-Hunedoara', *Sociologie Românească*, 3, 232–9.

Groza, O. (1994), 'Paşcani: Ville Industrielle de Roumanie: Années de Transition', *L'Espace Géographique*, 23, 329–41.

Gruescu, I.S. (1972), *Gruparea Industrială Hunedoara-Valea Jiului: Studiu de Geografie Economică*, Editura Academiei RSR, Bucharest.

Guran-Nica, L. (1997), 'Zonele Economice Libere ale Dunării: favorabilităţi şi perspective', *Geographica Timisiensis*, 6, 89–103.

Guran-Nica, L. (2002), *Investiţii Străine Directe: Dezvoltarea sistemului de Aşezări din România*, Editura Tehnică, Bucharest.

Guran-Nica, L. and Turnock, D. (2000), 'A Preliminary Assessment of Social Risk in Romania', *GeoJournal*, 50, 139–50.

Guvernul Romaniei (1997), *Carta Verde: Politica de Dezvoltare Regională în România*, Guvernul României şi Comisia Europeană, Bucharest.

Hillinger, N. (2000), 'Câteva consideraţii asupra Zonelor Defavorizate în Judeţul Caraş-Severin', *Geographica Timisiensis*, 8–9, 281–6.

Hillinger, N., Olaru, M. and Turnock, D. (2001), 'The Role of Industrial Archaeology in Conservation: The Reşiţa area of the Romanian Carpathians', *GeoJournal*, 55, 607–30.

Hillinger, N. and Turnock, D. (2000), 'The Reşiţa Industrial Complex: Restructuring and Diversification in the Post-Communist Period', in D. Light and D. Phinnemore (eds), *Post-Communist Romania: Geographical Perspectives*, Liverpool Hope Press, Liverpool, 7–27.

Ianoş, I. (2000), 'Less-Favoured Areas and Regional Development in Romania', in G. Horvath (ed), *Regions and Cities in the Global World*, Centre for Regional Studies, Pecs, 176–91.

Ianoş, I. (2001), 'Ariile Profund Dezavantajate din România: Consideraţii Preliminare', *Terra*, 31, 30–8.

Ianoş, I., Popescu, C. and Talanga, C. (1996), 'Repartiţia geografică a unor grupuri sociale marginale în România', *Studii şi Cercetări de Geografie*, 43, 13–22.

Kideckel, D.A., Botea, B.E, Nahorniac, R. and Soflan, V. (2000), 'A New "Cult of Labor": Stress and Crisis among Romanian Workers', *Sociologie Românească*, 2(1), 142–61.

Marga, A. (1993), 'Cultural and Political Trends in Romania before and after 1989', *East European Politics and Societies*, 7, 14–32.

Ministerul Dezvoltării şi Prognozei 2003, *Programul PHARE 2001: Coeziune Economică şi Socială – Asistenţa pentru IMM – ghidul soliicitanţului*, MDP, Bucharest.

Nadejde, S. (1999), 'Methodologie d'Analyse de l'Espace Rural selon les critères de l'Amenagement du Territoire', in V. Surd (ed), *Rural Space and Regional Development*, Editura Studia, Cluj-Napoca, 228–30.

Nica, N-A. (1993), *Criterii de Definire a Delimitare a Spaţiilor Rurale Defavorizate*, Urbanproiect, Bucharest.

Nica, N-A. (1999), 'Disadvantaged Areas in the Context of the Rural Development Policy in Romania', in V. Surd (ed), *Rural Space and Regional Development*, Editura Studia, Cluj-Napoca, 219–22.

Olaru, M. (2000), 'Calea Fierului Banatean şi Hunedorean: produs turistic unicat în cadrul ofertei turistice româneşti', in E. Sabiel (ed), *Turism Integrat: Banat şi Maramureş*, Editura InterGraf, Reşiţa, 83–104.

Pascariu, G. (1993), *A New Framework concerning the Sustainable Development of Human Settlements*, Urbanproiect, Bucharest.

Popescu, C.R., Negut, S., Roznovietchi. I. et al. (2003), *Zonele Miniere Defavorizate din România: abordare geografică*, Editura ASE, Bucharest.

Primarie Municipiului Petroşani (2003), *Plan Urbanistic Zonal: Zona de Agrement Masivul Părâng*, Getrix, Craiova.

Puwak, H. (1992), *Poverty in Romania: Territorial Distribution and the Intensity of Poverty Level*, Romanian Academy, Institute for the Quality of Life, Bucharest.

Radu, N. (2000), 'Anina: O Veche Regiune Industrială cu potenţial de viitor', in E. Sabiel (ed), *Turism Integrat: Banat şi Maramureş*, Editura InterGraf, Reşiţa, 147–56.

Ramboll Group (1996), *Disparităşi Regionale în România 1990–1994*, PHARE/Ramboll Grup de Consultanţa, Bucharest.

Ramboll Group (1997), *Profiles of Romanian Development Regions*, RCG for PHARE Programme, Regional Development Policy, Bucharest.

Săgeată, R. (2003), 'L'Organisation Administrative et Territoriale de la Roumanie entre le Modèle Traditionnel et les Réalités Contemporaines', in R. Creţan (ed), *Fifth Edition of*

the Regional Conference of Geography: Geographical Researches on the Carpathian-
Danubian Space: Reconsideration of the Geographical Approaches in the Context of
Globalisation, Editura Mirton, Timişoara, 601–10.

Săuleanu, D. and Pârlea, D. (1997), 'Restructurarea în Industria Minieră: disponibilizarea
forţei de muncă în bazinul carbonifer al Văii Jiului', *Revista Română de Sociologie*,
8(5–6), 505–18.

Timofticiuc, D. (2001), 'Zonele Defavorizate sub Monitorizarea Guvernului', *Adevarul
Economic*, 10(37), 10.

Timofticiuc, D. (2003), 'In Zonele Defavorizate s-au creat 54,000 de locuri de munca: cu ce
preţ?', *Adevarul Economic*, 12(31), 14.

Tismaneanu, V. (1997), 'Romanian Exceptionalism?: Democracy Ethnocracy and Uncertain
Pluralism in Post-Ceauşescu Romania', in K. Dawisha and B. Parrott (eds), *Politics
Power and the Struggle for Democracy in SEE*, Cambridge University Press, Cambridge,
403–51.

Turland, R. (2002), *Roşia Montană Project: Project Description*, RMGC, Roşia Montană.

Voiculescu, S. (1999), 'The Village of Izvin: A Case Study of Prospective Geography', in
V. Surd (ed), *Rural Space and Regional Development*, Editura Studia, Cluj-Napoca,
102–5.

Index